NEUROMETHODS ☐ 18

Animal Models in Psychiatry, I

NEUROMETHODS

Program Editors: Alan A. Boulton and Glen B. Baker

17. **Neuropsychology**
    Edited by *Alan A. Boulton, Glen B. Baker, and Merrill Hiscock,* 1990
16. **Molecular Neurobiological Techniques**
    Edited by *Alan A. Boulton, Glen B. Baker, and Anthony T. Campagnoni,* 1990
15. **Neurophysiological Techniques: Applications to Neural Systems**
    Edited by *Alan A. Boulton, Glen B. Baker, and Case H. Vanderwolf,* 1990
14. **Neurophysiological Techniques: Basic Methods and Concepts**
    Edited by *Alan A. Boulton, Glen B. Baker, and Case H. Vanderwolf,* 1990
13. **Psychopharmacology**
    Edited by *Alan A. Boulton, Glen B. Baker, and Andrew J. Greenshaw,* 1989
12. **Drugs as Tools in Neurotransmitter Research**
    Edited by *Alan A. Boulton, Glen B. Baker, and Augusto V. Juorio,* 1989
11. **Carbohydrates and Energy Metabolism**
    Edited by *Alan A. Boulton, Glen B. Baker, and Roger F. Butterworth,* 1989
10. **Analysis of Psychiatric Drugs**
    Edited by *Alan A. Boulton, Glen B. Baker, and Ronald T. Coutts,* 1988
9. **The Neuronal Microenvironment**
    Edited by *Alan A. Boulton, Glen B. Baker, and Wolfgang Walz,* 1988
8. **Imaging and Correlative Physicochemical Techniques**
    Edited by *Alan A. Boulton, Glen B. Baker, and Donald P. Boisvert,* 1988
7. **Lipids and Related Compounds**
    Edited by *Alan A. Boulton, Glen B. Baker, and Lloyd A. Horrocks,* 1988
6. **Peptides**
    Edited by *Alan A. Boulton, Glen B. Baker, and Quentin Pittman,* 1987
5. **Neurotransmitter Enzymes**
    Edited by *Alan A. Boulton, Glen B. Baker, and Peter H. Yu,* 1986
4. **Receptor Binding Techniques**
    Edited by *Alan A. Boulton, Glen B. Baker, and Pavel D. Hrdina,* 1986
3. **Amino Acids**
    Edited by *Alan A. Boulton, Glen B. Baker, and James D. Wood,* 1985
2. **Amines and Their Metabolites**
    Edited by *Alan A. Boulton, Glen B. Baker, and Judith M. Baker,* 1985
1. **General Neurochemical Techniques**
    Edited by *Alan A. Boulton and Glen B. Baker,* 1985

NEUROMETHODS

Program Editors: Alan A. Boulton and Glen B. Baker

NEUROMETHODS □ 18

Animal Models in Psychiatry, I

Edited by

Alan A. Boulton
University of Saskatchewan, Saskatoon, Canada

Glen B. Baker
University of Alberta, Edmonton, Canada

and

Mathew T. Martin-Iverson
University of Alberta, Edmonton, Canada

Springer Science+Business Media, LLC

Originally published by The Humana Press Inc in 1991.
Crescent Manor
PO Box 2148
Clifton, NJ 07015

ISBN 978-1-4899-3922-7 ISBN 978-1-59259-623-2 (eBook)
DOI 10.1007/978-1-59259-623-2
Softcover reprint of the hardcover 1st edition 1991

Library of Congress Cataloging-in-Publication Data

Animal models in psychiatry / edited by Alan A. Boulton, Glen B. Baker, and Mathew T. Martin-Iverson.
p. cm. — (Neuromethods ; 18)
Includes index.

1. Mental illness—Animal models. 2. Psychiatry—Research--Methodology. I. Boulton, A. A. (Alan A.) II. Baker, Glen B., 1947– III. Martin-Iverson, Mathew Thomas, 1956– IV. Series.
[DNLM: 1. Disease Models, Animal. 2. Mental Disorders. W1 NE337G v. 18 / WM 100 A5975]
RC455.4.A54A55 1991
616.89'027—dc20
DNLM/DLC
for Library of Congress 91-7076
CIP

Preface to the Series

When the President of Humana Press first suggested that a series on methods in the neurosciences might be useful, one of us (AAB) was quite skeptical; only after discussions with GBB and some searching both of memory and library shelves did it seem that perhaps the publisher was right. Although some excellent methods books have recently appeared, notably in neuroanatomy, it is a fact that there is a dearth in this particular field, a fact attested to by the alacrity and enthusiasm with which most of the contributors to this series accepted our invitations and suggested additional topics and areas. After a somewhat hesitant start, essentially in the neurochemistry section, the series has grown and will encompass neurochemistry, neuropsychiatry, neurology, neuropathology, neurogenetics, neuroethology, molecular neurobiology, animal models of nervous disease, and no doubt many more "neuros." Although we have tried to include adequate methodological detail and in many cases detailed protocols, we have also tried to include wherever possible a short introductory review of the methods and/or related substances, comparisons with other methods, and the relationship of the substances being analyzed to neurological and psychiatric disorders. Recognizing our own limitations, we have invited a guest editor to join with us on most volumes in order to ensure complete coverage of the field. These editors will add their specialized knowledge and competencies. We anticipate that this series will fill a gap; we can only hope that it will be filled appropriately and with the right amount of expertise with respect to each method, substance or group of substances, and area treated.

Alan A. Boulton
Glen B. Baker

Preface to the Animal Models in Neuropsychiatry Volumes

This and several subsequent volumes in the Neuromethods series will describe a number of animal models of neuropsychiatric disorders. Because of increasing public concern over the ethical treatment of animals in research, we felt it incumbent upon us to include this general preface to these volumes in order to indicate why we think further research using animals is necessary and why animal models of psychiatric disorders, in particular, are so important. We recognize that animals should only be used when suitable alternatives are not available. We think it self-evident, however, that humans can only be experimented upon in severely proscribed circumstances and alternative procedures using cell or tissue culture are inadequate in any models requiring assessments of behavioral change or of complex in vivo processes. However, when the distress, discomfort, or pain to the animals outweighs the anticipated gains for human welfare, then the research is not ethical and should not be carried out. It is imperative that each individual researcher examine his/her own research from a critical moral standpoint before engaging in it, taking into consideration the animals' welfare as well as the anticipated gains. Furthermore, once a decision to proceed with research is made, it is the researcher's responsibility to ensure that the animals' welfare is of prime concern in terms of appropriate housing, feeding, and maximum reduction of any uncomfortable or distressing effects of the experimental conditions, *and that these conditions undergo frequent formalized monitoring*. In the third of these volumes on animal models, we have included a chapter on the ethics of animal models by Dr. E. Olfert, a veterinarian who also directs a laboratory animal care facility. As indicated in Dr. Olfert's chapter, it is essential to conform to national and local animal welfare regulations, whether codified in law or by self-regulatory bodies. We urge readers who wish to adopt any of the procedures described to follow closely not only the letter of their own national and local regulations, but also the spirit of these guidelines.

The Editors

Preface

The two *Animal Models in Psychiatry* volumes are loosely organized by subject. The first volume contains a number of chapters concerned with schizophrenia, psychoses, neuroleptic-induced tardive dyskinesias, and other disorders that may involve dopamine, such as attention deficit disorder and mania. The second volume deals with affective and anxiety disorders, but also includes chapters on subjects not easily classified as either psychotic, or affective, or anxiety-related, such as aggression, mental retardation, and memory disorders. Four chapters on animal models of schizophrenia or psychoses are included in the present volume because of the importance of these disorders in psychiatry. Likewise, three chapters in the subsequent volume deal with depression.

The first of the two volumes begins with an introduction by Paul Willner reviewing the criteria for assessing the validity of animal models in psychiatry. He has written extensively on this subject, and his thorough description of the issues of various forms of validity provides a framework in which to evaluate the subsequent chapters. As will be seen, the remaining chapters in both volumes will refer frequently to these issues. The second chapter, by Melvin Lyon, describes a large number of different procedures that have been proposed as potential animal models of schizophrenia. This is a departure from the usual format, consisting of detailed descriptions of specific models. It was felt that the importance of schizophrenia in psychiatry required a more general overview of the models that are used for this mental illness. Nestor Schmajuk (Chapter 3) describes a hippocampal lesion model of schizophrenia that particularly addresses the cognitive and attentional dysfunctions of schizophrenia. One of the strongest supports for the dopamine hypothesis of schizophrenia is the observation that chronic administration of psychomotor stimulants can induce a psychosis similar in many respects to that experienced by schizophrenics. Chapter 4 by Mathew Martin-Iverson describes a behavioral procedure involving chronic stimulant treatment that examines the effects on the organization of behavior over time and explores the roles of dopamine receptor subtypes in the disorganization produced by stimulants. Gaylord Ellison continues this theme of stimulant-induced psychosis with his description of the development of hallucination-like behavior with chronic continuous infusions of stimulants (Chapter 5). An exciting aspect of this model of stimulant-induced hallucinations is the disruption of normal social behaviors that are likely to be especially relevant to schizophrenia. The appendix to this chapter describing the slow-release amphetamine pellets provides an inexpensive alternative to the rather costly osmotic minipumps that can make a large hole in research grants. Melvin Lyon (Chapter 6) points out the importance of interactions between dopamine and glutamate in his review of models of mania. A major problem with the treatment of schizophrenia is the iatrogenic development of tardive dyskinesias. Two chapters describe animal models of this disorder. Helen Rosengarten's rapid jaw movement model (Chapter 7) indicates the importance of considering both D1 and D2 dopamine receptor subtypes in understanding the development of tardive dyskinesias. The appearance of clozapine as a drug that binds to D1 receptors and that may have a decreased propensity for inducing tardive dyskinesias, as well as the existence of selective D1 antagonists currently undergoing clinical trials to determine their efficacy in treating schizophrenia demonstrate the significance of this model. A computerized methodology developed by Gaylord Ellison and Ronald See (Chapter 10) for measuring oral dyskinesias induced by neuroleptics presents a reliable system that may resolve the controversy surrounding animal models of tardive dyskinesia that quite possibly arose from differences in relatively subjective assessments of mouth movements across different labs. Anorexia nervosa is a psychiatric disorder that has received considerable media attention lately. David Pierce and Frank Epling's behavioral model of activity anorexia (Chapter 8) discloses the critical relationship between exercise and anorexia, providing an ethological framework within which to understand this disorder. I anticipate that application of neurobiological and pharmacological techniques within this model may well reveal the relevant neurochemical systems, such as the endogenous opiates described by the authors, that underlie anorexia and may well suggest future treatments. Joram Feldon and Ina Weiner's elegant work (Chapter 9) with the latent inhibition model of attention deficit disorders provides a well-characterized model of certain kinds of deficits that occur in both attention-deficit hyperactivity disorder and schizophrenia. This work indicates the utility of modeling symptoms, knowledge of which can lead to greater understanding of a number of different disorders that overlap in their symptoms. The observation that many (if not most) psychiatric patients suffer coexisting disorders indicates that this may be the most appropriate tack to take, at least at present.

Mathew T. Martin-Iverson

Contents

THE HIPPOCAMPAL-LESION MODEL OF SCHIZOPHRENIA
Nestor A. Schmajuk and Mabel Tyberg

AN ANIMAL MODEL OF STIMULANT PSYCHOSES
Mathew T. Martin-Iverson

ANIMAL MODELS OF HALLUCINATIONS: CONTINUOUS STIMULANTS

Gaylord D. Ellison

ANIMAL MODELS FOR THE SYMPTOMS OF MANIA

Melvin Lyon

Contributors

GAYLORD D. ELLISON • *Department of Psychology, University of California, Los Angeles, CA*

W. FRANK EPLING • *Department of Psychology, University of Alberta, Edmonton, Alberta, Canada*

JORAM FELDON • *Department of Psychology, Tel Aviv University, Tel Aviv, Israel*

ARNOLD J. FRIEDHOFF • *Department of Psychiatry, New York University School of Medicine, New York, NY*

MELVIN LYON • *Department of Psychiatry/Behavioral Science, University of Arkansas for Medical Sciences, Little Rock, AR*

MATHEW T. MARTIN-IVERSON • *Department of Psychiatry, University of Alberta, Edmonton, Alberta, Canada*

W. DAVID PIERCE • *Department of Sociology, University of Alberta, Edmonton, Alberta, Canada*

HELEN ROSENGARTEN • *Department of Psychiatry, New York University School of Medicine, New York, NY*

NESTOR A. SCHMAJUK • *Department of Psychology, Northwestern University, Evanston, IL*

JACK W. SCHWEITZER • *Department of Psychiatry, New York University School of Medicine, New York, NY*

RONALD E. SEE • *Department of Psychology, University of California, Los Angeles, CA*

MABEL TYBERG • *Department of Psychology, Northwestern University, Evanston, IL*

INA WEINER • *Department of Psychology, Tel Aviv University, Tel Aviv, Israel*

PAUL WILLNER • *Department of Psychology, City of London Polytechnic, London, UK*

Methods for Assessing the Validity of Animal Models of Human Psychopathology

Paul Willner

1. Background Considerations

Animal models are used very widely to investigate or illuminate aspects of human psychopathology. However, the extent to which it is possible to extrapolate from animals to people, and, therefore, the value of information derived from an animal model, will depend to a large extent on the validity of the model. This chapter outlines some methods that may be used to assess the validity of an animal model of psychopathology. It must be emphasized that these methods are primarily conceptual: They concern ways of evaluating experimental data and weighing different sources of evidence. The methods described may be used prescriptively, to indicate where a data base stands in need of expansion, or to indicate critical experiments. However, the validation exercise is more usually applied to form a view of the adequacy of a model on the basis of existing data.

Animal models are tools for our use: They are not developed as part of a beauty contest, with a prize for the most convincing. If a model cannot readily be used, it is of little value, however elegant. Thus, the successful construction of a valid model should not be seen as an end in itself, but as a useful step in the investigation of a scientific problem. As such, validity can be assessed only in relation to the broader objectives of the research program.

From: *Neuromethods, Vol. 18: Animal Models in Psychiatry I*
Eds: A. Boulton, G. Baker, and M. Martin-Iverson ©1991 The Humana Press Inc.

The two major uses of animal models of psychopathology are as screening tests for the development of new treatments, and as simulations within which to study aspects of the disorder (*see* Willner, 1991). As described below, these two objectives place very different requirements on an animal model. This can mean that different models are appropriate for different purposes; alternatively, the same model may be used for both purposes, but different considerations apply when assessing its suitability. The important point is that the assessment of validity takes place within a practical and scientific context. As an example, consider the use of animal models in antidepressant research. Practical considerations dictate that a screening test should be completed in the shortest possible time, and a response to acute drug treatment is a highly desirable feature (provided that the test accurately predicts clinical efficacy). However, if we are interested in using a model to study the physiological mechanisms of the clinical action of antidepressants, a valid model must involve chronic drug treatment; furthermore, this should be administered within a context of abnormal behavior, rather than prophylactically (*see,* e.g., Muscat et al., 1990; Sampson et al., 1990). These considerations of chronicity are larely irrelevant to the validation of a screening test; in the screening context, these features are highly undesirable on practical grounds.

In addition to a need for clarity as to the scientific objectives, it should also be recognized that conclusions arising from the use of a model are essentially hypotheses, which must eventually be tested against the clinical state. An assessment of the validity of a simulation gives no more than an indication of the degree of confidence that we can place in the hypotheses arising from its use. However, the fact that a simulation appears to be valid carries no guarantee that such predictions will be fulfilled; conversely, predictions derived from a manifestly invalid model may prove successful. The validity of a model is a matter of judgment, rather than measurement. The assessment of validity is therefore an interim and ongoing activity.

Against this background, there are a number of yardsticks on which a judgment may be based. Validating an animal model of psychopathology is, in principle, no different from validating any other psychological device, such as a psychometric test (Vernon, 1963) or a psychiatric diagnosis (Carroll, 1989), and the same generic approaches to validation are applicable:

1. *Predictive validity* means that performance in the test predicts performance in the condition being modeled;
2. *Face validity* means that there are phenomenological similarities between the two; and
3. *Construct validity* means that the model has a sound theoretical rationale (Willner, 1984a,1986,1991).

These three perspectives address three broad and different aspects of a model, from which a picture of its overall validity may be built.

Earlier attempts to develop criteria for validating animal models of human behavior have tended to concentrate largely on the assessment of face validity (Abramson and Seligman, 1977; McKinney and Bunney, 1969). The identification of two further categories reflects two ways in which the literature has developed in recent years. First, there has been a considerable expansion in the literature dealing with the pharmacological exploitation of animal models, much of which contributes to the assessment of predictive validity. Second, there has been significant growth in our understanding of the psychological mechanisms underlying psychopathological states, and examination of construct validity provides a convenient way of bringing animal models into contact with this very relevant literature. The exercise of distinguishing different types of validity has practical value: It allows ready identification of areas in which information about a particular model is weak or missing, and ensures that comparisons between different models are made on the basis of comparable data.

2. Assessment of Predictive Validity

The concept of predictive validity implies that manipulations known to influence the pathological state should have similar effects in the model: Manipulations known to precipitate or exacerbate the disorder should precipitate or exacerbate the abnormalties displayed in the animal model, and manipulations known to relieve the disorder should normalize behavior in the model. In principle, therefore, questions of predictive validity can address a number of features of simulations, including their etiology and physiological basis. In practice, the predictive validity of the animal models relevant to psychiatry is determined largely by their response to therapeutic drugs.

The assessment of predictive validity is based primarily on considerations of sensitivity and specificity. The detailed application of these criteria differs according to whether a model is being used as a screening test or as a simulation.

2.1. Pharmacological Sensitivity

Sensitivity to therapeutic drugs is the essential criterion for a screening test. Drugs known to be effective in the disorder should also be active in the model, and these effects should be achieved at reasonable doses that produce tissue concentrations of the drug comparable to those found clinically.

In assessing a test, attention should be paid not only to the number of effective agents shown to act in the model, but also to their chemical heterogeneity. This principle is most strikingly illustrated by the failure of many traditional animal models of anxiety to respond to the new generation of putative anxiolytics that act through serotonergic mechanisms (*see* Stephens and Andrews, 1991). The shock waves generated by this upset derive at least in part from a false sense of security in the traditional models, based on their response to over twenty traditional anxiolytics—all of which act through the GABA–BZ receptor complex, and the majority of which are closely related chemically. The problem of serotonergic anxiolytics has now been addressed in relation to the validity of animal models of anxiety; nevertheless, there is still considerable resistance to accepting that, in many types of anxiety, antidepressants are the drugs of choice (Lader, 1991).

This example of serotonergic anxiolytics illustrates one way in which the operation of validation criteria differs between the assessment of screening tests and simulations. For example, the failure of the Geller-Seifter conflict test to detect the prototypical serotonergic anxiolytic, buspirone, clearly detracts from the validity of this model as a screening test. However, in the clinic, buspirone acts more slowly than benzodiazepines (Tyrer et al., 1985). It is possible that buspirone would show anxiolytic activity in animal conflict tests after chronic administration. Such a similarity between the model and the clinical condition in the spectrum of pharmacological responsiveness would increase its validity as a simulation; however, the model would still be invalid (on the sensitivity criterion) as a screening test, unless the protocol were changed to one using chronic drug administration.

Although a screening test is required only to identify drugs that improve the clinical condition, a simulation should also respond to treatments that make the clinical condition worse. It should be possible to show, for example, that the behavioral abnormalities apparent in a simulation of schizophrenia are increased by acute administration of amphetamine or dopa, drugs that exacerbate schizophrenic symptoms (Angrist et al., 1973); this demonstration would have no relevance in a screening test for antischizophrenic drugs. A second divergence is that a simulation should be responsive to all the classes of drugs that are useful in treating the disorder, whereas a screening test might embody a strategic decision to look for drugs with particular chemical properties; this would then be recognized as an explicit limitation of the test. At the present time, for example, despite the possibility of serotonergic anxiolytics, it still makes good strategic sense to search for anxiolytic agents among the benzodiazepines and related compounds, and to design screening tests to minimize their undesirable side effects. However, as noted above, a simulation of anxiety should respond not only to benzodiazepines, but also to the various serotonergic anxiolytics, and perhaps also to beta blockers (*see* Lader, 1991).

In addition to pharmacological effects, a host of other factors may be brought into consideration. Their relevance will vary according to what is known about the disorder, but potential sources of information include nonpharmacological treatments (e.g., electroconvulsive shock or sleep therapies for depression, behavioral therapies for phobic anxiety), effects of brain damage, endocrinological influences, sex differences, and genetic contributions.

2.1.1. False Negatives

The failure to respond to a clinically effective agent should in principle detract unequivocally from the predictive validity of a model. However, before drawing this conclusion, it should be considered whether there might be some simple explanation. For example, false negatives can arise from species differences in drug kinetics. In particular, species differences in the rate or the route of drug metabolism can have profound effects on the concentration of the drug reaching its site of action; discrepancies arising in this way are especially likely in the case of a drug that has active metabolites. Returning to the example of buspirone, this agent is ineffective in conflict tests for anxiolytic activity in rats, but, in pigeons, buspirone shows anxiolytic activity comparable to chlordiazepoxide (Barrett et al., 1986); the reason for this species difference has not yet been established.

As noted above, another source of variability is the regime of drug administration. Clinical trials almost always assess the efficacy of drugs after a period of chronic treatment, and experiments carried out using acute drug administration may not accurately reproduce the appropriate conditions. It is well recognized that depression does not respond to acute treatment with antidepressants, and it is therefore routine for simulations of depression to use chronic treatment (Willner, 1989a); the same requirement for chronic neuroleptic administration in the treatment of schizophrenia has been less well assimilated in the field of animal models (*see* Cookson, 1991).

It is also important to consider that a "false negative" could reflect the fact that a drug that is considered to be clinically effective actually is not. A false impression of clinical effectiveness can easily arise in early, uncontrolled clinical trials. What is less frequently recognized is that this impression can continue through into controlled studies, if they compare the new drug against a reference drug that is known to be active. In such a trial, it is possible that the performance of the new agent might be worse than that of the reference drug, but for a variety of reasons, the difference might not be statistically significant. The new drug is then said to be "as good as drug X." However, if a placebo condition had been included, we might have observed that, in addition to being "no worse than drug X," the new drug was also "no better than no drug at all." This seems to be what happened in the case of the antidepressant iprindole, which has caused havoc by challenging antidepressant researchers to explain how it works while it actually appears not to be clinically effective in properly designed, placebo-controlled tests (Zis and Goodwin, 1979).

One further point to bear in mind is that false negatives raise the interesting possibility of clinical subgroups, distinguishable by their differential drug response. Unlike other antidepressants, for example, monoamine oxidase inhibitors fail to normalize behavior in olfactory-bulbectomized rats; some workers have suggested that this may be related to the fact that these drugs are clinically useful only in an atypical group of patients (Jesberger and Richardson, 1985).

2.2. Correlation of Potencies

In some circumstances, it may be possible to demonstrate that the relative potencies of different agents in a model correlate positively with their potencies in clinical use. This is potentially a very powerful test: Witness, for example, the almost perfect correlations among the clinical potencies of neuroleptic drugs in schizophrenia, their abilities to inhibit apomorphine-induced stereotyped behavior, and their affinities for dopamine D2 receptors (Creese and Snyder, 1978). However, this test should be applied cautiously, having regard to species differences in drug absorption and metabolism. Furthermore, it can be applied only if the chosen drugs vary widely in their clinical potencies. This condition is satisfied for the neuroleptics, the clinical potencies of which vary over four orders of magnitude. It is much less satisfactory for antidepressants, for example, most of which are used clinically within a very narrow dosage range; nevertheless, even in this case, there are sufficient "outliers" to allow a correlation of potencies to provide useful information (Willner, 1984b). It is also important to note that the use of this technique should, ideally, sample a range of chemically distinct compounds. The positive correlation between the clinical potency of benzodiazepines and their performance in several animal models of anxiety serves only to confirm that these drugs act at the same receptor.

2.3. Pharmacological Specificity

In addition to testing for false negatives, by investigating the effects of a wide range of drugs (and other manipulations) known to influence the clinical state, the specificity of a model should also be investigated; that is, its ability to respond appropriately to manipulations known to have no influence on the clinical condition. In assessing the validity of a model as a simulation of the disorder, false positives and false negatives carry equal weight. However, in a screening test, these two types of error are not of equal importance. If a screening test accepts an ineffective compound (false positive), the error will eventually come to light in further testing and no permanent damage will have been done. However, if the test wrongly rejects an effective compound (false negative), a potentially beneficial drug will be lost irretrievably.

Ideally, specificity should be tested by examining two potential sources of error. The more obvious test is to examine the response to agents from chemical classes unrelated to the known therapeutic agents, for example, psychomotor stimulants in animal models of depression, or neuroleptics in animal models of anxiety. Another strategy, which is rarely employed but is potentially powerful, is to examine the response to ineffective agents from the same chemical class as the effective drugs. The demonstration of specificity is particularly compelling if a test fails to respond to an agent that was developed as far as the clinic and rejected after formal clinical trials.

2.3.1. False Positives

As in the case of false negatives, before drawing conclusions from a false positive response, consideration sould be given as to whether it might have arisen for valid reasons. Questions of possible species differences in drug kinetics should be considered, as should the possibility that tolerance to a false positive response may develop after chronic drug treatment. For example, the Porsolt forced-swim test (an antidepressant-screening procedure) shows a false positive response to antihistamines after acute treatment, but not after chronic treatment (Kitada et al., 1981). It is also important to consider whether a drug that appears to be clinically ineffective might have revealed a positive effect if higher doses had been used. In practice, clinical trials at an adequate dose might prove impossible to carry out, owing to the emergence of unacceptable side effects, which in an animal-model simulation would be far less of a deterrent. It is possible, for example, that the very poor performance of cholinergic agonists in dementia, in contrast to their positive effects in many animal models (*see* Dunnett and Barth, 1991), reflects a difficulty in achieving an appropriate dose level in the clinical studies (Summers et al., 1986).

As in the case of false negatives, it should be borne in mind that a failure to identify inactive agents correctly may not necessarily be the fault of the model: It is always possible that the clinical effects of a drug have been incorrectly classified. Many animal models of depression, for example, admit anticholinergic drugs as "false positives" (*see* Willner, 1984a,1989a). But are these responses really false? Anticholinergics have never been subjected to properly controlled clinical trials (and, almost certainly, never will be), but there are numerous reports from the 1940s and 1950s that they have antidepressant activity (Janowsky and Risch, 1984), and there is little evidence that antidepressants are more efficaceous in depression than atropine (Thomson, 1982). The general principle here is that the validity of a model is absolutely limited by the quality of the information available to describe the condition modeled (*see also* Carroll, 1989).

3. Assessment of Face Validity

Face validity refers to a phenomenological similarity between the model and the disorder modeled: On one hand, the model should resemble the disorder; on the other, there should be no major dissimilarities. A modicum of face validity is essential in a model that is to be used experimentally as a surrogate for the disorder. However, it should be apparent that face validity is not, in principle, a requirement in a screening test: If a model succeeds in accurately predicting clinical efficacy, its face validity should be of little interest. In practice however, face validity is also desirable in a screening test. The main reason for this is that the assessment of predictive validity is so inexact, and the time-scale for testing clinical predictions so extended, that positive judgments of predictive validity must always be tentative. This being so, the decision to rely on a particular screening test needs collateral support, and face validity is one obvious area from which this may derive.

In a seminal paper that endowed the topic of modeling psychopathology in animals with a degree of scientific respectability, McKinney and Bunney (1969) suggested that to be valid, a model should resemble the condition it models in four respects: etiology, symptomatology, treatment, and physiological basis. Though admirable in principle, this requirement is too stringent. In practice, we cannot require a simulation to correspond in every particular to the disorder it simulates, for the very good reason that there are major gaps in our understanding of the disorders themselves. It is not reasonable, for example, to require similarity of etiology in a simulation of schizophrenia, when the etiology of schizophrenia is virtually a closed book. Indeed, the main objective in setting up a simulation is precisely to fill the missing pages.

We can, however, reasonably ask that a simulation be compared with the disorder it models in those areas in which we do understand something of the disorder. A corollary to this position is that in areas in which more is known about the disorder, we should examine the model more carefully for its degree of phenomenological similarity. A second corollary is that, as the understanding of a disorder develops, criteria for evaluating the face validity of an animal model will automatically change. When facts have been established about the etiology of a disorder or its physiological basis, they should obviously be included in an assessment of face validity. In general, however, the etiology and the physiological basis of psychiatric disorders are poorly understood; assessment of face validity will usually be based primarily on symptomatology, about which most is known, and, to some extent, on treatment.

3.1. Phenomenology

The central activity in an assessment of race validity is likely to be a comparison of symptoms displayed in the disorder with behavioral and other abnormalities apparent in the model. Since most disorders are syndromes that include a number of different symptoms, it should usually be possible to make multiple comparisons between the disorder and the model; the symptom-checklist approach adopted in the DSM-III system of diagnosis provides a useful starting point for this exercise (American Psychiatric Association, 1987).

In practice, models tend to focus on a single behavior, making the significant question that of its centrality in the disorder; again, the DSM-III distinction between essential symptoms and "optional extras" may be helpful. However, the question of centrality cannot fully be answered at the level of phenomenology. Even though a reduction in food intake is the essential feature of anorexia nervosa, it remains a moot point whether anorexia is properly described as a disorder of feeding or a disorder of personality (*see* Coscina and Garfinkel, 1991); if the latter, then simulations based on a reduction in food intake may be missing the target (*see* Montgomery, 1991). A related question is the specificity of symptoms for a particular disorder (i.e., their value in differential diagnosis); clearly, a model will be less valid if the symptoms it simulates are common to a variety of disorders (Abramson and Seligman, 1977).

The question of the feasibility of extrapolation from species to species lies dormant at the heart of all animal models. This problem does not intrude into the assessment of predictive validity, but it positively erupts in a consideration of face validity. If we are dealing with, for example, an endocrinological disorder (or an endocrinological correlate or symptom of a psychological disorder), then there is little difficulty in evaluating whether a symptom is present in both the disorder and the model. However, the difficulties of making this evaluation in relation to psychological symptoms are immense. There is no sure way out of this quandary. However, it helps considerably if the behavior evinced in the model is at a level of complexity that permits investigation of the underlying behavioral mechanisms. Such investigations are not readily undertaken in models that rely exclusively on a change in locomotor behavior. Many animal models of depression, for example, are based on a decrease in locomotor activity (*see* Willner, 1984a,1989a). It is certainly possible that a decrease in locomotor activity might simulate psychomotor retardation or even loss of motivation, but these remain unsupportable analogies. If the intention is to model a loss of motivation, then this is achieved far better in a paradigm that actually involves motivated behavior; the question of whether a decrease in behavior results from motivational or motor causes can thus be asked directly. Some models of depression, such as learned helplessness (*see* Willner, 1986,1989a) or the reduction of rewarded behavior by chronic mild stress (*see* Willner et al., 1991), have generated research aimed at answering this type of question.

As a general rule, the less sophisticated the behavior displayed in the model (in the sense that its interpretation is less open to experimental investigation and analysis), the lower the possibility of making a judgment of face validity. Specious claims for face validity are frequently advanced on the basis of unsupportable interpretations of behaviorally unsophisticated models.

Even more basic than the interpretation of behavioral changes is the language used to describe them. The language of behavioral change requires careful scrutiny to ensure that like is compared with like in different species. Animals subjected to uncontrollable electric shocks are undoubtedly stressed, but if severe electric shock is the only stressor that can be shown to generate a particular set of behavioral or neurochemical changes, then the model has little relevance to the concept of stress as used in human contexts. On similar lines, the fact that the term amphetamine anorexia is used to describe the reduction of food intake that follows amphetamine administration to animals does not necessarily mean that this paradigm can serve as a model of anorexia nervosa. In fact, the two behaviors are clearly different: Amphetamine anorexia is accompanied by an increase in the rate of food ingestion, and this phenomenological dissimilarity to anorexia nervosa is sufficient to invalidate the model (*see* Montgomery, 1991).

As Abramson and Seligman (1977) wisely observed, similarity between a model and a disorder must be demonstrated, rather than assumed. The demonstration of similarity between model and disorder can proceed only on the basis of a thorough conceptual and experimental analysis of both.

3.2. Coherence

When a number of points of similarity may be adduced between a model and the disorder it simulates, it becomes necessary to ask whether the identified cluster of symptoms forms a coherent grouping that might realistically be seen in a single patient, or whether are they drawn from a variety of diagnostic subgroupings (Abramson and Seligman, 1977). However, when asking this question, it is important not to rely too heavily on the diagnostic categories defined in DSM-III and related systems, since these categories were derived consensually, rather than empirically, and themselves stand in need of validation (*see*, e.g., Carroll, 1983,1989). Simulations offer a way of relating clusters of symptoms to specific etiological factors, specific treatment modalities, and specific physiological abnormalities. It remains an optimistic possibility that, through this process, valid simulations of psychiatric disorders might themselves contribute to the clearer definition of diagnostic boundaries.

3.3. Treatment Considerations

In addition to the issues raised earlier under the heading of "predictive validity," some aspects of treatment are relevant to the assessment of face validity. The most important of these questions concerns the treatment regime. Most prescriptions of psychotropic drugs envisage that the drug will be taken regularly over an extended period of weeks or months; consequently, the drugs must continue to exert their effects after chronic treatment. It is therefore essential to establish that drug effects in the model are also demonstrable after a period of chronic administration; tolerance must not develop to the point at which the effects disappear. It is also necessary to establish that therapeutic effects are actually present while the drug is being administered: It is often expedient, following chronic drug treatment, to precede testing by a period of drug "wash-out;" in these circumstances, confusion between drug effects and withdrawal effects can easily arise (*see,* e.g., Noreika et al., 1981; Towell et al., 1986).

A related issue is less straightforward. In clinical use, not only must drugs continue to act after chronic treatment, but also their onset of action is frequently delayed. This act has been used to argue against the validity of models in which therapeutic effects can be demonstrated using acute drug treatment. The problem with accepting this argument is that delayed onsets of clinical action are poorly understood, and may in part reflect a delay in delivering an adequate concentration of drug to the brain (*see* Willner, 1989b). As with so many of the guidelines for validation animal models, an acute drug effect in a model of a disorder that appears to require chronic treatment may or may not indicate a lack of face validity.

4. Assessment of Construct Validity

There is one important natural limitation to the attempt to establish face validity by mapping a point-to-point correspondence between a disorder and an animal model: There is no good reason to suppose that a given condition will manifest itself in identical ways in different species (Hinde, 1976); in fact, there are many obvious instances of divergence. Patterns of maternal behavior vary widely across species, for example, as do behavioral responses to psychotropic drugs: For example, rearing on the hind legs is a prominent component of stimulant-induced stereotyped behavior in rats, but not in primates, whereas the reverse is true of scratching (Randrup and Munkvad, 1970). The physical topography of these behaviors is quite different; nevertheless, we are able to say that they are homologous across species. In the case of stimulant-induced rearing and scratching, our judgment of homology arises from an understanding that the two behaviors have the same physiological substrate. In the case of a human mother cradling her baby and a female rat retrieving her pups, the judgment of homology arises primarily from an understanding of the psychosocial environment in which the behaviors take place. In both cases, the decision that different behaviors in different species reflect different manifestations of a similar underlying process is based on a theoretical rationale that derives from consideration of factors other than the behaviors themselves.

The theoretical rationale behind a model requires evaluation, and a satisfactory outcome endows the model with construct validity. In advancing this concept it is implicitly assumed that it is possible to construct theories of psychopathology that have some application to species other than human. Since this has in the past been a source of some confusion (exemplified by the extreme difficulty of defining what might or might not be meant by "a schizophrenic rat"), it may be helpful to consider briefly the general shape of such theories.

I have argued elsewhere (Willner, 1984c,1985) that it is inadequate for a biological theory of psychopathology simply to describe a link between a set of biochemical events (e.g., antagonism of benzodiazepine receptors) and a change in human experience (relief of anxiety). An adequate psychopharmacological theory must also interpose an account at two intermediate levels. At the first, it must explain the consequences of the identified biochemical changes for functional activity within the brain (which pathways are affected, and how); at the second, it must explain how these changes in functional activity affect the brain's ability to process information (which requires some understanding of the role of the affected systems in the cognitive activity of the normal brain). It then becomes necessary to consider, in addition, the manner in which those underlying cognitive changes are incorporated into subjective experience.

Having distinguished between these various levels of analysis, we are now in a position to consider which aspects of a theory of psychopathology may be addressed in animals, and which may not. For most practical purposes, we do not possess experimental tools that will allow us to address queries to animals about their subjective state; the questions we can ask them are strictly limited by the fact that their replies must be behavioral rather than verbal. Even in the case of a drug-discrimination procedure, the data from human experiments caution against equating internal states with subjective states (Chait et al., 1986).

We do not, therefore, attempt to simulate the experiential aspects of drug action or of psychopathology. What we are attempting in a simulation is to model the constraints on information processing and behavior that underly the experiential changes: the cognitive foundations of psychopathology (*see* Brewin, 1988), rather than its overt experiential manifestations.

The assessment of construct validity therefore involves not only an evaluation of the theoretical status of the model, but also an evaluation of the theoretical status of the condition modeled, which must involve an adequate conceptualization of the disorder using concepts appropriate to the context of animal models. From this perspective, the assessment of construct validity is a three-stage process (Willner, 1986). The first step is to identify the behavioral variable that is being modeled. The second is to assess the degree of homology between the identified variable and behavior in the simulation. The third is to assess the significance of the identified variable in the clinical picture.

Each of these states is fraught with difficulty. It might, for example, seem a straightforward matter to decide what is being modeled, and this may indeed be so in those cases in which the behavioral objectives of the model are specified in advance. Many models of dementia, for example, embody an explicit choice to simulate specific disorders of learning and memory (*see* Dunnett and Barth, 1991). However, there are many other starting points for a simulation, such as the application of a supposed etiological factor or the use of drugs to induce a supposedly relevant physiological state; the behavioral state simulated may then be shrouded in obscurity. It remains a total mystery what aspect of depression was being addressed by experiments in which monkeys were isolated in the dark, in a vertical cylinder called "the well of despair" (Harlow and Suomi, 1971).

Having established what is being modeled, the next step is to assess whether the two behaviors are homologous. However, this can be done only if both behaviors are well understood, and frequently they are not. The learned-helplessness model of depression provides an excellent example of this problem. In this model, prior exposure to uncontrollable aversive events causes a subsequent learning deficit; this effect may be demonstrated very readily in both rats and people (Miller et al., 1977). However, despite a very substantial research effort, the nature of the underlying mechanisms remains unclear. The explanation originally proposed was that subjects exposed to uncontrollable events learn that events are uncontrollable—a homologous process in rats and people (Seligman, 1975). However, for rats there is evidence that motor disablilties and stress-induced analgesia contribute to the behavioral deficits (Jackson et al., 1978,1979), whereas in people helplessness effects are sensitive to very minor changes in the experimental procedure (Buchwald et al., 1978). Returning to the original examples of maternal behavior and drug-induced stereotypies, we are able to see that for two behaviors to be homologous, they should share a similar physiological basis and occur in a similar behavioral context. In the case of learned helplessness, the growth of information about the two behaviors has, if anything, reduced the likelihood of homology (Willner, 1986).

The final step in the assessment of construct validity is an evaluation of the significance in the overall clinical picture of the behavior modeled by the simulation. Many animal models of drug dependence, for example, focus on the rewarding properties of drugs. However, although rewarding effects are of undoubted importance in the early stages of drug use, some theoretical formulations emhasize the fear of withdrawal as a major factor in dependence (*see* Goudie, 1991; Hartnoll, 1991). If this is correct, then models that simulate rewarding effects may be of only limited relevance. A similar problem concerning the significance of low food intake in anorexia nervosa has already been remarked: The anorectic's low food intake should not be mistaken for a reduced appetite for food (*see* Montgomery, 1991). Because debates of this kind take place entirely within the human literature, these problems do not figure prominently in discussions of animal models. However, it is obviously of some considerable importance to know whether a simulation is asking the correct question.

Of the three roads to validity, construct validity is the most fundamental: Although a model may, for a variety of reasons, fail to meet criterial or predictive or face validity and still survive, it would be difficult to retain confidence in a model whose theoretical rationale had been exploded. Construct validity is the most difficult aspect of a model to establish, but also the most challenging.

5. Theory into Practice

The methods described provide a framework within which to carry out a broad assessment of an animal model from three rather different perspectives. The weighting accorded to each area of validity will remain a matter of individual judgment and preference; unfortunately, there are no methods for combining the three relatively independent blocks of information, which makes the attempt to produce "best-buy" lists rather unrewarding. However, it is important to be aware that disputes between the protagonists of rival models often arise because different sources of evidence are being weighted differently. This situation will probably continue until the question of the validity of animal models is addressed more systematically and the appropriate studies are carried out to evaluate a wide range of models on all three sets of criteria.

It must be emphasized that, at present, few models, if any, in any area of psychopathology perform well on all three sets of criteria. To some extent, the shortfall reflects an incomplete set of clinical data. Some important disorders—for example, the dementias—respond poorly, if at all, to pharmacotherapy; inevitably, therefore, the predictive validity of models in these areas is poorly characterized. In other areas, predictive validity is difficult to determine because of the sheer range of unproven experimental therapies currently undergoing clinical trials. This excess of riches makes it difficult, at present, to determine the predictive validity of animal models of depression: The problem that the development of putative serotonergic anxlolytics brings to models of anxiety has already been mentioned. The poverty of the clinical data is particularly apparent when we attempt to assess construct validity, for this requires a good theoretical understanding of the disorder, and well-grounded theories of psychopathology are few and far between.

In other instances, however, the "validity gap" reflects a lack of awareness of the importance of understanding the subject of the model. A deep familiarity with the disorder and with current experimental approaches to it (particularly those of cognitive psychologists) is almost a prerequisite for the construction of a model possessing face and construct validity. The historical accident through which the modeling of psychopathology in animals has developed in close contact with, and often as an integral part of, psychopharmacology has led to a situation in which the pharmacological sophistication of the field cannot be faulted. However, with notable exceptions, the psychological basis of most models is unacceptably simplistic.

This rather sweeping criticism of the field at large must be tempered by the recognition that practical considerations are paramount. The overriding question is whether a particular model is well-suited to address a specific scientific problem. If not, it is of little value (for that purpose), no matter how elegant or how valid. The considerations that can stand in the way of a model being adopted range from the scientific (e.g., wrong time-course) through the ethical (e.g., levels of stress unjustifiable in relation to the potential outcome) to the mundane (e.g., animal-husbandry problems). The practical constraints on the choice of an experimental tool, combined with the limited validity of the tools available to choose from, provides a working environment that is far from ideal. In these circumstances, it is particularly important to ensure, whenever possible, that the work carried out constitutes good science that can be justified in its own right, without reference to the status of the paradigm used as an animal model of psychopathology.

In effect, an animal model constitutes a theory of some aspect(s) of the psychopathology it models. Its purpose, therefore, is to generate predictions that can be tested against the clinical condition. Assessment of the validity of the model provides us with an indication of the degree of confidence that we can place in those predictions. However, a prediction derived from an invalid model (and therefore advanced somewhat tentatively) may prove successful, whereas a prediction derived from a model judged to be more valid (and therefore advanced wlth some confidence) may fail to be upheld. The development of scientific knowledge is an iterative process: Models are used to expand our understanding of the disorder, and the results feed back to refine the validity of the models. Ultimately, we may reach a situation in which the model is fully validated; we may then find that the model is no longer needed because the disorder is perfectly understood.

References

Abramson L. Y. and Seligman M. E. P. (1977) Modelling psychopathology in the laboratory. History and rationale, in *Psychopathology: Animal Models* (Maser J. D. and Seligman M. E. P., eds.), Freeman, San Francisco, pp. 1–26.

American Psychiatric Association (1987) Diagnostic Criteria from DSM-III-R. APA, Washington, DC.

Angrist B., Sathananthan G. S., and Gershon S. (1973) Behavioral effects of L-dopa in schizophrenic patients. *Psychopharmacologia* **31,** 1–12.

Barrett J. E., Witkin J. M., Mansbach R. S., Skolnick P., and Weissman B. A. (1986) Behavioral studies with anxiolytic drugs. III. Antipunishment actions of buspirone do not involve benzodiazepine receptor mechanisms. *J. Pharmacol. Exp. Ther.* **238,** 1009–1013.

Brewin C. R. (1988) *Cognitive Foundations of Clinical Psychology.* Erlbaum, Hillsdale, NJ.

Buchwald A. M., Coyne J. C., and Cole C. S. (1978) A critical evaluation of the learned helplessness model of depression. *J. Abnorm. Psychol.* **87,** 180–193.

Carroll B. J. (1983) Neurobiologic dimensions of depression and mania, in *The Origins of Depression: Current Concepts and Approaches* (Angst J., ed.), Springer, Berlin, pp. 163–186.

Carroll B. J. (1989) Diagnostic validity and laboratory studies: The rules of the game, in *The Validity of Psychiatric Diagnosis* (Robins L. N. and Barrett J. E., eds.), Raven, New York, pp. 229–244.

Chait L. D., Uhlenluth E. H., and Johanson C. E. (1986) The discriminative stimulus and subjective effects of phenylpropanoamine, mazindol and *d*-amphetamine in humans. *Pharmacol. Biochem. Behav.* **24,** 1665–1672.

Cookson J. (1991) Animal models of schizophrenia and mania: Clinical perspectives, in *Behavioural Models in Psychopharmacology: Theoretical, Industrial and Clinical Perspectives* (Willner P., ed.), Cambridge University Press, Cambridge, pp. 331–356.

Coscina D. V. and Garfinkel P. E. (1991) Animal models of eating disorders: A clinical perspective, in *Behavioural Models in Psychopharmacology: Theoretical, Industrial and Clinical Perspectives* (Willner P., ed.), Cambridge University Press, Cambridge, pp. 237–250.

Creese I. and Snyder S. H. (1978) Behavioral and biochemical properties of the dopamine receptor, in *Psychopharmacology: A Generation of Progress* (Lipton M. A., DiMaschio A., and Killam, K. F., eds.), Raven, New York, pp. 377–388.

Dunnett S. B. and Barth T. M. (1991) Animal models of Alzheimer's disease and dementia (with an emphasis on cortical cholinergic systems), in *Behavioural Models in Psychopharmacology: Theoretical, Industrial and Clinical Perspectives* (Willner P. ed.), Cambridge University Press, Cambridge, pp. 359–418.

Goudie A. J. (1991) Animal models of drug abuse and dependence, in *Behavioural Models in Psychopharmacology: Theoretical, Industrial and Clinical Perspectives* (Willner P., ed.), Cambridge University Press, Cambridge, pp. 453–484.

Harlow H. F. and Suomi S. J. (1971) Production of depressive behaviors in young monkeys. *J. Autism Child Schiz.* **1,** 246–255.

Hartnoll R. (1991) The relevance of behavioral models of drug abuse and dependence liabilities to the understanding of drug misuse in humans, in *Behavioural Models in Psychopharmacology: Theoretical, Industrial and Clinical Perspectives* (Willner P., ed.), Cambridge University Press, Cambridge, pp. 503–519.

Hinde R. A. (1976) The uses of similarities and differences in comparative psychopathology, in *Animal Models in Human Psychobiology* (Serban G. and Kling A., eds.), Plenum, New York, pp. 187–220.

Jackson R. L., Maier S. F., and Coon D. J. (1979) Long-term analgesic effect of inescapable shock and learned helplessness. *Science* **206,** 91–93.

Jackson R. L., Maier S. F., and Rapoport, P. M. (1978) Exposure to inescapable shock produces both activity and associative deficits in rats. *Learn. Motiv.* **9,** 69–98.

Janowsky D. S. and Risch S. C. (1984) Cholinomimetic and anticholinergic drugs used to investigate an acetylcholine hypothesis of affective disorders and stress. *Drug Dev. Res.* **4,** 125–142.

Jesberger J. A. and Richardson J. S. (1985) Animal models of depression: Parallels and correlates to severe depression in humans. *Biol. Psychiatr.* **20,** 764–786.

Kitada Y., Miyauchi T., Satoh A., and Satoh S. (1981) Effects of antidepressants in the rat forced swim test. *Eur. J. Pharmacol.* **72,** 145–152.

Lader M. (1991) Animal models of anxiety: A clinical perspective, in *Behavioural Models in Psychopharmacology: Theoretical, Industrial and Clinical Perspectives* (Willner P., ed.), Cambridge University Press, Cambridge, pp. 76–88.

McKinney W. T. and Bunney W. E. (1969) Animal model of depression: Review of evidence and implications for research. *Arch. Gen. Psychiat.* **21,** 240–248.

Miller W. R., Rosellini R. A., and Seligman M. E. P. (1977) Learned helplessness and depression, in *Psychopathology: Animal Models* (Maser J. D. and Seligman M. E. P., eds.), Freeman, San Francisco, pp. 104–130.

Montgomery A. M. J. (1991) Animal models of eating disorders, in *Behavioural Models in Psychopharmacology: Theoretical, Industrial and Clinical Perspectives* (Willner P., ed.), Cambridge University Press, Cambridge, pp. 177–214.

Muscat R., Sampson D., and Willner P. (1990) Dopaminergic mechanism of imipramine action in an animal model of depression. *Biol. Psychiat.* **28,** 223–330

Noreika L., Pastor G., and Liebman J. (1981) Delayed emergence of antidepressant efficacy following withdrawal in olfactory bulbectomized rats. *Pharmacol. Biochem. Behav.* **15,** 393-398.

Randrup A. and Munkvad I. (1970) Biochemical and psychological investigations of stereotyped behavior, in *Amphetamines and Related Compounds* (Costa E. and Garattini S., eds.), Raven, New York, pp. 695–713.

Sampson D., Muscat R., and Willner P. (1990) Reversal of antidepressant action by dopamine antagonists in an animal model of depression. *Psychopharmacology* (in press).

Seligman M. E. P. (1975) *Helplessness: On Depression, Development and Death.* Freeman, San Francisco.

Stephens D. N. and Andrews J. S. (1991) Screening for anxiolytic drugs, in *Behavioural Models in Psychopharmacology: Theoretical, Industrial and Clinical Perspectives* (Willner P., ed.), Cambridge University Press, Cambridge, pp. 50–75.

Summers W. K., Majovski L. V., Marsh G. M., Tachiki K., and Kling V. (1986) Oral tetrahydroaminoacridine in long-term treatment of senile dementia. *New Engl. J. Med.* **315,** 1241–1245.

Thomson R. (1982) Side effects and placebo amplification. *Brit J. Psychiat.* **140,** 64–68.

Towell A., Willner P., and Muscat R. (1986) Behavioural evidence for autoreceptor subsensitivity in the mesolimbic dopamine system during withdrawal from antidepressant drugs. *Psychopharmacology* **90,** 64–71.

Tyrer P., Murphy S., and Owen R. T. (1985) The risk of pharmacological dependence with buspirone. *Br. J. Clin. Pract.* **39(Suppl. 38),** 91–93.

Vernon R. E. (1963) *Personality Assessment: A Critical Survey.* Methuen, London.

Willner P. (1984a) The validity of animal models of depression. *Psychopharmacology* **83,** 1–16.

Willner P. (1984b) The ability of antidepressant drugs to desensitize beta-adrenergic receptors is not correlated with their clinical potency. *J. Affect. Dis.* **7,** 53–58.

Willner P. (1984c) Drugs, biochemistry and subjective experience: Towards a theory of psychopharmacology. *Perspect. Biol. Med.* **28,** 49–64.

Willner P. (1985) *Depression: A Psychobiological Synthesis.* Wiley, New York.

Willner P. (1986) Animal models of human mental disorders: Learned helplessness as a paradigm case. *Prog. Neuropsychopharmacol. Biol. Psychiat.* **10,** 677–690.

Willner P. (1989a) Animal models of depression: An overview. *Pharmacol. Ther.* **45,** 425–455.

Willner P. (1989b) Sensitization to the actions of antidepressant drugs, in *Psychoactive Drugs: Tolerance and Sensitization* (Emmett-Oglesby, M. W. and Goudie A. J., eds.), Humana, Clifton, NJ, pp. 407–459.

Willner P. (1991) Behavioural models in psychopharmacology, in *Behavioural Models in Psychopharmacology: Theoretical, Industrial and Clinical Perspectives* (Willner P., ed.), Cambridge University Press, Cambridge, UK, pp. 3–18.

Willner P., Muscat R., Papp M., and Sampson D. (1991) Dopamine, depression and antidepressant drugs, in *The Mesolimbic Dopamine System: From Motivation to Action* (Willner P. and Scheel-Kruger J., eds.), Wiley, Chichester, UK, pp. 387–410.

Zis A. P. and Goodwin F. K. (1979) Novel antidepressants and the biogenic amine hypothesis of depression. The case for iprindole and mianserin. *Arch. Gen. Psychiat.* **36,** 1097–1107.

Animal Models with Parallels to Schizophrenia

Melvin Lyon

1. Introduction

Schizophrenia is frequently regarded in psychiatry as a purely human disorder, with only vague parallels to certain forms of animal behavior. Although monkeys, cats, and dogs, and even rats, may be described as having certain psychotic symptoms, the existence of full-blown schizophrenia has been difficult to model in animals. It is the purpose of this chapter to review critically some of the methods used to induce in animals symptoms thought to be highly similar to those of schizophrenia. An attempt will also be made to indicate promising areas for future research.

Since the dopamine (DA) overactivation hypothesis has been the leading theoretical view in schizophrenia research (Carlsson, 1988), much of what follows will consider how well this hypothesis holds. An important characteristic of any model must be that the symptoms produced are ameliorated by neuroleptic drugs, which, in the main, are antidopaminergic in function.

Clearly, there are obvious problems with the acceptance of a simple dopaminergic overactivation theory, in view of conflicting findings regarding DA-receptor increases in the brains of schizophrenic patients (Wong et al., 1986; Farde et al., 1990). As a result of these disagreements, and also because of the significant modulating effects on DA systems caused by other neurotransmitters and neuromodulators, it is necessary to consider multiple causative factors in the neurochemistry of schizophrenia.

From: *Neuromethods, Vol. 18: Animal Models in Psychiatry I*
Eds: A. Boulton, G. Baker, and M. Martin-Iverson

In the following, the presently accepted definition and clinical manifestations of schizophrenic symptoms (American Psychiatric Association, *DSM-III*, 1980) will be compared with parallels from animal behavior. This will be followed by a review of extant models and more detailed descriptions of a few selected methods.

1.1. Definition of Schizophrenia

Part of the difficulty in describing animal behaviors as parallels to schizophrenia lies in the limitations of the *DSM-III* nomenclature. The basic emphasis in *DSM-III* is on the classical symptoms of thought disorder, delusions, hallucinations, and flat or inappropriate affect. These symptoms are more central to the Schneiderian emphasis in classification than to that of Bleuler (1950). The major point of difference, and one that is considered central to the present article (*see also* Lyon, 1990), is that the powerful evidence for increased response switching, followed by perseveration and stereotypy, is totally omitted in the *DSM-III* characterization of schizophrenia. Yet, as Bleuler pointed out and as experiments with both human patients and animals treated with dopaminergic agents demonstrate, switching and perseverative or stereotyped responding are increased as the illness or the drug effect progresses. Furthermore, the general symptomatology of many schizophrenic patients worsens under the influence of DA stimulation (Angrist, 1983).

The type of forced perseveration that Bleuler described is also exemplified in the extreme perseveration of the licking response in the rat under heavy doses (10 mg/kg) of *d*-amphetamine, as reported by Teitelbaum and Derks in 1958. In that study, drugged animals showed such a strong tendency to continue the conditioned licking response that they returned to the licking tube even after the session had ended and they had been placed outside the experimental box. The significance of the similarity of this type of stereotypy of responding in both humans and animals was first made clear by Randrup and Munkvad (1967) when they demonstrated that excitation of the DA systems in the brain was essential to the stereotyped behavior.

The fact that increased switching between behaviors will also be a necessary concomitant of the initial stages of amphetamine intoxication, hence should also be found in schizophrenia, was predicted from an animal model by Robbins and Watson (1981). Evidence that both the initially increased switching of response categories and the later stereotypy of single responses exist in schizophrenic populations has recently been seen in a series of human studies (Frith and Done, 1983; Lyon et al., 1986; Lyon and Gerlach, 1988). Hence, these behavioral changes, first observed in animal models, have now been directly correlated with the behavioral symptoms of schizophrenic patients. Because of the above evidence, the present review of animal methods for modeling aspects of schizophrenia includes both increased switching and stereotypy of responding as basic elements in the symptomatology to be investigated. This will become more obvious in the listing of symptoms and animal parallels in Table 1. The present question is whether changes in DA stimulation in the brain are the sole cause of these alterations.

Obviously, if the parallel between amphetamine dose level and increasing schizophrenia holds, one would expect that increased switching in behaviors would dominate the early symptomatic phases of schizophrenia, whereas stereotyped, single-response, perseveration would be more common in later stages. Antipsychotic medication, which is almost always antidopaminergic in its basic effects, would be expected to reduce the stereotypy or enhanced switching to more moderate levels. These expectations are also confirmed in the human studies mentioned above.

All of this is consistent, in the present view, with the fact that chronic schizophrenia leads to a different set of symptoms than does recent-onset schizophrenia. This difference is seen mostly in terms of the increasing dominance of certain major symptoms, such as thought disorder, delusions, hallucinations, and stereotyped forms of behavior and thought.

There can also be a subdivision of schizophrenic symptoms into "positive" and "negative" types, with "positive" referring to active, overtly recognizable symptoms, and "negative" refer-

Table 1
Suggested Parallels Between Animal Behaviors and the Symptoms of Schizophrenia

Human symptoms	Potential animal parallels
Early phase—positive symptoms; characterized by excessive switching.	
Digressive speech, overelaboration, repetitions, many puns	Inappropriate switching, increased repetition of selected responses with exclusion of others; excessively repeated vocalizations
Odd ideation, bizarre or unusual associations; inappropriate affect	Incorrect vocalizations, including danger, threat, or aggression signals (NB: such signals may exceed human frequency thresholds); inappropriate affect to social contact or provocation; inappropriate sexual or aggressive actions
Peculiar or bizarre behavior	Close contact with others despite being attacked or bitten; inappropriate, awkward, fragmented, or dissociated actions. Self-destructive behaviors. Head or paw shakes, ataxia, obstinate progression (festination), abortive grooming, postural imbalance, hyperextension and hyperflexion of limbs, akathisia
Unusual perceptual experiences	Concentrated examination of minute details, especially near objects, own paws and feet; hyperstartle to humanly indetectible stimuli; crouching, hiding, or fearful responses to innocuous stimuli or familiar individuals

(continued)

Table 1 (cont.)
Suggested Parallels Between Animal Behaviors and the Symptoms of Schizophrenia

Human symptoms	Potential animal parallels
Early phase—negative symptoms; characterized by excessive perseveration and stereotypy.	
Lack of personal hygiene or grooming	Lacks normal grooming, hair in disorder, does not wash or clean body; overly attentive to specific points on body, excluding the rest
Flat affect, lack of normal emotional responses	Staring into space, no emotional response to others
Social isolation (self-imposed)	Loss of bodily contact with others, failure to respond to others, poor eye contact, hiding, avoiding others; loss of social grooming
Late phase—positive symptoms; characterized by extreme shifting or stereotyped repetition.	
Delusional thinking	May relate to failures to observe social hierarchy, including those involving a normally dominant animal
Hallucinations	Behavior seemingly directed at invisible objects and having a sequential nature, with attention remaining concentrated on one spot during the activity; "fly-catching" with coordinated eye, tongue, and mouth movements (cats); attack, flight, or feeding behaviors associated with invisible stimuli (monkeys); excessive parasitotic-like grooming and biting at specific points on the skin (monkeys) (NB: hallucinatory behavior not readily measurable in rodents)

(continued)

Table 1 (cont.)
Suggested Parallels Between Animal Behaviors and the Symptoms of Schizophrenia

Human symptoms	Potential animal parallels
Late phase—negative symptoms; characterized by fragmentation and/or total lack of responding.	
Incoherence or extreme poverty of speech	Atypical, fragmented calls or noises; muteness
Flat affect	Flat, or absent response to social contact or provocation; staring and lack of eye contact; emotionally unresponsive; remains isolated and passive

[a]Modified from Lyon, M. (1990) Animal models of mania and schizophrenia, in *Behavioral Models in Psychopharmacology: Theoretical, Industrial and Clinical Perspectives.* (Willner, P., ed.) Cambridge University Press, Campbridge, UK, Table 11-2, pp. 262–263.

ring to symptoms of neglect or omission, such as poverty of speech, lack of personal hygiene, and flat affect. Since the positive/negative dimension is sometimes visible from the very beginning of the illness, it is possible that two subgroups, with different brain defects, are to be considered. However, it is equally possible that the major positive and negative symptom categories find their most explicit basis in the switching vs stereotypy changes in behavior that differentiate the earlier, more active phase of schizophrenia from the most stereotyped and behaviorally passive late, chronic stage of the illness. Those individuals showing negative symptoms from the start would be viewed as having more serious brain defects, which seems to be the case (Andreasen et al., 1982; Weinberger et al., 1979).

In any case, both the "recent-onset/chronic" and "positive/negative" symptom dimensions will also be considered in relation to certain animal models. Animal models provide parallels to all these types of symptoms, but, just as with the human cases themselves, the negative symptoms are more difficult to assess, since failure to act may have many reasons.

1.2. Clinical Signs of Schizophrenia and Animal Parallels

On the above basis, Table 1 compares human symptoms of schizophrenia with the animal behaviors that are thought to parallel these symptoms. The Table is divided into early and late phases of schizophrenic symptomatology, since these tend to differ in patients with recent-onset schizophrenia compared with chronically ill, hospitalized patients. Table 1 also includes reference to the "positive" and "negative" aspects of symptoms. The Table is specifically intended to suggest *structural* elements of behavior that characterize the symptom categories, rather than to judge symptoms by their content alone. For example, perseveration in a certain behavior, whether it is motor or cognitive in function, is considered a prime characteristic of schizophrenic behavior. This is in accordance with the view of Bleuler (1950) that highly repetitious and stereotyped patterns are found in all types of schizophrenic activity, including both thought and action. This consideration is particularly important in animal models, in which the *content* of behaviors may differ, although their inherent *structure* remains the same. This adds a new, more inclusive dimension to the *DSM-III* analysis of schizophrenic behavior.

2. Types of Animal Models

In the following review of animal models, an arbitrary grouping has been imposed, mainly in order to aid identification of specific fields of interest. The subdivisions chosen are not mutually exclusive, but indicate the initial methodological impetus. These groupings are:

1. Pharmacologically based models;
2. Brain-lesion models;
3. Genetic and fetal-development models;
4. Behaviorally based models; and
5. Cerebral-asymmetry and laterality models.

Each grouping will be discussed with special attention to the methods used and the intended reference to parallels in human behavior.

2.1. Pharmacologically Based Models

2.1.1. DA-Related Models

Because of the importance of the DA-overactivation theory of schizophrenia, which led successfully to the development of the majority of neuroleptic agents, animal models involving changes in DA systems are the most common of all animal models of schizophrenia (Carlsson, 1988). In the development of behavioral tests for DA overstimulation, Randrup and Munkvad (1967) emphasized the significance of stereotyped behavior induced by amphetamines in both animals and humans. The perseverative and stereotyped nature of the behavior was shown to be related specifically to DA. The highly stereotyped motor behavior induced in the rat by dopaminergic agonists, such as amphetamine, became the cornerstone of behavioral tests for antipsychotic agents (Janssen, 1967a,b).

The theoretical basis for the relationship of amphetamine effects to behavioral changes in perseveration and stereotypy was clarified by Lyon and Robbins (1975), who showed that the end behaviors (stereotypies) could be predicted as the result of the increasingly rapid behavior elicited by the drug, which resulted in more and more frequent responses being made in fewer and fewer categories of behavior. This interpretation was further strengthened by the demonstration of Robbins and his colleagues (Robbins and Watson, 1981; Evenden and Robbins, 1983) that, under lower doses of amphetamine (1.0–5.0 mg/kg), all behavioral categories were stimulated, so that stereotyped response sequences were not obvious. The lower-dose drug effects induced, instead, an increased behavioral *switching*, which mimics that aspect of schizophrenic behavior that is commonly attributed to a lack of continued attention (attentional shift). In extreme cases, this would lead to the continual changing in behavior that was earlier characterized as "hebephrenia."

However, as one progresses within the 1.0–5.0-mg/kg dose range of *d*-amphetamine, some differentiation of behaviors does appear. At 2.0–3.0 mg/kg, animals begin to show abnormally increased frequency and duration of rearing on the hind legs, accompanied by less locomotion. These changes are strong enough to affect significantly even conditioned shock-avoidance behavior (Lyon and Randrup, 1972). At about 5.0 mg/kg, the rearing becomes less and less frequent, and the animals begin to orient toward the floor of the cage, with nasal contact, and sniffing and licking of the cage floor (especially in wire-net cages). At 10.0 mg/kg, locomotion is almost nonexistent, and the animal stays in one spot, licking and sniffing at the floor with only slight side-to-side motions of the head. Occasionally, "backward locomotion" is seen, and self-mutilation by excessive licking and biting at one spot on the body or tail. The behavior has become extremely stereotyped and localized.

Using such behavioral measures as these, antiamphetamine tests were a major tool in the discovery of most of the so-called "typical" neuroleptics among the phenothiazines and butyrophenones (Deniker, 1966; Janssen, 1967a,b). It was largely because of this type of research that animal models of psychosis began to be accepted as directly related to the human symptoms of schizophrenia.

Randrup et al. (1981) summarized the work relating DA overstimulation and its behavioral effects in animals to the symptoms of human psychosis. However, in the intervening time, it has become evident that, although DA is central to the behavioral effects, there are many modulating factors, including other neurotransmitters (Scheel-Kruger, 1986), the localization of the stimulant effect (Stevens, 1975), and the action of neuromodulators, such as certain peptide chains (de Wied, 1979). Nevertheless, the dopaminergic theory remains the focal point of nearly all extant models of schizophrenia (Carlsson, 1988), and this will be reflected in the following review, which often refers back to the DA-system relevance, and sensitivity to DA receptor-blockers as the benchmarks for antischizophrenic effects.

2.1.1.1. Single- and Repeated-Injection Methods Using DA Agonists

These methods are perhaps best characterized by the extensive studies of Ellinwood and his colleagues (Ellinwood et al., 1972; Ellinwood and Sudilovsky, 1973; Ellinwood and Kilbey, 1977). These studies include single-injection methods with varying doses of *d*- and methamphetamines and chronic daily-injection routines with amphetamines and cocaine. Most studies were done on cats and monkeys, and a very careful description of the behavioral results has been given in the articles mentioned.

Chronic amphetamine or cocaine intoxication in cats (Ellinwood and Kilbey, 1977) led to symptoms of excessive startle, abortive and fragmented behaviors, awkward postures, akathisia, "obstinate progression" (festination), "hallucinatory-like" behaviors, and "fearful hyperreactivity" (crouching with ears flattened, hair raised, and pupils enlarged). It is important to note that there is an increasing sensitization to the amphetamine effect over days, and that the given symptoms will not all appear on the first day of treatment.

All of these symptoms are improved by treatment with neuroleptic drugs, but improvement begins within an hour of injection with haloperidol, for example, whereas human symptoms of schizophrenia seem to require several weeks of neuroleptic treatment. The reasons for this are still not understood, but may involve learning as well as long-term drug effects.

2.1.1.2. Minipumps vs Implanted Capsules

There is some suggestion that the maintenance of a stable dose level over days during chronic treatment with stimulant drugs may alter their behavioral effect (Eison et al., 1983). Two basic methods are considered here. The first method uses the Alzet® minipump, which is a device that may be implanted under the skin and releases its contents over a period of 7 or 14 d by allowing osmotic diffusion across the pump's outer shell to force the drug contents out into the tissue. This system yields a relatively uniform pumping rate, with a drop of <25% from the initial peak level across the recommended duration of pump use.

The second method was developed by Huberman et al. (1977), initially for use in the rat. As used by Nielsen et al. (1983) in monkeys, it consisted of a Silastic® tube fitted with two inner, diffusion-resistant, polyethylene plastic tubes, which leave only a small gap (5 mm) at the center, through which the outer Silastic® tube has contact with the drug contents. The device is filled with an amphetamine base and sealed at both ends with a thick layer of silicone cement, through which relatively little diffusion is thought to occur. Depending on the size of the exposed Silastic®

surface, the amphetamine can diffuse into the surrounding tissue over a period of more than 2 wk, but by that time the amount released becomes erratic and is reduced by more than 60% from its original release rate.

Another form of this capsule was used for long-term release of the drug by Nielsen and Lyon (1982). In that experiment, the inner polyethylene tubes were replaced by a single diffusion-resistant polystyrene tube with only one (1-mm diameter) hole allowing diffusion to the outer Silastic® layer. This capsule continues to release amphetamine for more than 80 d, although the rate of release becomes minimal after about 60 d.

Eison et al. (1983) compared the capsule and minipump methods in rats and found that capsules produced both a more peaked release rate, as noted above, and a change in behavior that more nearly paralleled the expected course of symptoms in schizophrenia. Apparently the more stable release rate from the minipumps led to a spreading of the behavioral effects over a longer period, with less spectacular symptomatic changes. Nielsen (1981), using only minipumps, found that stereotyped behavior was rapidly lost over a period of days, even though the dose level was constant. Since this effect also occurs in monkeys implanted with capsules (Nielsen et al., 1983), the loss of the stereotypy is not a consequence of the peaking in release of amphetamine in the capsule-implanted animals. However, the loss of overt stereotypy in normal *cage* behaviors is not a reliable sign of the loss of this effect on behavior in *test* situations (Fischman and Schuster, 1974). Users of minipump and capsule methods should distinguish between test and home-cage situations in generalizing their test results.

2.1.1.3. Direct Injections into the Nucleus Accumbens (NAcc)

Eison et al. (1983) found the only real difference between minipumps and capsules in distribution of DA activity: Although both methods increased the mean DA levels within the nucleus accumbens (NAcc), only the pump method also significantly increased DA activity in the caudate nucleus. Since the mesolimbic DA system, with its connections to the NAcc, is suggested to be of prime importance in schizophrenia (Stevens, 1975), this finding suggests that manipulation of DA content within the NAcc may provide a better model.

DA injections into the NAcc have behavioral effects that are at least partly related to placement along the dorso/ventral axis of the structure. Injections into the ventral aspect of the accumbens in squirrel monkeys induce mainly hyperactivity accompanied by increased self-grooming, exaggerated startle responses, and increases in exploratory behavior (Dill et al., 1979; Jones et al., 1981a). Loss of social grooming and increased visual scanning of the environment were seen following DA injections into the *dorsal* aspect of the NAcc (Dubach and Bowden, 1983). However, the total symptom picture is not convincing as an overall model for schizophrenia.

It should be remembered that the NAcc is functionally related to the rate of locomotor activities, whereas the caudate seems more involved in drug-induced stereotypy (Creese and Iversen, 1975). Furthermore, Robbins et al. (1986) have shown that attentional switching is potentially affected by DA stimulation of the mesolimbic region. How this is related to the switching between motor response tendencies is still not clear, but since both stereotypy and increased switching are characteristics of schizophrenia, clearly the separation of NAcc and caudate nucleus effects will be increasingly important. *See* Lyon's Chapter 6 on mania in this volume for details on the methods for intraaccumbens injections.

Most other references to localized injections of DA agonists and antagonists in various nigrostriatal and mesolimbic regions produce either limited or diffuse behavioral effects that are not specific models for schizophrenia. One of these more limited models does deserve attention. Swerdlow et al. (1986) used mesolimbic injections of very small amounts of 6-hydroxydopamine (6-OHDA) to induce DA supersensitivity in specific regions, arguing that such effects would locally model the DA overactivation that supposedly occurs in schizophrenia. Their test for increased sensitivity was the acoustic startle-response paradigm, with measurement of prepulse inhibition (PPI) effects. In this paradigm, a warning signal precedes the auditory pulse stimulus; in normal individuals, this induces a partial inhibition of the central nervous system response to the pulse. The point of interest is that schizophrenic patients show a reduced PPI effect; they continue to respond fully to the pulse even after several repetitions of the warning stimulus.

Swerdlow et al. (1986) found that NAcc injections of small amounts of 6-OHDA induced a similar loss of PPI effects in the rat. In support of a DA role in this behavior, increases have been found in DA and in polyamine concentrations in the mesolimbic system during similar experiments.

2.1.1.4. Phenylethylamine (PEA)

PEA is the best-known of the endogenous amphetamine-like compounds found in the CNS. It was hoped that models using PEA would be superior to those using exogenous drugs, but this has not been the case so far. Injection of PEA, even in large doses, does not mirror the range of symptoms produced by *d*-amphetamine. For example, Dourish (1982) reports that mice injected with PEA do not show the same progressive changes produced by *d*-amphetamine, but show a hyperactive early phase, followed by a middle phase of depressed and fragmented motor responses and a late phase characterized by a return to hyperreactivity. Such changes are not typically seen in schizophrenia, especially the extreme neurological effects of the middle phase, in which Dourish describes splayed hindfeet and forepaw "padding." The abortive grooming said to be most characteristic of the PEA effect, and the rearing and perseverative hyperactivity, are familiar signs in amphetamine studies also (Ellinwood and Kilbey, 1977).

2.1.2. Serotonin-Related Models

The hallucinogenic lysergic acid diethylamide (LSD) has often been suggested to produce symptoms that mimic schizophrenia better than dopaminergic agonists (e.g., Claridge, 1978). Claridge summarizes evidence that LSD increases stimulus generalization, which has been suggested as a characteristic of schizophrenia (Mednick, 1958), and disrupts hippocampal theta rhythms, which may be important for suppression of behavior when the environment is altered.

Although LSD also has an effect on DA, its major influence is probably exerted on serotonin (5-HT) raphe neurons. However, very high doses of *d*-amphetamine and similar agonists also produce a marked imbalance in 5-HT systems, and at such high dose levels it can be shown that the 5-HT effect is important for producing the observed behavioral changes (Sloviter et al., 1980). It remains to be determined whether strong hallucinations, which can also occur with amphetamine abuse in humans and probably in monkeys (Nielsen et al., 1983), are always dependent on 5-HT system changes.

Attempts to manipulate directly the available serotonin in the brain, for example, by treating animals with the 5-HT antagonist parachloro-phenylalanine (PCPA), can result in hyperactivity and increased hypersensitivity to external stimuli. The end result often includes irritability and aggression (Fibiger and Campbell, 1971) in a syndrome resembling mania (*see* Lyon's chapter on mania in this volume). However, the marked perseveration and stereotyped behavior characteristic of schizophrenia are absent, and there are no clear signs of thought disorder.

All in all, alterations in 5-HT activity, except for those related to LSD hallucinations, do not provide convincing parallels to schizophrenia, even though they do indicate psychotic behavior.

2.1.3. Opioid-Related Models

Although opioid effects have been specifically tied to manic symptoms (*see* Lyon's chapter on mania), they have been less convincing as models for schizophrenic behavior until the more recent discovery of the endogenous opiates, enkephalins, and endorphins. It was noted that excessive amounts of β-endorphin could induce a catatonic-like state (Bloom et al., 1976). Then De Wied (1978,1979) suggested a number of ways in which β-endorphin peptide chains could operate as neuromodulators leading to different types of psychosis.

The evidence from animal models testing this hypothesis is not yet very strong. De Wied and colleagues presented evidence that extinction of avoidance responding was delayed by α- and β-endorphin and met-enkephalin; in another study, it was shown that des-tyr-γ-endorphin and the morphine antagonist naltrexone tended to facilitate extinction. Since facilitation of extinction also occurs with neuroleptics, it was suggested that peptide fragments such as β-endorphin and des-tyr-γ-endorphin had opposing effects on the brain systems involved in psychosis.

Moreover, the extended avoidance responding during extinction may be explained as a direct influence of the endorphins on DA systems (Wahlstrom and Terenius, 1981). This suggests that the endorphin release activates DA systems, with resulting stimulatory effects.

This seems to make it difficult to explain why Van Ree et al. (Verebey, 1982, p. 492) found some *antipsychotic* effects with γ-endorphins in early-phase schizophrenic patients who had received little previous neuroleptic treatment. However, it should be remembered that even DA agonists can temporarily produce clinical improvement in a small subgroup of schizophrenic patients. Such effects could be attributable either to massive DA overstimulation, with resulting decreases in overtly active ("positive") symptoms, or to DA stimulation in individuals with severely depleted DA reserves, which may be the case in the "negative" symptomatology in schizophrenia. This is clearly an area in which animal models are needed, particularly with techniques that are less influenced by DA systems than is shock avoidance.

In essence, certain peptide chains may be capable of mimicking both positive and negative schizophrenic symptom complexes, depending on local dose levels. However, if opioid receptors were the major element in producing schizophrenic reactions, then blocking their effect with such antagonists as naloxone and naltrexone should also strongly reduce the symptoms, but this is not the case (*see* Verebey, 1982). Very few adequate behavioral measures exist for these models, and the fact that they are minimally affected by naloxone suggests that other, perhaps DA-related, paths of influence may be essential.

2.1.4. Acetylcholine-Related Models

Some early studies seemed to indicate a cholinergic effect related to schizophrenia. The anticholinergic drug scopolamine, in doses >10 μg, is known to produce hallucinations, and some cases of catatonia are temporarily improved by intraventricular injection of cholinesterase (ChE) (Schuberth, 1978). The cholinesterase inhibitor physostigmine reportedly can convert mania to depression (Davis et al., 1978), yet it produces no improvement in schizophrenia. The latter researchers also noted that physostigmine seemed to affect behavioral rate measures more than ideational content or emotional tone.

Interestingly, physostigmine *does* block the exacerbating effect of methylphenidate on schizophrenia. The reasons for this are not clear, since physostigmine tends to have a behavioral stimulant effect, which can even resemble the effects of methylphenidate on locomotor behavior in rats. However, highly increased stimulant effects have been suggested to lead to *decreased* overt behavior despite intense brain activity (Lyon and Robbins, 1975; Grilly, 1977).

The behavioral effects on animals of reduced ChE levels are similar to those of amphetamine, with lower doses seeming to hasten maze-learning behavior, and higher doses leading to response perseveration and exponentially increasing numbers of errors (Russell, 1978). Lesions of cholinergic neurons in the corpus striatum prevent the normal acquisition and retention of a passive avoidance task in rats (Sandberg et al., 1984).

For animal models of schizophrenia, the important anatomical configuration of cholinergic neurons is probably that found in the caudate/putamen and NAcc as well as that in the rostrally projecting system from the medial surface of the lateral lemniscus, along the periaquectal gray matter, to a medial position near the substantia nigra (Cuello and Sofroniew, 1984). These two systems run somewhat parallel to, and definitely are affected by, the nigrostriatal and mesolimbic DA systems that have already been implicated in models of schizophrenia (Ladinsky et al., 1978). Nonetheless, it has been shown that the muscarinic receptor blocker atropine does not directly block DA stimulant effects, nor is there a direct cholinergic interneuron in the DA pathways related to locomotion (Jones et al., 1981b). This suggests a modulatory role for acetylcholine (ACh) systems in schizophrenia, but not a principal role, since cholinergic stimulation or inhibition alone is not sufficient to produce visibly schizophrenic symptoms. Nevertheless, Tandon and Greden (1989) have suggested that cholinergic "overdrive" may result in a negative schizophrenic symptomatology, and this possibility remains to be tested with animal models.

An as-yet-unstudied effect of cholinergic agents is the potential relationship of muscarinic transmission to potentiation of other noncholinergic pathways that may stimulate alerting or attentional responses (Brown, 1983) that are so frequently reported to be dysfunctional in schizophrenia.

2.1.5. Gamma-Amino-Butyric Acid Related Models

Scheel-Kruger and his colleagues (Scheel-Kruger et al., 1978, 1980; Arnt and Scheel-Kruger, 1979; Scheel-Kruger, 1986) have shown that γ-amino-butyric acid (GABA) systems are intimately related to the DA systems that are known to have some relationship with schizophrenia. The general findings have emerged from studies involving localized cerebral microinjections of GABA agonists (such as muscimol) and GABA antagonists (such as picrotoxin and bicucculine) into the corpus striatum, substantia nigra, and ventral tegmental area of the midbrain. Much of this work has been done in comparison with the effects of systemic amphetamine, apomorphine, and neuroleptic drugs.

Many of the DA stimulation effects on behavior, plus additional symptoms, such as sedation and catalepsy, can be mimicked by the injection of the GABA-ergic agents into the appropriate brain regions. Scheel-Kruger (1986) depicts these effects, which usually are considered the sole result of DA stimulation, as the consequence of the combined stimulation of GABA and DA systems. He has suggested that instead of the "dopamine hypothesis" of schizophrenia, one should speak of the "dopamine–GABA hypothesis," since many, if not all, of the critical behaviors associated with the DA hypothesis are dependent in part on a GABA–DA interaction.

Two particular aspects of this research are of interest here. The first is related to the topographic organization of GABA systems within the corpus striatum and the nucleus accumbens, and the second relates to the effects of GABA within the ventral tegmental area of the mesencephalon. In both regions, it has been demonstrated that there exists a topographical separation of regions with differing behavioral effects following localized GABA stimulation (Scheel-Kruger et al., 1980).

2.1.5.1. GABA Effects in the Corpus Striatum and Globus Pallidus

Muscimol injected directly into the basal ganglia of the rat in small doses (25 ng) produces catalepsy only when placed in the ventromedial portion of the intermediate striatum or in the intermediate ventral portion of the rostral globus pallidus (Scheel-Kruger, 1986).

If the GABA agonist is given in higher doses (100–250 ng) in the same localized regions, it first produces a cataleptic reaction, followed by nearly continuous locomotion, rearing, and sniffing. This strong localization of effect may be responsible for some earlier negative reports on the behavioral effects of GABA agonists within the striatum.

Similar localizations of function were reported for the ventral tegmental area (VTA), with injections (10–25 ng muscimol; 100–500 ng THIP) into the caudal portion producing much locomotion, aggressive behavior, and increased food intake. There were no strongly stereotyped responses following this caudal VTA injection, which separates the effect from that of striatal stimulation. The finding of aggressive and attack behavior corresponds to the strong emotional responses reported following GABA stimulation in the cat (Stevens, 1979). Injections of muscimol into the rostral VTA caused a loss of activity, with sedative effects, a hunched-back posture, and some reflexive rigidity. As with the striatum, there was a clear differentiation of symptoms related to localization within the VTA (Arnt and Scheel-Kruger, 1979).

Since it is possible to produce many of the behavioral effects, including the stereotyped behaviors usually related to the nigrostriatal DA systems, by intracerebral muscimol alone (Scheel-Kruger et al., 1980), these data suggest that the interaction of GABA and DA systems is very intimate. The older notion of the GABA "feedback loop" controlling the nigrostriatal system should probably be altered to include a nigrothalamic circuit (Scheel-Kruger, 1986).

2.1.5.2. Suggestions for Further GABA-Related Models

It appears that DA overstimulation in the brain cannot reasonably be proposed as the major cause of schizophrenia without consideration of neurotransmitters, such as GABA. As Carlsson (1988) has pointed out, even reserpine-like agents, which almost totally remove available DA, do not entirely prevent the animal symptoms associated with schizophrenia.

The best method for continuing these studies would be to use intracerebral injections accompanied by more sophisticated behavioral tests and more precise definitions for such terms as "stereotypy," "emotionality," and "aggression." It is also important to establish whether the GABA-related behaviors can be induced *only* by direct injection into the brain, or whether GABA functions can be altered by systemic treatment.

2.1.6. Glutamic Acid Related Models

Glutamate and aspartate receptor functions need to be considered both in relation to excitatory neurotransmission and for their possible role in schizophrenia. The *n*-methyl-*d*-aspartate (NMDA) receptor site has come in for special attention because of its sensitivity to phencyclidine (PCP), which may produce an even better model of schizophrenic symptoms than many DA agonists. However, it has been reported recently that only quisqualate and kainate receptors were found to be reduced in number in the left hippocampus of brains from schizophrenic patients (Kerwin, 1990).

To complicate the picture further, all the excitatory amino acid receptor sites are also potentially involved in the neurotoxic effects of these putative neurotransmitters. There is also no clear separation of behavioral actions related specifically to the stimulatory vs the neurotoxic effects of these agents. It has been suggested that DA exerts a modulatory effect on the release of L-glutamate in the neostriatum, in such a manner that high levels of L-glutamate invoke antagonistic action, and very low levels result in dopaminergic stimulation of L-glutamate release (Godukhin et al., 1984). This implies a strict regulatory function by DA of the glutamic acid (Glu) activity in the striatum, with the possibility that failure to control high Glu levels might result in cell death, but the behavioral consequences of this relationship are still unknown. In fact, the problem of identifying schizophrenia-like behavioral effects resulting from Glu dysregulation in animals has been difficult.

The only potential animal model suggested so far is based on the work of Kornhuber and Fischer (1982), who injected the glutamate reducing agent glutamate diethyl esterase (GDEE) into the lateral ventricles of rats and reported behavioral changes resembling the *negative* symptoms of schizophrenia. The doses of GDEE used were 1.1, 2.0, and 3.0 mg, all being injected as GDEE-HCl dissolved in 5 μL of distilled water with an NaOH-adjusted pH value of 6.0. However, the only behavioral measures obtained were photocell–cage activity counts and catalepsy, as measured by the latency of removal of at least one forepaw from a rod placed 6 cm above the floor. The 1.1-mg dose produced a very strong catalepsy and a reduced locomotion lasting more than 2 h, and the 2.0-mg dose frequently resulted in strong catalepsy in the first 20 min, but with increased locomotion peaking about 1 h after injection, and greatly increased salivation. The animals receiving a 3.0-mg dose were severely affected, and some died, with signs of pulmonary edema.

These results are similar to those reported for localized GABA stimulation by Scheel-Kruger (1986). Furthermore, the excessive salivation may be a sign of high body temperature, which rats attempt to alleviate by generating excessive amounts of saliva, which they spread around the body. In any case, having only catalepsy and locomotor stimulation to use as symptoms, this is not a convincing model of schizophrenia, and the great increases in locomotor activity do not fit well with the negative symptoms of schizophrenia.

However, despite the lack of immediate evidence, there is much to suggest some general link between Glu-controlled functions and schizophrenia. Singh and Kay (1976) have reported that increased dietary gluten intake exacerbates schizophrenic symptoms, implying that increased Glu intake may affect the brain. However, Glu is normally synthesized in the brain in such quantities that it is transported out of, rather than into, the brain

(Pardridge, 1979). Therefore, the findings of Singh and Kay suggest a defect in the transport or synthesizing mechanisms for Glu in schizophrenia. These mechanisms may, nevertheless, be amenable to change under normal circumstances.

Both Williams (1979) and Lyon (unpublished) have found that increases in dietary wheat gluten content tend to increase sensitivity to DA stimulation in rats. However, at least one small-peptide fraction of the gluten produces behavioral effects that can be blocked by naloxone, although this is not true for the larger protein component of the diet (Williams, 1980). The analysis of the effects of these protein components is another obvious area in which animal models can be of great future value.

2.2. Brain-Lesion and/or Stimulation Models

2.2.1. Hippocampal-Lesion Models

Since these models are thoroughly evaluated in the chapter by Schmajuk and Tyberg in this volume, this section will consist mainly of a historical note on the changing types of models that have referred to the hippocampus. Following the early suggestion of Papez (1937), lesions of the hippocampus were thought to produce hyperemotionality, but this soon gave way to the implication of the hippocampus in short-term memory (Milner, 1965). More recent developments point to an important role in spatial working memory (Olton, 1983) and to the long-term potentiation (LTP) effects of hippocampal stimulation. The LTP effects have a known close relationship to kindling phenomena and the generalization of epileptic seizures, but may also be the source of highly perseverative thought, fixed delusions, and so on (Post et al., 1984). Post et al. (1977) suggested some time ago that kindling effects from cocaine stimulation may result in a recurring seizure-like state that has similarities to mania and schizophrenia (*see* Lyon's chapter on mania models in this volume). However, changes in behavior resulting from seizure activity are not considered typical of schizophrenia, and it would appear that the more general functions that are disturbed in the hippocampus also have cortical and mesolimbic system side effects.

Attempts to lesion the hippocampus have been frustrated by its curvilinear structure and by the fact that it lies beneath important cortical and callosal fiber systems. It is very difficult to control for the effects of the incidental lesions to overlying structures, since these structures are complex enough to be involved in many systems. Only one example will be given here. Devenport et al. (1981) reported that electrolytic lesions made simultaneously at eight locations in the hippocampus of rats produced an animal with excessively stereotyped behavior and other signs similar to those of catecholaminergic overstimulation. They likened this response to amphetamine psychosis, and referred to the loss of passive avoidance, increases in active avoidance, lengthened extinction responding, and loss of ability to delay responding as similar to the effects of catecholamine overstimulation. Their interpretation was that the hippocampus normally opposed the catecholaminergic stimulation, and this suggestion was bolstered by evidence that pretreatment with haloperidol reversed these effects.

The problem of this interpretation is that control animals were given large lesions of the overlying cortex and corpus callosum, and such lesions may interfere with catecholaminergic systems and behavior by a set of pathways separate from those involved in the hippocampal damage itself (Goldman-Rakic, 1987). Thus, the effects could equally well result from a combined lesion effect. Although the volume of the hippocampus may be reduced in some schizophrenic patients (Bogerts et al., 1985), there are still many schizophrenic patients who do not show a measurable structural difference, and, even when they do, the loss of tissue is about 1–2% (Bogerts, in press), which hardly resembles the large lesions used by Devenport et al. (1981).

Perhaps more important for models of schizophrenia is the growing evidence that defects in cell migration during fetal development may lead not so much to volume changes, but to a changed hippocampal structure (Kovelman and Scheibel, 1984). These structural changes will have to be modeled by inducing alterations in early fetal development and then tracing behavioral changes in the offspring. This seems to be a very promising area for future models of schizophrenia in animals.

2.2.2. Ventral Tegmental Lesion and Stimulation Models

Stevens and Livermore (1978) suggested that some form of kindling within the mesolimbic system could be a relevant model of schizophrenia. This concept was derived, in part, from Stevens' (1975) demonstration that GABA blockade in the VTA led to a state of apparently extreme "anxiety" in the cat (*see also* Arnt and Scheel-Kruger, 1979). However, as noted earlier, the VTA is not homogeneous with respect to behavioral effects. Since the GABA receptor agonist muscimol, when injected into the rostral VTA, caused inactivity and abnormal passivity (Scheel-Kruger, 1986), it is likely that Stevens' (1975) injections with the GABA antagonist bicucculine would cause an opposite effect if given in this same region. Arnt and Scheel-Kruger (1979) reported extreme hypermotility and much rearing on the hindlegs in rats after the same treatment with bicucculine. However, neither Stevens nor Arnt and Scheel-Kruger examined conditioned behavior in any detail.

Several lesion studies of the rat VTA have provided evidence that cognitive-learning capacities are reduced after lesions in this area. Oades (1982) produced lesions by mechanical damage with a metal stylet, and Simon et al. (1980) used 6-OHDA injections into the VTA. In each study, the lesions were extensive enough to include both the rostral and caudal regions mentioned by Arnt and Scheel-Kruger (1979). The lesions produced significant losses in both retention of delayed alternation performance and in a complex, hole-board, foraging task. Based on performance in the complex task, Oades suggested that VTA-lesion effects were a potential model for thought disorder in schizophrenia. Indeed, such tasks as the hole board (Oades, 1982), radial maze (Olton et al., 1977), and Morris water maze (Morris et al., 1982) all require the retention of a spatial-orientation plan (cognitive map) within a complex environment. Behavioral methods having these qualities are very much needed if animal models are to demonstrate any exact parallel to the subtleties of human thought disorder.

2.2.3. Nucleus Accumbens Lesion and Stimulation Models

At the other end of the mesolimbic system from the VTA lie the nucleus accumbens (NAcc) and the medial frontal cortex.

These regions have also been explored with some animal models. Stevens (1975) injected the GABA antagonist bicuculline into the NAcc of the cat and noted the same type of excitability that she had found in the VTA, in one case with a strong aggressive and attack component. In the rat, it has been found that 6-OHDA lesions of the NAcc potentiate the locomotor-excitatory effects of apomorphine in activity cages. Such lesions also increase a "place preference" conditioned to the presentation of apomorphine, which suggests that positive reinforcing effects of the drug are increased by the 6-OHDA lesions of the NAcc (Van Der Kooy et al., 1983).

These experiments suggest that overstimulation of the NAcc leads to excessive excitability with accompanying hypersensitivity to stimuli, which leads to aggressive behavior and attack. When the DA system within the NAcc is largely inoperative (92% depletion of DA), the effect of systemic apomorphine stimulation (0.1 mg/kg/mL dissolved in saline with 0.1 mg/mL ascorbic acid and injected sc in the neck region) was to increase the positively reinforcing effects of locomotor activity and of place preference for the location previously associated with apomorphine treatment.

However, in assessing models that involve the NAcc, it is important to consider the close relationship of the NAcc to the *initiation* of repeated actions, although the choice of action *type* is probably more closely related to the caudate nucleus/cerebellum/thalamus complex (Iversen, 1977). The choice of a plan of action undoubtedly involves cortical functions, including the dopaminergically relevant areas of the frontal cortex. Recent evidence suggests that not only the orbitofrontal and medial prefrontal cortex are involved, but also a large part of the dorsolateral frontal cortex (Goldman-Rakic, 1987). Since the frontal cortex has a close topographical connection with the basal ganglia, both of these regions have potential for changing the plan of action.

A further important complication resides in the capacity of the NAcc to control certain circadian-rhythm effects related to the DA systems. Costall et al. (1984a,b) have shown that local injections of L-dopa into the NAcc of rats with a demonstrable sensitivity to systemic injections of (-)*N*-*n*-propylnorapomorphine [(-)NPA] can produce lasting cyclic variations in sensitivity to DA challenge. This cyclic variation was demonstrated over a period of days, even with continuous DA infusions; following (-)NPA challenge, the sensitivity changes could be demonstrated even after a year without intervening treatment. These effects are altered, but notably not abolished, by treatment with haloperidol and sulpiride. Potentially, such experiments may lead to a model for certain aspects of schizophrenia, but unfortunately no proper behavioral analyses have yet been made.

Injection of cholecystokinin octapeptide (CCK-8) into the NAcc appears to work against the locomotor-stimulating effects of low-dose amphetamine, and to exacerbate the stereotyped responding seen with high doses of that drug (Weiss et al., 1988). This suggests that CCK-8 strengthens the amphetamine effect, since both of these changes could be in the same direction (Lyon and Robbins, 1975).

2.2.4. Periventricular-Lesion Models

A large number of human studies in schizophrenia have demonstrated enlargement of the lateral and third ventricles. This suggests the plausibility of an animal model using agents that affect the periventricular tissue. Petty and Sherman (1981) infused 6-OHDA into the ventricles and found that it markedly increased motor activity. They interpreted this as a possible model for mania, but other behavioral tests were not used, and the clinical relevance is obscure. Kline and Reid (1985) used a more general agent for intraventricular injection, lysophosphatidyl choline (LPC), which causes acute demyelination.

A further advantage of the Kline and Reid study was the large number of behavioral tests used, including a neurological examination, hole-board test (Makanjuola et al., 1977), open-field measures, activity counts, a catalepsy test (seconds to remove a forepaw from a 2.5-cm block) and a "step-down" test (without aversive stimulus training). Counts of defecation revealed an apparently decreased emotionality, and there was increased catalepsy, excessive aggression during handling, and impaired grooming.

These response changes were likened by Kline and Reid (1985) to the symptoms of schizophrenia, including flat affect, inappropriate aggression, increased cataleptic responding, and lack of interest in physical appearance. There was histological confirmation of increased lateral and third-ventricle size and subcallosal demyelination in these animals, all of which are seen in the brains of schizophrenic patients (Weinberger et al., 1979; Andreasen et al., 1982; Bogerts et al., 1985). These close parallels suggest that this particular animal model has much to recommend it from both anatomical and behavioral views. This model may be especially important because the structures involved in the periventricular damage include those of the rostral mesolimbic and neostriatal systems. These are also the regions thought to have greater vulnerability during prenatal development (Lyon et al., 1989; Lyon and Barr, in press), which will be discussed in the following section.

2.3. Genetic and Fetal Development Models

There seems little doubt that schizophrenia is frequently, if not always, a result of changes begun in fetal development or at birth (Mednick et al., in press). Brain damage caused by both pre- and perinatal disturbances are thought to be secondary factors that act together with a genetic predisposition and result in adult schizophrenia.

The only genetically relevant animal models of schizophrenia developed so far are quite broad. For instance, Nowakowski (1987) has shown that particular strains of mice have brain structure deficits similar to those found in schizophrenia (Kovelman and Scheibel, 1984; Jakob and Beckmann, 1986). However, behavioral measurements directly related to these structural changes are still primitive. On the other hand, the sequence of structural growth in the human fetal brain has been well studied (Gilles et al., 1983), and the effects of growth disturbances and of other perinatal factors is also available (Freeman, 1985). Furthermore, it is known that disturbances resulting from viral infection, radiation exposure, and emotional stress during the second trimester of pregnancy is probably critical to the fetal brain changes leading to schizophrenia (Mednick et al., 1988;

Otake and Shull, 1984; Huttunen and Niskanen, 1978). It is necessary to produce animal models of the critical changes in brain structure during the second trimester that are most likely to be involved in schizophrenia. These changes are presently thought to be most closely related to such periventricular structures as the basal ganglia, hippocampus, and mesolimbic systems, and there are reasons to believe that exactly these regions are especially sensitive to disturbance during the second trimester of development (Lyon and Barr, in press).

It has long been known that haloperidol and other neuroleptic substances, when given during pregnancy, can cause nervous malfunction (Golub and Kornetsky, 1974; Spear et al., 1980). More recently, an animal model has been tested using the DA agonist cocaine to treat pregnant rats (Spear et al., 1989). Although this procedure resulted in some cognitive deficits, the effect seems to involve reduced DA influence, and the exact parallels with schizophrenia remain to be worked out.

One of the unknown features of such animal models is whether or not there is any similar, naturally occurring genetic deficit in the rat, or even in the monkey, which resembles that responsible for human schizophrenia. Because of the relatively rare expected occurrence of this disorder, animal-population experiments have never been done, but it would be premature to state, as some researchers have (Kornetsky, 1977), that no exact animal parallel exists. In fact, it is probably the lack of animal models using a fetal-development premise that has led to this attitude toward animal models in general. It is predicted that this will be a major field of endeavor for animal research into schizophrenia in the near future.

2.4. Behaviorally Based Models

2.4.1. Social Isolation and Isolation Stress Models

It has long been noticed that, when animals tend to become abnormal in captivity, one of the first and most persistent signs is the lack of response to others, with accompanying social isolation. This is frequently referred to as "isolation stress." Einon and Morgan (1977) showed that social isolation of young rats shortly after weaning, even if for only a week or so, was sufficient to induce changes in the dopaminergic system. Increased sensitivity to haloperidol in previously isolated animals has been demonstrated, and DA overactivity in social isolates has been directly confirmed by Jones et al. (1987,1989) using in vivo dialysis of the corpus striatum.

The results of social isolation can be seen in neurotransmitter functions, hormonal disturbances, and in abnormal response to stressful or punishing stimulation. This indicates that comparisons of behavior in rat models of schizophrenia should include appropriate, socially caged controls. In passing, it may be noted that handling by the experimenter is also a form of social stress, and ought never to be done by a person not normally engaged in the experiment (no matter how knowledgeable he or she may be about animal handling).

2.4.1.1. Ethologically Based Models of Isolation Stress

To further approach the human condition in an animal model requires looking at the overtly different, but structurally parallel natural patterns of behavior for each species. A valuable ethological viewpoint has been added to the concept of social dysfunction as a sign of psychotic abnormality by descriptions of how social isolation in young monkeys can lead to severe emotional disturbance (Mason, 1968; Suomi et al., 1973; Goosen, 1981). In cases of mother deprivation and early loss of peer social contact, rhesus monkeys develop habits of social isolation, depression, loss of bodily hygiene, and even self-mutilation that closely resemble the symptoms of schizophrenia. In examining the behavior of singly caged female rhesus monkeys, Goosen found that many of the self-directed behaviors (hugging, grooming, and so on) could be interpreted as attempts to restore the missing social contact.

Strong parallels are seen between these effects, produced by social isolation, and those following amphetamine treatment (Haber et al., 1977). The latter researchers saw a distinct likeness between the symptoms of paranoid schizophrenia and the social isolation, intense staring, and heightened agonistic behaviors in amphetamine-treated rhesus monkeys. Both Haber et al. (1977) and Nielsen et al. (1983) have commented on the gradual limiting of the sphere of attention by the drugged animals. In the end, they are almost totally isolated socially, do not interact with others, and stare at or fixate often on points in space near them or on their own body. Especially during the early stages of the amphetamine effect, the monkeys appear to be tense, watchful, and may resort to overly aggressive or overly submissive behavior, depending on their colony status (Haber et al., 1977).

2.4.1.2. Social Isolation Plus Amphetamine Challenge

The most interesting model used with socially isolated monkeys is probably that of Kraemer et al. (1983), who elected to challenge previously isolated rhesus monkeys with amphetamine. They compared the effects of isolation for 1 mo, at 6 mo of age, with 11 mo of isolation beginning 1 mo after birth. Both groups were then raised in social groups until 30 mo of age.

The interesting feature of this procedure is that the long intervening period of social grouping, after the early isolation, removed any obvious signs of difference between these animals and nonisolated control animals. Upon amphetamine challenge (by intubation at doses of 0.5, 1.0, 2.0, and 3.0 mg/kg/d), there were obvious differences between the previously isolated and nonisolated groups. Nonisolated animals showed the usual reduced social body contact, and increased stereotypic exploration of their surroundings, ending in almost total removal from the social context. Monkeys with a history of early social isolation showed a remarkable conflict in behaviors, with increased facial grimaces and screeching usually indicative of fear together with increased "contact cling" to other animals. This led to their being constantly attacked, yet they persisted in this behavior in spite of even serious wounding. Overall, the nonisolated animals directed their behavior under amphetamine into more and more stereotyped activity directed at the environment, and the previously isolated animals directed their repetitive behavior toward cagemates who explicitly tried to rebuff them. Both types of behavior have parallels in schizophrenia, and both lead to a loss of normal social contact, in the one case also inducing strong agonistic reactions from conspecific cagemates.

2.4.2. Anhedonia Models

Schizophrenic patients treated with neuroleptics frequently complain that, although their psychotic symptoms are reduced, they lack the positive emotional tone previously associated with things as diverse as playing the guitar or sexual activity. Following this implication that positive motivation (hedonia) might be reduced during neuroleptic treatment, Wise et al. (1978) predicted that haloperidol and other neuroleptics should produce anhedonia in animal models.

However, if a major effect of neuroleptics is anhedonia, then schizophrenics without neuroleptic treatment should be overly responsive to the hedonic effects of stimuli. If true, such hyperresponsiveness would allow minor environmental stimuli to acquire enormous secondary reinforcing properties. The problem in proving this point with animal models is that the typical neuroleptics, such as haloperidol, also have effects on motor responses via the frontal cortex/basal ganglia interaction (Ettenberg et al., 1981). However, Ettenberg and Horvitz (1987) did not find evidence of any simple motor disturbance as an explanation. Fantie and Nakajima (1987), using conditioned bursts of hippocampal theta waves as the basic measure of response, showed that this responding continued during neuroleptic treatment.

On the other hand, the anhedonia hypothesis does not apply well to punishment and conditioned suppression studies, in which the evidence is strongly suggestive of an impairment in *response initiation,* rather than in associative learning (Beninger et al., 1980).

In summary, the anhedonia models appear to provide limited evidence on the negative-symptom side of the disease, yet they do suggest that increased secondary reinforcement may be a prime feature of the schizophrenic disturbance. This would be in accord with the evidence from DA-agonist models of schizophrenia advanced earlier in this chapter, and with the fact that the DA-agonist models are most closely related to the "positive" symptoms of schizophrenia.

2.5. Cerebral Asymmetry and Laterality Models

2.5.1. Handedness and Bodily Rotation Measures

There is an increasing amount of data supporting hemispheric-dominance and side-preference changes in schizophrenia (Gur, 1978; Flor-Henry, 1976). In handedness, the major change seems to be a tendency to lose the usual human right-hand preference. Sometimes this results in a shift to left-hand use, but most commonly shows as a weakness or partial loss in the previous right-hand dominance in schizophrenia (Green et al., 1989).

Since animals do not share the very strong human preference for right-handedness, attention has been paid here to the general issue of lateral preference, regardless of side chosen. Ungerstedt and Arbuthnott (1970), using rotational behavior as a measure, showed that DA agonists, such as amphetamine and apomorphine, could influence the direction of turning in rats, both with and without various lesions of the substantia nigra and other DA-system lesions. Glick and colleagues (Glick et al., 1977; Glick and Cox, 1978) have gone one step further and demonstrated that there are hemispheric differences in DA activity in a majority of rats and that turning behavior, at least in male rats, is directly related to the side with the highest DA activity.

Although human hand preference cannot be directly equated with rotational-preference measures in animals, recent experiments have shown an important relationship between cerebral hemispheric dominance (presumably of a dopaminergic nature) and the choice of the side to which schizophrenic patients rotate (Bracha, 1987; Lyon and Satz, in press). Measured in full 360° turns, normal individuals rotate approximately equally to both sides during a sample of daily bodily activity, whereas schizophrenic patients rotate significantly more to the left than to the right. This suggests a relationship to altered hemispheric dominance in DA activity, either with the right side becoming overactive (Bracha, 1989) or with a reduction of influence from the left side in such a way that the right side now becomes dominant. There is some evidence from positron emission tomography (PET) of a lateralized change in activity in dopaminergically controlled brain regions (Buchsbaum and Haier, 1987). These issues could be addressed by an animal model that would first determine the side preference for rotation or induce a preference by conditioning, and then manipulate the levels of DA activity in the manner suggested by the human data.

From a methodological viewpoint, it should be pointed out that there is also a strong sex-linked influence in lateral preference. This is true of both humans and animals; in both cases, it is the females that show a stronger lateral preference. By accident or design, Glick's early experiments used exclusively female rats. Robinson et al. (1980) and others, in later experiments, have shown that male animals much more frequently show neither significant lateral preference nor an increased amount of DA on one side of the brain. Therefore, any animal models to be developed for lateral preference must take sex differences into account.

2.5.2. Conditioned Lateral Preference Measures

One of the more interesting of animal models presently available in relation to lateral preference is that of Castellano et al. (1989). These researchers have shown that *conditioned* side preferences are, at least in part, related to DA activity. The combination of conditioned behavior and measures of hemispheric DA activity appears to be the next logical step in finding animal models that come closer to approximating the changes in schizophrenia. In a similar vein, Lyon and Magnusson (1982) have also reported changes in paw preference for problem solving by vervet monkeys *(cercopithecus Aethiops)* when they were given methylphenidate.

Both of these studies indicate a close relationship between lateral preference and DA stimulation. Further studies are needed, but it appears that even a rat model of these behaviors could have significance for the study of schizophrenia.

3. Summary and Conclusions

It appears certain that extremely useful animal models of schizophrenia are within reach of present methods. However, at this time, insufficient notice has been paid to the importance of fetal development, during which the genetic expression occurs that will later result in the schizophrenic syndrome. Most present models are based on disturbances of the DA–GABA systems in adult animals only, and have demonstrated some very clear examples of schizophrenia-like behavior. Models based on other neurotransmitters or neuromodulators, such as the endogenous opioids or 5-HT imbalances, or on single factors, such as social isolation or anhedonia, have not yet been demonstrated to mimic broadly the symptoms of schizophrenia.

Acknowledgments

This chapter was completed while the author was on leave of absence from Copenhagen University and a Research Associate at the Social Science Research Institute, University of Southern California. Thanks are expressed to Director Ward Edwards and his staff for their unfailing support and assistance. Special thanks to W. O. McClure, S. A. Mednick, and C. E. Barr for important theoretical discussions.

References

American Psychiatric Association (1980) *DSM-III: Diagnostic and Statistical Manual of Mental Disorders,* 3rd Ed. (American Psychiatric Association, Washington, DC).

Andreasen N. C., Olsen S. A., Dennert J. W., and Smith M. R. (1982) Ventricular enlargement in schizophrenia: Relationship to positive and negative symptoms. *Am. J. Psychiat.* **139,** 297–302.

Angrist B. (1983) Psychoses induced by central nervous system stimulants and related drugs, in *Stimulants: Neurochemical, Behavioral, and Clinical Perspectives* (Creese I., ed.) Raven, New York, pp. 1–30.

Arnt J. and Scheel-Kruger J. (1979) GABA in the ventral tegmental area: Differential regional effects on locomotion, aggression and food intake after microinjection of GABA agonists and antagonists. *Life Sci.* **25,** 1351–1360.

Beninger R. J., Mason S. T., Phillips A. G., and Fibiger H. C. (1980) The use of conditioned suppression to evaluate the nature of neuroleptic-induced avoidance deficits. *J. Pharm. Exp. Ther.* **213,** 623–627.

Bleuler E. (1950) *Dementia Praecox* (International University Press, New York).

Bloom F., Segal D., Ling N., and Guillemin R. (1976) Endorphins: Profound behavioral effects in rats suggest new etiological factors in mental illness. *Science* **194,** 630–632.

Bogerts B., Meertz E., and Schonfeldt-Bausch R. (1985) Basal ganglia and limbic system pathology in schizophrenia. *Arch. Gen. Psychiat.* **42,** 784–791.

Bogerts B. (in press) The neuropathology of schizophrenia: Pathophysiological and neurodevelopmental implications, in *Fetal Neural Development and Adult Schizophrenia* (Mednick S. A., Cannon T. D., Barr C. E., and Lyon M., eds.) Cambridge University Press, Cambridge, UK.

Bracha H. S. (1987) Asymmetric rotational (circling) behavior, a dopamine-related asymmetry: Preliminary findings in unmedicated and never-medicated schizophrenic patients. *Biol. Psychiat.* **22,** 995–1003.

Bracha H. S. (1989) Is there a right hemi-hyper-dopaminergic psychosis? *Schiz. Res.* **2,** 317–324.

Brown D. A. (1983) Slow cholinergic excitation—a mechanism for increasing neuronal excitability. *Trends Neurol. Sci.* **12,** 302–307.

Buchsbaum M. S. and Haier R. J. (1987) Functional and anatomical brain imaging: Impact on schizophrenia research. *Schiz. Bull.* **13(1),** 115–132.

Carlsson A. (1988) The current status of the dopamine hypothesis of schizophrenia. *Neuropsychopharmacology* **1(31),** 179–186.

Castellano M. A., Diaz-Palarea M. D., Barroso J., and Rodriguez M. (1989) Behavioral lateralization in rats and dopaminergic system: individual and population laterality. *Behav. Neurosci.* **103,** 46–53.

Claridge G. (1978) Animal models of schizophrenia: The case for LSD-25. *Schizophr. Bull.* **4,** 186–209.

Costall B., Domeney A. M., and Naylor R. J. (1984a) Locomotor hyperactivity caused by dopamine infusion into the nucleus accumbens of rat brain: Specificity of action. *Psychopharmacology* **82,** 174–180.

Costall B., Domeney A. M., and Naylor R. J. (1984b) Long-term consequences of antagonism by neuroleptics of behavioural events occurring during mesolimbic dopamine infusion. *Neuropharmacology* **23,** 287–294.

Creese I. and Iversen S. D. (1975) The pharmacological and anatomical substrates of the amphetamine response in the rat. *Brain Res.* **83,** 242–248.

Cuello A. C. and Sofroniew M. V. (1984) The anatomy of the CNS cholinergic neurons. *Trends Neurol. Sci.* **13,** 74–78.

Davis J. M., Janowsky D., Tamminga C., and Smith R. C. (1978) Cholinergic mechanisms in schizophrenia, mania and depression, in *Cholinergic Mechanisms and Psychopharmacology* (Jenden D. J., ed.) Plenum, New York, pp. 805–815.

Deniker P. (1966) *La Psychopharmacologie* (Presses Universitaires de France, Paris).

Devenport L. D., Devenport J., and Holloway F. A. (1981) Reward-induced stereotypy: Modulation by the hippocampus. *Science* **212,** 1288–1289.

De Wied D. (1978) Psychopathology as a neuropeptide dysfunction, in *Characteristics and Function of Opioids* (Van Ree J. M. and Terenius L., eds.), Elsevier/North-Holland, Amsterdam, pp. 113–122.

De Wied D. (1979) Schizophrenia as an inborn error in the degradation of beta-endorphin—a hypothesis. *Trends Neurosci.* **2,** 79–82.

Dill R. E., Jones D. L., Gillin C., and Murphy G. (1979) Comparison of behavioral effects of systemic L-DOPA and intracranial dopamine in mesolimbic forebrain of nonhuman primates. *Pharmacol. Biochem. Behav.* **10,** 711–716.

Dourish C. T. (1982) A pharmacological analysis of the hyperactivity syndrome induced by beta-phenylethylamine in the mouse. *Br. J. Pharmacol.* **77,** 129–139.

Dubach M. F. and Bowden D. M. (1983) Response to intracerebral dopamine injection as a model of schizophrenic symptomatology, in *Ethopharmacology: Primate Models of Neuropsychiatric Disorders* (Miczek K. A., ed.), Alan R. Liss, New York, pp. 157–184.

Einon D. F. and Morgan M. J. (1977) A critical period for social isolation in the rat. *Dev. Psychobiol.* **10,** 123–132.

Eison M. S., Eison A. S., and Iversen S. D. (1983) Two routes of continuous amphetamine administration induce different behavioral and neurochemical effects in the rat. *Neurosci. Lett.* **39,** 313–319.

Ellinwood E. H., Jr. and Kilbey M. M. (1977) Chronic stimulant intoxication models of psychosis, in *Animal Models in Psychiatry and Neurology* (Hanin I. and Usdin E., eds.), Pergamon, Oxford, UK, pp. 61–74.

Ellinwood E. H., Jr. and Sudilovsky A. (1973) Chronic amphetamine intoxication: Behavioral model of psychoses, in *Psychopathology and Psychopharmacology* (Cole J. O., Freedman A. O., and Friedhoff A. J., eds.), Johns Hopkins University Press, Baltimore, pp. 51–70.

Ellinwood E. H., Jr., Sudilovsky A., and Nelson L. M. (1972) Behavioral analysis of chronic amphetamine intoxication. *Biol. Psychiat.* **4,** 215–225.

Ettenberg A. and Horvitz J. C. (1987) Haloperidol blocks the incentive motivational properties of food reinforcement. *Soc. Neurosci. Abst.* **13,** 219.

Ettenberg A., Koob G. F., and Bloom F. E. (1981) Response artifact in the measurement of neuroleptic-induced anhedonia. *Science* **213,** 357–359.

Evenden J. L. and Robbins T. W. (1983) increased response switching, perseveration and perseveration switching following *d*-amphetamine in the rat. *Psychopharmacology* **80,** 67–73.

Fantie B. D. and Nakajima S. (1987) Operant conditioning of hippocampal theta: Dissociating reward from performance deficits. *Behav. Neurosci.* **101,** 626–633.

Farde L., Wiesel F-A., Stone-Elander S., Halldin C., Nordstrom A-L., Hall H., and Sedvall G. (1990) D2 dopamine receptors in neuroleptic-naive schizophrenic patients. *Arch. Gen. Psychiat.* **47,** 213–219.

Fibiger H. C. and Campbell B. A. (1971) The effect of parachlorophenylalanine on spontaneous locomotor activity in the rat. *Neuropharmacology* **10,** 25–32.

Fischman M. W. and Schuster C. R. (1974) Tolerance development to chronic methamphetamine intoxication in the rhesus monkey. *Pharmacol. Biochem. Behav.* **2,** 503–508.

Flor-Henry P. (1976) Lateralized temporal-limbic dysfunction and psychopathology. *Ann. NY Acad. Sci.* **280,** 777–795.

Freeman J. M., ed. (1985) *Prenatal and Perinatal Factors Associated with Brain Disorders* (Publ. No. 85–1149, NIMH, Rockville, MD) April, 1985.
Frith C. D. and Done D. J. (1983) Stereotyped responding by schizophrenic patients on a two-choice guessing task. *Psychol. Med.* **13**, 779–786.
Gilles F. H., Leviton A., and Dooling E. C., eds. (1983) *The Developing Human Brain. Growth and Epidemiological Neuropathology* (Wright PSG, Boston).
Glick S. D. and Cox R. D. (1978) Nocturnal rotation in normal rats: Correlation with amphetamine-induced rotation and effects of nigrostriatal lesions. *Brain Res.* **150**, 149–161.
Glick S. D., Zimmerberg B., and Jerussi T. P. (1977) Adaptive significance of laterality in the rodent. *Ann. NY Acad. Sci.* **299**, 180–185.
Godukhin O. V., Zharikova A. D., and Budantsev A. Yu. (1984) Role of presynaptic dopamine receptors in regulation of the glutamatergic neurotransmission in rat neostriatum. *Neuroscience* **12(2)**, 377–383.
Goldman-Rakic P. S. (1987) Development of cortical circuitry and cognitive function. *Child Dev.* **58**, 601–622.
Golub M. and Kornetsky C. (1974) Seizure susceptibility and avoidance conditioning in adult rats treated prenatally with chlorpromazine. *Dev. Psychobiol.* **7**, 79–88.
Goosen C. (1981) Abnormal behavior patterns in rhesus monkeys: Symptoms of mental disease? *Biol. Psychiat.* **16**, 697–716.
Green M., Satz P., Smith C., and Nelson L. (1989) Is there a typical handedness in schizophrenia? *J. Abnorm. Psychol.* **98(1)**, 57–61.
Grilly D. M. (1977) Rate-dependent effects of amphetamine resulting from behavioural competition. *Biobehav. Rev.* **1**, 87–93.
Gur R. E. (1978) Left hemisphere dysfunction and left hemisphere overactivation in schizophrenia. *J. Abnorm. Psychol.* **87**, 226–238.
Haber S., Barchas P. R., and Barchas J. D. (1977) Effects of amphetamine on social behaviors of rhesus macaques: An animal model of paranoia, in *Animal Models in Psychiatry and Neurology* (Hanin I and Usdin E., eds.), Pergamon, New York, pp. 107–114.
Huberman H. S., Eison M. S., Bryan K., and Ellison G. (1977) A slow-release pellet for chronic amphetamine administration. *Eur. J. Pharmacol.* **45**, 237–240.
Huttunen M. O. and Niskanen P. (1978) Prenatal loss of father and psychiatric disorders. *Arch. Gen. Psychiat.* **35**, 429–431.
Iversen S. D. (1977) Striatal function and stereotyped behaviour, in *The Psychobiology of the Striatum* (Cools A. R., Lohman A. H. M., and van der Bercken J. H. L., eds.), North-Holland, Amsterdam, pp. 99–118.
Jakob H. and Beckmann H. (1986) Prenatal developmental disturbances in the limbic allocortex in schizophrenia. *J. Neurotrans.* **65**, 303–326.
Janssen P. A. J. (1967a) The pharmacology of haloperidol. *Int. J. Neuropsychiat.* **3**, S10–S18.
Janssen P. A. J. (1967b) Haloperidol and related butyrophenones, in *Psychopharmacological Agents, Vol. 4* (Gordon M., ed.) Academic, New York, pp. 69–72.

Jones D. L., Berg S. L., Dorris R. L., and Dill R. E. (1981a) Biphasic locomotor response to intra-accumbens dopamine in a nonhuman primate. *Pharmacol. Biochem. Behav.* **15,** 243–246.
Jones D. L., Mogenson G. J., and Wu M. (1981b) injections of dopaminergic, cholinergic, serotoninergic and gabaergic drugs into the nucleus accumbens: Effects on locomotor activity in the rat. *Neuropharmacology* **20,** 29–37.
Jones G. H., Hernandez, T. D., and Robbins, T. W. (1987) Isolation-rearing impairs the acquisition of schedule–induced polydipsia. *Soc. Neurosci. Abst.* **13(1),** 405.
Jones G. H., Marsden C. A., and Robbins T. W. (1989) Hypersensitivity to reward-related stimuli and to intraaccumbens *d*-amphetamine following social deprivation in rats. *Soc. Neurosci. Abst.* **15,** 412.
Kerwin R. (1990) The neurochemical anatomy of the hippocampus in postmortem schizophrenic brain. *Schizophr. Res.* **3,** 33,34.
Kline J., Jr. and Reid K. H. (1985) The acute periventricular injury syndrome: A possible animal model for psychotic disease. *Psychopharmacology* **87,** 292–297.
Kornetsky C. (1977) Animal models: promises and problems, in *Animal Models in Psychiatry and Neurology* (Hanin I. and Usdin E., eds.), Pergamon, New York, pp. 1–8 .
Kornhuber J. and Fischer E. G. (1982) Glutamic acid diethyl ester induces catalepsy in rats. A new model for schizophrenia? *Neurosci. Lett.* **34,** 325–329.
Kovelman J. A. and Scheibel A. B. (1984) A neurohistological correlate of schizophrenia. *Biol. Psychiat.* **19,** 1601–1621.
Kraemer G. W., Ebert M. H., Lake C. R., and McKinney W. T. (1983) Amphetamine challenge: Effects in previously isolated rhesus monkeys and implications for animal models of schizophrenia, in *Ethopharmacology: Primate Models of Neuropsychiatric Disorders* (Miczek K., ed.), Liss, New York, pp. 199–218.
Ladinsky H., Consolo S., Peri G., Crunelli V., and Samanin R. (1978) Pharmacological evidence for a serotoninergic–cholinergic link in the striatum, in *Cholinergic Mechanisms and Psychopharmacology* (Jenden D. J., ed.), Plenum, New York, pp. 615–627.
Lyon M. (1990) Animal models of mania and schizophrenia, in *Behavioural Models in Psychopharmacology* (Willner P., ed.), Cambridge University Press, Cambridge, UK.
Lyon M. and Barr C. E. (in press) Possible interactions of obstetrical complications and abnormal fetal brain development in schizophrenia, in *Fetal Neural Development and Adult Schizophrenia* (Mednick S. A., Cannon T. D., Barr C. E., and Lyon M., eds.), Cambridge University Press, New York.
Lyon M., Barr C. E., Cannon T. D., Mednick S. A, and Shore D. (1989) Fetal neural development and schizophrenia. *Schiz. Bull.* **15,** 149–161.
Lyon M. and Magnusson M. S. (1982) Central stimulant drugs and the learning of abnormal behavioral sequences, in *Behavioral Models and the*

Analysis of Drug Action (Spiegelstein, M. Y. and Levy, A., eds.), Elsevier Scientific, Amsterdam, pp. 135–153.

Lyon M. and Randrup A. (1972) The dose-response effect of amphetamine on avoidance behavior in the rat seen as a function of increasing stereotypy. *Psychopharmacologia (Berl.)* **23,** 324–347.

Lyon M. and Robbins T. W. (1975) The action of central nervous system stimulant drugs: A general theory concerning amphetamine effects, in *Current Developments in Psychopharmacology,* Vol. 2. (Essman W. and Valzelli L., eds.), Spectrum, New York, pp. 79–163.

Lyon N. and Gerlach J. (1988) Perseverative structuring of responses by schizophrenic and affective disorder patients. *J. Psychiat. Res.* **23,** 261–277.

Lyon N. and Satz P. (1991) Left-turning (swivel) in medicated chronic schizophrenia patients. *Schiz. Res.*

Lyon N., Mejsholm B., and Lyon M. (1986) Stereotyped responding in schizophrenic outpatients: Cross-cultural confirmation of perseverative switching on a two-choice task. *J. Psychiat. Res.* **20,** 137–150.

Makanjuola R. O. A., Hill G., Dow R. C., Campbell G., and Ashcroft G. W. (1977) The effects of psychotropic drugs on exploratory and stereotyped behaviour of rats studied on a hole-board. *Psychopharmacology* **55,** 67–74.

Mason W. A. (1968) Early social deprivation in the nonhuman primates: Implications for human behavior, in *Environmental Influences* (Glass, D. C., ed.), Rockefeller University Press and Russell Sage Foundation, New York, pp. 70–100.

Mednick, S. A. (1958) A learning theory approach to research in schizophrenia. *Psychological Bulletin* **55,** 316–327.

Mednick S. A., Cannon T. D., Barr C. E., and Lyon M., eds. (in press) *Fetal Neural Development and Adult Schizophrenia.* Cambridge University Press, Cambridge, UK.

Mednick S. A., Machon R. A., Huttunen M. O., and Bonett D. (1988) Adult schizophrenia following prenatal exposure to an influenza epidemic. *Arch. Gen. Psychiat.* **45,** 189–192.

Milner B. (1965) Memory disturbance after bilateral hippocampal lesions, in *Cognitive Processes and the Brain* (Milner P. and Glickman S., eds.), Van Nostrand, Princeton, NJ.

Morris R. G. M., Garrud P., Rawlins J. N. P., and O'Keefe J. (1982) Place navigation impaired in rats with hippocampal lesions. *Nature* **297,** 681–683.

Nielsen E. B. (1981) Rapid decline of stereotyped behavior in rats during constant one week administration of amphetamine via implanted ALZET osmotic minipumps. *Pharmacol. Biochem. Behav.* **15,** 161–165.

Nielsen E. B. and Lyon M. (1982) Behavioral alterations during prolonged low level continuous amphetamine administration in a monkey family group (Cercopithecus aethiops). *Biol. Psychiat.* **17,** 423–434.

Nielsen E. B., Lyon M., and Ellison G. (1983) Apparent hallucinations in monkeys during around-the-clock amphetamine for seven to fourteen

days. Possible relevance to amphetamine psychosis. *J. Nerv. Ment. Dis.* **171,** 222–233.
Nowakowski R. S. (1987) Basic concepts of CNS development. *Child Dev.* **58,** 568–595.
Oades R. D. (1982) Search strategies on a hole-board are impaired in rats with ventral tegmental damage: Animal model for tests of thought disorder. *Biol. Psychiat.* **17,** 243–258.
Olton D. S. (1983) Memory functions and the hippocampus, in *Neurobiology of the Hippocampus* (Siefert W., ed.), Academic, New York.
Olton D. S., Collison C., and Werz M. A. (1977) spatial memory and radial arm maze performance in rats. *Learn. Motiv.* **8,** 289–314.
Otake M. and Shull W. J. (1984) in utero exposure to A-bomb radiation and mental retardation: A reassessment. *Br. J. Radiol.* **57,** 409–414.
Papez J. W. (1937) A proposed mechanism of emotion. *Arch. Neurol. Psychiat.* **38,** 725–744.
Pardridge W. M. (1979) Regulation of amino acid availability to brain: Selective control mechanisms for glutamate, in *Glutamic Acid: Advances in Biochemistry and Physiology* (Filer L. J., Jr. et al., eds.), Raven, New York, pp. 125–137.
Petty F. and Sherman A. D. (1981) A pharmacologically pertinent animal model of mania. *J. Affective Disord.* **3,** 381–387.
Post R. M., Rubinow D., and Ballenger J. (1984) Conditioning, sensitization, and kindling: Implications for the course of affective illness, in *Neurobiology of Mood Disorders* (Post R. M. and Ballenger J., eds.), Williams & Wilkins Co., Baltimore, pp. 432–466.
Post R. M., Squillace K. M., Sass W., and Pert A. (1977) Drug sensitization and electrical kindling, in *Society for Neuroscience Abstracts* (Society for Neuroscience Meeting, Anaheim, CA).
Randrup A. and Munkvad I. (1967) Stereotyped activities produced by amphetamine in several animal species and Man. *Psychopharmacologia (Berl.)* **11,** 300–310.
Randrup A., Munkvad I., and Fog R. (1981) Mental and behavioural stereotypies elicited by stimulant drugs. Relation to the dopamine hypothesis of schizophrenia, mania, and depression, in *Recent Advances in Neuropsychopharmacology* (Angrist B., Burrows G. D., Lader M., Lingjaerde O., Sedvall G., and Wheatley D., eds.), Pergamon, Oxford, UK, pp. 63–74.
Robbins T. W. and Watson B. A. (1981) Effects of *d*-amphetamine on response repetition and win–stay behaviour in the rat, in *Quantification of Steady-State Operant Behaviour* (Bradshaw C. M., Szabadi W., and Lowe C. F., eds.), Elsevier/North-Holland, Amsterdam, pp. 441–444.
Robbins T. W., Evenden J. L., Ksir C., Reading P., Wood S., and Carli M. (1986) The effects of *d*-amphetamine, alpha flupenthixol and mesolimbic dopamine depletion on a test of attentional switching in the rat. *Psychopharmacology* **90,** 72–78.

Robinson T. E., Becker J. B., and Ramirez V. D. (1980) Sex differences in amphetamine-elicited rotational behavior and the lateralization of striatal dopamine in rats. *Brain Res. Bull.* **5,** 539–545.

Russell R. W. (1978) Cholinergic substrates of behavior, in *Cholinergic Mechanisms and Psychopharmacology* (Jenden D. J., ed.), Plenum, New York, pp. 709–731.

Sandberg K., Sanberg P. R., Hanin I., Fisher A., and Coyle J. T. (1984) Cholinergic lesion of the striatum impairs acquisition and retention of a passive avoidance response. *Behav. Neurosci.* **98,** 162–165.

Scheel-Kruger J. (1986) Dopamine–GABA interactions: Evidence that GABA transmits, modulates and mediates dopaminergic functions in the basal ganglia and the limbic system. *Acta Neurol. Scand. (Suppl.)* **107,** 1–54.

Scheel-Kruger J., Christensen A. V., and Arnt J. (1978) Muscimol differentially facilitates stereotypy but antagonizes motility induced by dopaminergic drugs: A complex GABA–dopamine interaction. *Life Sci.* **22,** 75–84.

Scheel-Kruger J., Arnt J., Magelund G., Olianas M., Przewlocka B., and Christensen A. V. (1980) Behavioural functions of GABA in basal ganglia and limbic system. *Brain Res. Bull.* **5 (Suppl. 2),** 261–267.

Schuberth J. (1978) Central cholinergic dysfunctions in man: Clinical manifestations and approaches to diagnosis and treatment, in *Cholinergic Mechanisms and Psychopharmacology* (Jenden D. J., ed.), Plenum, New York, pp. 733–745.

Simon H., Scatton B., and Le Moal M. (1980) Dopaminergic A10 neurones are involved in cognitive functions. *Nature* **286,** 150,151.

Singh M. M. and Kay S. R. (1976) Wheat gluten as a pathogenic factor in schizophrenia. *Science* **191,** 401–404.

Sloviter R. S., Damiano B. P., and Connor J. D. (1980) Relative potency of amphetamine isomers in causing the serotonin behavioral syndrome in rats. *Biol. Psychiat.* **15,** 789–796.

Spear L., Shalaby I. A., and Brick J. (1980) Chronic administration of haloperidol during development: Behavioral and psychopharmacological effects. *Psychopharmacology* **70,** 47–58.

Spear L. P., Kirstein C. L., Frambes N. A., Moody C. A., Miller J., and Spear, N. E. (1989) Neurobehavioral teratogenic effects of gestational cocaine exposure. *Proc. Am. Psychol. Soc.* **1,** 20.

Stevens J. R. (1975) GABA blockade, dopamine and schizophrenia: Experimental activation of the mesolimbic system. *Int. J. Neurol.* **10,** 115–127.

Stevens J. R. (1979): Schizophrenia and dopamine regulation in the mesolimbic system. *Trends Neurosci.* **2,** 102–105.

Stevens J. R. and Livermore A. L., Jr. (1978) Kindling of the mesolimbic dopamine system: Animal model of psychosis. *Neurology* **28,** 36–46.

Suomi S. J., Collins M. L., and Harlow H. F. (1973) Effects of permanent separation from mother on infant monkeys. *Dev. Psychol.* **9,** 376–384.

Swerdlow N. R., Braff D. L., Geyer M. A., and Koob G. F. (1986) Central dopamine hyperactivity in rats mimics abnormal acoustic startle response in schizophrenics. *Biol. Psychiat.* **21,** 23–33.

Tandon R. and Greden J. F. (1989) Cholinergic hyperactivity and negative schizophrenic symptoms. A model of cholinergic/dopaminergic interactions in schizophrenia. *Arch. Gen. Psychiat.* **46,** 745–753.

Teitelbaum P., and Derks P. (1958) The effect of amphetamine on forced drinking in the rat. *J. Comp. Physiol. Psychol.* **51,** 801–810.

Ungerstedt, U. and Arbuthnott, G. W. (1970) Quantitative recording of behavior in rats after 6-hydroxy-dopamine lesions of the nigrostriatal dopamine system. *Brain Res.* **24,** 85–493.

Van der Kooy D., Swerdlow N. R., and Koob G. F. (1983) Paradoxical reinforcing properties of apomorphine: Effects of nucleus accumbens and area postrema lesions. *Brain Res.* **259,** 111–118.

Van Ree J. M., Verhoeven W. M. A., De Wied D., and Van Praag H. M. (1982) The use of the synthetic peptides gamma-type endorphins in mentally ill patients, in *Opioids in Mental Illness,* Vol. 398, *Ann. NY Acad. Sci.* (Verebey K., ed.) New York Academy of Sciences, New York, pp. 478–495.

Verebey K., ed. (1982) Opioids in Mental Illness: Theories, Clinical Observations, and Treatment Possibilities. Vol. 398, *Ann. NY Acad. Sci.* New York Academy of Sciences, New York.

Wahlstrom A. and Terenius L. (1981) Endorphin hypothesis of schizophrenia. *Mod. Probl. Pharmacopsychiat.* **17,** 181–191.

Weinberger D. R., Torrey E. F., Neophytides A. N., and Wyatt R. J. (1979) Lateral cerebral ventricular enlargement in chronic schizophrenia. *Arch. Gen. Psychiat.* **36,** 735–739.

Weiss F., Tanzer D. J., and Ettenberg A. (1988) Opposite actions of CCK-8 on amphetamine-induced hyperlocomotion and stereotypy following intracerebroventricular and intra-accumbens injections in rats. *Pharmacol. Biochem. Behav.* **30,** 29–33.

Williams E. W. (1979) The effect of dietary wheat protein on rat behaviour. *J. Orthomol. Psychiat.* **8(2),** 113–117.

Williams E. W. (1980) *d*-Amphetamine behaviour in rats after a wheat protein challenge, and its reversal by naloxone hydrochloride. *J. Orthomol. Psychiat.* **14(4),** 281–292.

Wise R. A., Spindler J., de Wit H., and Gerber G. J. (1978) Neuroleptic-induced "anhedonia" in rats: Pimozide blocks the reward quality of food. *Science* **201,** 262–264.

Wong D. F., Wagner H. N., Jr., Tune L. E., Dannals R. F., Pearlson G. D., Links J. M., Tamminga C. A., Broussole E. P., Ravert H. T., Wilson A. A., Toung J. K. T., Malat J., Williams J. A., O'Tuama L. A., Snyder S. H., Kuhar M. J., and Gjedde, A. (1986) Positron emission tomography reveals elevated D2 dopamine receptors in drug-naive schizophrenics. *Science* **234,** 1558–1563.

The Hippocampal-Lesion Model of Schizophrenia

Nestor A. Schmajuk and Mabel Tyberg

1. Introduction

1.1. Summary

Animals with hippocampal lesions are evaluated as models for schizophrenia according to the criteria of McKinney and Bunney (1969): similarity of inducing conditions, similarity of behavioral states, similarity of underlying neurobiological mechanisms, and reversibility by usual pharmacological treatment. Hippocampal-lesioned animals seem to comply adequately with McKinney and Bunney's criteria because (a) schizophrenia might be the consequence of hippocampal damage resulting from anoxia or of hippocampal structural abnormalities induced by viral infection during pregnancy; (b) animals with hippocampal lesions share many of the characteristics of schizophrenics in both cognitive and psychophysiological processes; (c) hippocampal dysfunction seems to be present in schizophrenia; and (d) the effects of hippocampal lesions might be reversed by neuroleptics. The model seems to be able to reproduce several of the cognitive and psychophysiological symptoms of the disorder, offering a good experimental tool for the analysis of its inducing conditions, impaired neurobiological mechanisms, and clinical treatment.

1.2. Limbic Dysfunction in Schizophrenia

Exhaustive reviews of the biological basis of schizophrenia point to abnormalities in the limbic system of schizophrenic

From: *Neuromethods, Vol. 18: Animal Models in Psychiatry I*
Eds: A. Boulton, G. Baker, and M. Martin-Iverson

patients. Davison and Bagley (1969) reviewed schizophrenia-like psychoses associated with brain disorders, concluding that lesions in the temporal lobe and diencephalon are linked to psychotic behavior. Torrey and Peterson (1974) argued that the limbic–mesolimbic system might be altered in schizophrenics because (a) schizophrenic patients with implanted electrodes in limbic areas produce electrical abnormalities coinciding with their delusions, agitation, or hallucinations; (b) humans with known limbic-system neuropathologies have schizophrenic symptoms; (c) many patients with limbic seizures have schizophrenic symptoms; (d) lesions of the limbic system render animals unable to screen out multiple visual stimuli; (e) lesions or stimulation of the limbic structures in humans causes schizophrenic symptoms (paranoia, depersonalization, distortion of perception, violence, and catatonia); (f) chronic schizophrenic patients have different electroencephalographic profiles; and (g) psychedelic drugs produce maximal electrical abnormalities in the limbic system of monkeys. Seidman's (1983) review on schizophrenia and brain dysfunction concludes that (a) brain impairment exists in a substantial number of schizophrenic patients; (b) the type of brain dysfunction varies from patient to patient, arguing against a unitary-disease concept of schizophrenia; (c) positive symptoms (delusions, hallucinations, stereotypy) originate from pathophysiological dysfunction in limbic, midbrain, and upper-brainstem regions; (d) negative symptoms (apathy, blunted effect, social withdrawal, and avolition) originate from a frontal lobe dysfunction. Similarly, Weinberger et al. (1983) reviewed neuropathological studies of schizophrenia and concluded that pathology of the limbic system is associated with schizophrenia. More recently, Meltzer (1987) reviewed current research concerning the biological contribution to the etiology of schizophrenia. He pointed out that (a) computed tomography reveals ventricular enlargement and cortical atrophy in a subgroup of schizophrenic patients; (b) neuropathological studies demonstrate decreased volume and cell loss in the limbic system, basal ganglia, and frontal cortex; and (c) postmortem and positron emission tomography suggest an increased number of D_2 dopamine (DA) receptors in some schizophrenics.

Some theories of schizophrenia suggest that the limbic-system abnormalities are concentrated in the hippocampus. Mednick (1974a) proposed that hippocampal damage is associated with schizophrenic symptoms. Mednick pointed out that rats are hyperactive after hippocampal lesions (Kimble, 1968) and related this behavioral change to the state of hyperarousal characteristic of schizophrenics (Venables, 1966). Poor habituation in rats with hippocampal lesions (Kimble, 1968) would be related to the poor habituation of the galvanic skin response shown by schizophrenic patients (Zahn, 1964). The resistance to extinction of conditioning shown by lesioned rats (Isaacson et al., 1961) corresponds to the unusually large number of trials shown to be needed by schizophrenics to extinguish a conditioned plethysmographic response (Vinogradova, 1962). These results, according to Mednick (1974a), would be related to two physiological changes observed after hippocampal lesions. First, the lack of inhibition of adrenocorticotropic hormone (ACTH) release (Knigge, 1966), which plays a role during states of stress, can contribute to prolonging the extinction of a conditioned response (DeWied and Bohus, 1966). The second change is the lack of inhibition on the reticular formation after hippocampal lesions (Redding, 1967), which would produce hyperarousal. Venables (1973) also suggested that schizophrenia might be the result of hippocampal dysfunction. He pointed out hippocampal vulnerability to disturbances, the action of several tranquilizing agents on hippocampal function, the high methionine uptake by the hippocampus as related to the worsening of schizophrenic symptoms after methionine administration, and the association of temporal lobe epilepsy with psychosis. Also relating limbic dysfunction and schizophrenia, Frith and Done (1988) suggested that negative symptoms of schizophrenia (flattening of affect, poverty of speech, cognitive impairment) reflect a defect in the initiation of spontaneous action and are associated with damage to pathways connecting the prefrontal lobes and the basal ganglia. Positive symptoms (delusions, hallucinations) reflect a deficit in the internal monitoring of action and are associated with damage to pathways connecting the prefrontal cortex and the hippocampus.

Recently, Schmajuk (1987) analyzed how the hippocampally lesioned animal reproduces cognitive, psychophysiological, and neurophysiological changes found in schizophrenic patients. This chapter presents the hippocampal-lesion model and discusses how it meets the requirements proposed by McKinney and Bunney (1969) for an animal model of schizophrenia. In addition, it presents a theory of hippocampal function that helps to understand the behavioral changes suffered by animals with hippocampal lesions.

2. The Hippocampal-Lesion Model

2.1. Surgical Procedures

Given the relatively large size of the hippocampus and its accessibility from the surface of the brain, removal by aspiration is the technique of choice (Bures et al., 1976). Isaacson and Woodruff (1975) described a method of removal by aspiration of the hippocampus in the rat. Rats are anesthetized with sodium pentobarbital (50 mg/kg, administered intraperitoneally), and atropine sulfate (1 mg/kg, also administered intraperitoneally) is injected to inhibit secretions in the bronchi that might block respiration.

Animals are assigned to one of three groups. The hippocampal group receives bilateral aspiration of the hippocampus and the overlying cortex. The cortical group receives bilateral aspiration of the same part of the neocortex that had been aspirated in the hippocampal group. A sham group receives only two burr holes in the skull at the same place as those made in the previous groups when surgery began.

After the rat is anesthetized, its head is shaved and fixed in a head-holder. With a scalpel, the skull is exposed by making an incision at the midline behind the level of the eyes. The skin is retracted laterally and the muscles and connective tissue overlying the skull are scraped. The skull is cleaned and dried to expose the cranial sutures.

The point at which the anterior skull suture crosses the midline is called bregma. The point at which the posterior suture intersects the midline is called lambda. With a drill, two small holes are made approx 2 mm posterior to bregma and 2 mm lateral to the midline suture. The holes are enlarged with rongeurs, keeping away from the sagittal sinus to avoid profuse bleeding. The final holes in the skull should extend laterally from 2 mm off the midline to the bony ridge at the dorsal-lateral edge of the skull and then 2–3 mm down the side of the skull. The holes should be made bilaterally.

After the holes are finished, the dura is cut with the scalpel and removed with scissors and forceps. Aspiration is performed using a hypodermic needle connected to a suction pump attached to a water faucet. The needle is slightly bent at the tip and its end is cut and filed. Aspiration of the cortex proceeds in two steps. The first one reaches the corpus callosum, which can be distinguished by its whiter color. The second step consists of the careful removal of the corpus callosum to expose the hippocampus. The lateral ventricles, two fluid-filled spaces, lie under the corpus callosum. The hippocampi can be seen lying inside the ventricles. The hippocampus has a shiny surface and a whiter color than the overlying white matter of the cortex. Moving the needle in the rostral and caudal directions, the hippocampus is removed. The hippocampus is aspirated until the thalamus is seen beneath.

After finishing the procedure, the skin is sutured and the animal stays in the operating room until it recuperates from anesthesia. Animals are allowed to recover postoperatively in their home cages for two weeks.

2.2. Behavioral Testing

Many experimental paradigms can be used to evaluate the effect of hippocampal lesions. General changes in locomotor activity, exploration, spontaneous alternation, and neuroendocrine regulation have been evaluated in hippocampal-lesioned animals (Isaacson, 1982). More specific changes in avoidance conditioning, operant conditioning, discrimination learning, classical conditioning, and spatial learning have also been studied. Summaries of the effects of hippocampal lesions in different paradigms can be found in Gray (1982), Isaacson (1982), O'Keefe and Nadel (1978), and Schmajuk and Moore (1988).

2.3. Histological Procedures

A complete description of the morphological procedures is presented in Bures et al. (1976). After behavioral testing is concluded, animals are sacrificed with an overdose of pentobarbital (100 mg/kg, administered intraperitoneally). The animal is placed on its back, the heart exposed, and a needle inserted into the left ventricle. The right ventricle is opened to allow the venous blood to flow out of the system. A volume of 10 to 20 mL of saline is injected, followed by the same volume of 10% formalin. After perfusion, the animal is decapitated, the skull exposed, and the bone overlying the brain is cut off in small pieces with rongeurs. When the brain is totally exposed, it is held and lifted by the spinal cord with the help of forceps while the cranial nerves are cut with a pair of small scissors. The brain is fixed in 10% formalin and frozen sections are cut at 30-µm intervals. Every sixth section is retained, mounted, stained with thionine, and examined for the extent and location of the lesions. Figure 1 shows a typical lesions in rats with cortical and hippocampal ablations. The lined and black areas show the maximal and the minimal extent of the lesions, respectively.

3. Evaluation of the Model

McKinney and Bunney (1969) proposed four criteria for the evaluation of animal models of psychiatric disorders:

1. Similarity of inducing conditions;
2. Similarity of behavioral states;
3. Common underlying neurobiological mechanisms; and
4. Reversal by clinically effective treatment techniques.

The following sections analyze how animals with bilateral lesions of the hippocampus meet the requirements proposed by McKinney and Bunney (1969) for a model of schizophrenia.

3.1. Similarity of Inducing Conditions

Hippocampal anoxia and viral infections during pregnancy have been proposed as inducing mechanisms for schizophrenia.

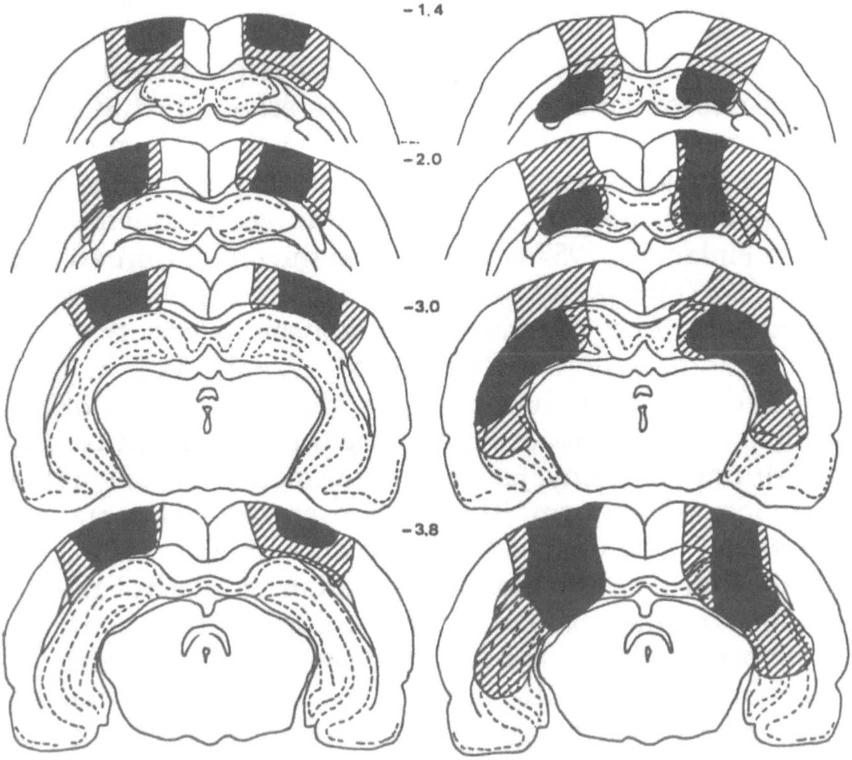

Fig. 1. Typical lesions in rats with cortical and hippocampal ablations. The lined areas show the maximal, and the black areas show the minimal, extent of the lesions. From Schmajuk et al. (1983). *Physiological Psychology* **11**, 59–62. Reprinted with permission.

3.1.1. Pregnancy or Birth Anoxia

Mednick (1974a) related the presence of schizophrenic symptoms to pregnancy or birth complications (PBCs), such as anoxia, prematurity, prolonged labor, placental difficulty, or mother's illness during pregnancy. These complications would trigger a genetically predisposed sensitivity of the hippocampus. In agreement with this view, Handford (1975) and Luchins et al. (1980) found that intrauterine or perinatal factors, such as anoxia, might cause brain damage that contributes to schizophrenia.

3.1.2. Viral Infections

Conrad and Scheibel (1987) suggested that the structural alterations found in the hippocampus of schizophrenic patients (*see* Section 4.3.2.) might result from maternal infection with one of several neurominidase-bearing viruses, especially during the second trimester of pregnancy. Mednick et al. (1987), reported that the incidence of schizophrenia in the offspring of mothers in their second trimester of pregnancy during an influenza epidemic in Finland in 1957 was increased threefold. Conrad and Scheibel (1987) proposed that genetic factors might have been involved in the form of reduced immunocompetence that rendered the mothers more susceptible to the viral infection.

In addition to the hippocampus, the organization of the entorhinal cortex (a major hippocampal input) might be disturbed during ontogeny. Jacob and Beckman (1986) suggested that structural abnormalities in the entorhinal cortex of schizophrenics presupposed a disturbance of neuronal migration in the fourth or fifth month of gestation. Similarly, Falkai et al. (1988) submitted that the reduced volume of the entorhinal cortex in brains of schizophrenics is developmental in origin.

In summary, if schizophrenia is induced by hippocampal damage from anoxia, hippocampal structural abnormalities induced by viral infection during pregnancy, or embryological alterations in the organization of the entorhinal cortex, the hippocampal-lesion model might be regarded as complying with the first criterion proposed by McKinney and Bunney (1969).

3.2. Similarity of Behavioral States

Schizophrenia is frequently characterized by disorders in thought, perception (delusions or hallucinations), affect (blunted and inappropriate), motor activity (rigidity or excitement), and interpersonal relationships (withdrawal) (Neale and Oltmanns, 1980). Schizophrenic individuals have been evaluated in different tests that determine cognitive abilities and psychophysiological manifestations. The results of these tests are assumed to reflect some of the above-mentioned schizophrenic symptoms. Many cognitive and psychophysiological characteristics of schizophrenic patients have been found after hippocampal lesions in animals. The following sections compare the performance of schizophrenic subjects and hippocampal-lesioned animals in various experimental paradigms.

3.2.1. Cognitive Processes

Schizophrenic subjects and hippocampal-lesioned animals have been assessed in different paradigms that explore their cognitive abilities.

3.2.1.1. Selective Attention

According to Payne et al. (1959), Weckowicz and Blewett (1959), and Neale and Cromwell (1968), schizophrenic *overinclusive thinking, loss of abstract reasoning,* and *thought disorder* are associated with a failure to selectively attend to relevant stimuli. Schizophrenic subjects are more distractible than normal subjects in a digit-span test in the presence of distracting stimuli (Lawson et al., 1967). Hemsley and Zawada (1976) found that schizophrenic (and depressive) patients have difficulties in distinguishing between relevant and irrelevant stimuli in a dichotic memory task.

Like schizophrenic patients, animals with hippocampal lesions show deficiencies in classical conditioning that have been interpreted as attentional deficits. For instance, Solomon and Moore (1975) found that lesioned animals fail to show latent inhibition. In latent inhibition, a conditioned stimulus (CS) is preexposed in the absence of the unconditioned stimulus (US), thereby making it irrelevant. When the CS is subsequently paired with the US, acquisition of the CS–US association is slower than when the CS has not been preexposed. Therefore, like schizophrenic patients, animals with hippocampal lesions have difficulties in distinguishing between relevant and irrelevant stimuli. As predicted by the hippocampal lesion model (Schmajuk, 1987), Baruch et al. (1988) reported that latent inhibition is absent in acute schizophrenics tested within the first week of the beginning of a schizophrenic episode.

Other classical conditioning paradigms affected by hippocampal lesions but not yet explored in schizophrenic patients are blocking and overshadowing. In blocking and overshadowing paradigms, animals selectively attend to the most relevant stimulus. In the first phase of a blocking paradigm, a CS, CS(A), is paired with the US. In the second phase, another CS, CS(B), is added and presented with CS(A) and the US. Under these circumstances, CS(B) accrues a weaker association with the US than that it would have attained had it been presented with only the US. It is assumed that animals do not learn the CS(B)–US association, because they attend to the most relevant CS(A). Overshadowing refers to the weaker conditioning of CS(B) when paired with a (more relevant) CS(A) than that it would have attained had it been presented with only the US. Rickert et al. (1978) and Solomon (1977) found that hippocampal ablation attenuates blocking. Rickert et al. (1979) and Schmajuk et al. (1983) found no overshadowing after hippocampal lesions.

3.2.1.2. Acquisition of Classical Conditioning

Pffafman and Schlosberg (1930) found that a knee-jerk response was conditioned faster in schizophrenic patients than in normal individuals. Mays (1934) and Shipley (1934) found that galvanic skin response was conditioned faster in schizophrenic patients. Correspondingly, Schmaltz and Theios (1972) and Schmajuk and Isaacson (1984) found that animals with hippocampal lesions showed facilitated classical conditioning.

3.2.1.3. Extinction of Classical Conditioning

The resistance to extinction shown by hippocampally lesioned rats under given conditions (Isaacson et al., 1961) corresponds to the unusually large number of trials required by schizophrenics to extinguish a conditioned plethysmographic response (Vinogradova, 1962).

3.2.1.4. Generalization

Bender and Schilder (1930) noted overgeneralization in conditioned withdrawal from shock in schizophrenic subjects. Cameron (1951) considered overinclusion and broadening of generalization as part of the schizophrenic syndrome.* Mastroiani (1979) and Solomon and Moore (1975) found that animals with hippocampal lesions showed augmented generalization to auditory stimuli.

*However, Harvey and Neale (1983) suggested that overinclusion is more common in manic than in schizophrenic patients.

3.2.1.5. Spatial Effects

In a size-constancy procedure, chronic schizophrenic subjects showed size underestimation (Weckowicz, 1957), a result that implies an impairment in evaluating the distance to the object. Correspondingly, animals with hippocampal lesions show important spatial deficits in place learning, i.e., when the distances to distal cues have to be correctly evaluated (Morris et al., 1982).

3.2.1.6. Contextual Effects

Normal individuals can take advantage of contextual information to retrieve adequate information from memory. If schizophrenic patients are deficient in this ability, their recall performance should not improve as a function of increasing contextual information. Lawson et al. (1964) found that, with noncontextual cues, schizophrenic and normal subjects did not differ, but that with more contextual information, the difference was significant. Chapman (1958) found that schizophrenic patients selected responses mediated by strong verbal associations, instead of conceptual relationships.* In correspondence with data showing a contextual deficit in schizophrenic patients, Hirsh et al. (1978) found that rats with hippocampal lesions are unable to retrieve information according to contextual cues (*see also* Hirsh, 1974).

3.2.1.7. Serial Position Curves

Oltmanns (1978) analyzed serial position curves for normal and schizophrenic individuals in the presence and absence of distraction, and found that schizophrenic individuals show a smaller primacy effect in the distraction, but not in the neutral, condition. In agreement with the results obtained in the distraction condition, Kesner and Novak (1982) reported that rats with hippocampal lesions show no primacy effect.

3.2.1.8. Recognition Memory

Bauman and Murray (1968) and Nachmani and Cohen (1969) reported that, although schizophrenic subjects recognized items as well as control subjects did, they were significantly impaired in their ability to recall stimulus items. In contrast, Gaffan

*Naficy and Willerman (1980) found that this problem was present in both schizophrenic and manic patients.

(1972) suggested that recognition memory, but not associative memory, is affected after hippocampal lesions.

3.2.1.9. Complex Learning

Hunt (1936) and Hunt and Cofer (1944) showed that schizophrenic patients performed poorly on complex tasks, mainly because they were retarded by irrelevant, incorrect responses. Similarly, animals with hippocampal lesions performed poorly in complex mazes, but not in simple ones (Kimble, 1963).

3.2.1.10. Stereotyped Behavior

Bleuler (1911/1950) pointed out the stereotyped behavior of schizophrenic patients. Devenport et al. (1980) found that rats with hippocampal lesions showed behavioral stereotypy after receiving signaled delivery of reward.

3.2.1.11. Superstitious Behavior

According to Matthysse (1981), superstitious behavior (operational responses maintained even if they are not contingent with reward) is a behavioral model for the failure in reality testing that is characteristic of schizophrenic individuals. Devenport (1979,1980) found that animals with hippocampal lesions behaved as if a relationship existed between their behavior and reinforcement, even when reward was noncontingent on their responses.

3.2.1.12. Discrimination Reversal

Kolb and Wishaw (1983) reported that schizophrenic patients had impaired performance in the Wisconsin Card-Sort Test. This test consists in sorting cards according to different stimuli: color, form, or number of elements. The correct stimulus changes without warning from one to another. Schizophrenic patients (and patients with frontal lesions) tend to continue responding to color when the correct stimulus has become form. An equivalent result has been found in hippocampal lesioned animals. Buchanan and Powell (1980), Berger and Orr (1983), Weikart and Berger (1986), and Port et al. (1986) found that animals with hippocampal lesions show normal acquisition of classical discrimination, but are impaired during discrimination reversal. As in the case of patients in the Wisconsin Card-Sort Test, the inferior performance of hippocampal-lesioned animals is attributable to an elevated response to the stimulus rewarded during the acquisition of discrimination.

3.2.2. Psychophysiological Processes

In addition to cognitive studies, schizophrenic patients and hippocampal-lesioned animals have been subject to different psychophysiological evaluations.

3.2.2.1. Hyperarousal

Chronic schizophrenic patients show spontaneous fluctuations in electrodermal activity (Thayer and Silber, 1971) and higher resting levels of skin conductance (Magaro, 1973; Zahn, 1964; Zahn et al., 1968). Both characteristics appear to be correlated with higher arousal. Mednick (1974a) related hyperarousal to the absence of corticosterone receptors subsequent to the destruction of the hippocampus, which would increase ACTH release and, as a consequence, increase the response to stress. The inhibitory control of ACTH by the hippocampus has been confirmed by Dunn and Orr (1984). Osborne and Seggie (1980) found increased corticosterone response to stress in rats with fornix lesions, and this increase is correlated with increased activity in the open field.* This hyperactivity in the open field had been previously reported by other authors (Kimble, 1968).

3.2.2.2. Habituation of the Orienting Response

As mentioned above, Zahn (1964) reported poor habituation of the galvanic skin response shown by schizophrenic patients. In addition, Zahn et al. (1968) found that schizophrenic patients were overreactive, showing a failure to habituate an orienting response. In a recent review, Berstein (1987) suggested that schizophrenics with negative symptoms show no orienting responses, whereas those with positive symptoms display heightened orienting responses with slow habituation.

*The adrenal cortex is stimulated to release corticosteroids by the ACTH carried by the blood from the adenohypophysis. The adenohypophysis is activated by the corticotropin-releasing factor (CRF) produced in the hypothalamus. Plasma corticosterone exerts a negative feedback on ACTH release, through corticosteroid receptors in the hippocampus.

Leaton (1981) found no difference between normal and hippocampal-lesioned animals in the habituation of the orienting response to neutral stimuli. Although animals with hippocampal lesions show no impairment in the orienting response to a neutral stimulus itself, their orienting response to neutral stimuli is attenuated when the novel stimulus is presented when the animal is engaged in a consummatory behavior (Crowne and Riddell, 1969; Hendrickson et al., 1969; Wickelgren and Isaacson, 1963). However, hippocampal-lesioned animals show an increased orienting response when biologically meaningful stimuli (a US or a CS associated with it) are presented during the acquisition of classical conditioning (Buchanan and Powell, 1980).

In summary, it seems that both schizophrenic individuals and hippocampal lesioned animals show impairment in habituation when biologically significant stimuli are presented, but they both habituate normally to less salient stimuli.

3.2.2.3. Skin Conductance Recovery

Mednick (1974b) and Venables (1974) have suggested that the faster skin conductance recovery (SCR) shown by schizophrenic patients might be linked to limbic factors. Bagshaw et al. (1965) had studied the galvanic skin response in monkeys with hippocampal ablation. When they reanalyzed the data for SCR, they found that it was faster in hippocampectomized monkeys than in amygdalectomized or control monkeys (Bagshaw and Kimble, 1972). Venables (1977) suggested that, since lesions of the hippocampus increase ACTH levels and adrenocortical steroids, faster sodium reabsorption and shorter SCR could be expected.

3.2.2.4. Event-Related Potentials

Shagass et al. (1978) reported that chronic schizophrenics show less variation and higher amplitudes in the early somatosensory-evoked potentials than normal subjects. They viewed these results as reflecting impaired (reticular) filtering. Baribeau-Braun et al. (1983) and Saitoh et al. (1984) found that P300 waves (event-related potentials typically elicited by unexpected stimuli) are present, but reduced in amplitude, in schizophrenic patients. A similar result was reported by Pritchard (1986). Paller et al.

(1988) found that monkeys with bilateral medial temporal lesions, which included amygdala and hippocampus, produced normal P300-like brain waves.

3.2.2.5. Heart Rate

Although Lang and Buss's (1965) review concluded that schizophrenics had a higher resting level than normal control individuals, more recent studies failed to find a difference between resting heart rates of schizophrenic patients and those of control individuals (Fowles et al., 1970; Goldstein and Acker, 1967; Gray, 1975). In agreement with the results in schizophrenics, Powell and Buchanan (1980) reported similar baseline heart rates for hippocampal-lesioned and control rabbits.

3.2.2.6. Polydipsia

Polydipsia has been consistently found in patients with psychiatric disorders, mainly schizophrenic patients (Goldman and Luchins, 1987). Correspondingly, Devenport (1978) reported that hippocampal lesions were followed by a rapid and uniform release of schedule-induced polydipsia, an unnecessarily frequent and intense drinking that accompanies the internal delivery of small pieces of food.* Devenport (1978) suggested that the hippocampus modulates schedule-induced polydipsia by inhibition of ACTH (*see* Section 3.2.2.1.).

Table 1 compares the performance of schizophrenic patients with the effect of hippocampal lesions in different experimental procedures. In almost 90% of the cases, the hippocampal-lesion model is able to reproduce the characteristics of schizophrenic patients in both cognitive and psychophysiological processes. Some of the symptoms listed in Table 1 are not exclusive to schizophrenic patients. However, it has been recognized that manic-depressive and schizophrenic disorders overlap in many cases (Swerdlow and Koob, 1987). Given the similarity of many (but not all) of the cognitive and psychophysiological attributes of schizophrenics and animals with hippocampal lesions, the hippocampal-lesion model seems, in general, to comply with the second of McKinney and Bunney's (1969) criteria.

*Interestingly, destruction of dopaminergic terminals in the nucleus accumbens attenuates polydipsia (Robbins and Koob, 1980).

Table 1
Comparison of Schizophrenic Symptoms, the Effect of Hippocampal Lesions, and the Predictions Generated by an Attentional Model of the Hippocampus in Different Experimental Paradigms

	Schizophrenia	Hippocampal lesion	Model
Selective attention	Deficit	Deficit	Deficit
Aquisition of classical conditioning	Facilitation	Facilitation	Facilitation
Extinction of classical conditioning	Deficit	Deficit	Deficit[b]
Generalization	Increased	Increased	Increased
Spatial effects	Deficit	Deficit	Deficit
Contextual effects	Deficit	Deficit	Deficit
Serial position curves	Deficit	Deficit	Deficit
Recognition memory	Normal	Deficit[a]	Deficit
Complex learning	Deficit	Deficit	Deficit
Stereotyped behavior	Increased	Increased	?[c]
Superstitious behavior	Increased	Increased	?
Discrimination reversal	Deficit	Deficit	Deficit
Arousal	Increased	Increased	Increased
Habituation of the orienting response	Deficit	Deficit	Deficit
Skin conductance recovery	Facilitated	Facilitated	?
Event-related potentials	Attenuated	Present	?
Heart rate	Normal	Normal	?
Polydipsia	Present	Present	?

[a]The animal model fails to accurately describe the schizophrenic symptom.
[b]The model predicts facilitation or deficit according to the duration of the conditioned stimulus.
[c]The theoretical model cannot describe the experimental result.

3.3. Similar Underlying Neurobiological Mechanisms

The next sections concentrate on the evidence showing that the hippocampus and several neurobiological systems linked to it seem to be affected in schizophrenia.

3.3.1. Reduced Volume of Limbic Areas

A common finding in the brain of schizophrenic patients has been an increased ventricular size, measured by pneumoencephalography (Huag, 1982) or by computed tomography

(Dewan et al., 1983). Because ventricular enlargement may result from degenerative processes of the different structures surrounding lateral ventricles, Lesch and Bogerts (1984) tried to establish the relationship between ventricular enlargement and the atrophy of periventricular regions. Lesch and Bogerts (1984) found no difference in the volume of the thalamus and large subthalamic nuclei between schizophrenic and control individuals. This result suggests that ventricle enlargement results from atrophy of the limbic periventricular areas. Bogerts et al. (1985) measured several parts of the basal ganglia and of the limbic system in postmortem brains of schizophrenic patients and control individuals. They found that the medial limbic structures of the temporal lobe (including amygdala, hippocampus, and parahippocampal gyrus) and the pallidum internum were significantly smaller in the schizophrenic group. In contrast, the volume of the putamen, nucleus caudatus, nucleus accumbens, and the bed nucleus of the stria terminalis did not differ between schizophrenic patients and control subjects.

Brown et al. (1986) compared the brains of patients with schizophrenia, affective disorders, Huntington's chorea, and Alzheimer's disease, and found that the brains of patients with schizophrenia had larger lateral ventricles and thinner parahippocampal cortices than the brains of patients with an affective disorder. However, the lateral ventricles and parahippocampal cortices of patients with schizophrenia were comparable to those of patients with Huntington's chorea or Alzheimer's disease. Schizophrenic patients differed from patients with Huntington's chorea in that the patients with Huntington's chorea showed reduced striatal mass. Schizophrenic patients also differed from the subjects with Alzheimer's disease, because the patients with Alzheimer's disease showed a generalized loss of temporal-lobe substance.

Recently, Suddath et al. (1989) analyzed magnetic resonance scans of schizophrenic patients' brains with computerized image processing. They found that (a) the volume of temporal gray matter was smaller in patients than in control subjects; (b) the lateral ventricular volume was larger in patients than in control subjects; and (c) the volume of prefrontal gray matter was comparable in patients and control subjects.

As in the case of reported reduced volumes of amygdala, hippocampus, and parahippocampal gyrus, Falkai et al. (1988) found that the volume of the entorhinal cortex, a main input to the hippocampus, is significantly reduced in postmortem brains of schizophrenic compared with control subjects. This reduction in volume is the result of a reduced number of neurons. Because there is no significant increase in the number of glial cells, Falkai et al. (1988) supported the hypothesis that structural changes in the medial temporal lobes of schizophrenic patients are developmental in origin.

3.3.2. Hippocampal, Entorhinal, and Cingulate Cell Disarray

Scheibel and Kovelman (1981) and Kovelman and Scheibel (1984) reported a change in the orientation of hippocampal pyramidal cells, mainly in the anterior and medial hippocampal regions. Pyramidal cells were rotated 35° with respect to the normal orientation. Conrad and Scheibel (1987) discarded the view of gliosis associated with neural cell loss as a mechanism capable of "tilting" neurons once they are in their final position and immersed in an extensive neuropil. They proposed, instead, that disturbed processes of cell migration and laminar organization during organization of the hippocampal primordium are the cause of neural disorientation in the hippocampus, and suggested that the disturbance might result from maternal viral infection during the second trimester of development. This suggestion is supported by data presented by Mednick et al. (1987), who reported that the incidence of schizophrenia in the offspring of mothers in their second trimester of pregnancy during an influenza epidemic in Finland in 1957 was increased threefold (*see* Section 3.1.1.). Altshuler et al. (1987) attempted to determine whether the disorientation in pyramidal cells found in brains of schizophrenic patients were present in patients never exposed, or briefly exposed, to neuroleptics (the Yakovlev collection). Although Altshuler et al. (1987) did not find a statistically significant greater amount of disorganization in the hippocampal pyramidal layer in the brain of schizophrenic individuals than in the brains of nonpsychotic control subjects, their data suggest a relationship between the degree of cell disarray and the severity of the psychosis.

In addition to changes in hippocampal organization, the entorhinal cortex also appears to be abnormal in the brains of schizophrenic patients. Jacob and Beckman (1986) found that brains of schizophrenic patients exhibit cytoarchitectonic abnormalities of the rostral entorhinal cortex in the parahippocampal gyrus and insular cortex. Another cortical region intimately connected to the hippocampus is the cingulate cortex. Benes and Bird (1987) found that the anterior cingulate cortex of the schizophrenic patient may contain aggregates of neurons in layer II that are smaller and more separated than those observed in control goups.

3.3.3. Regional Cerebral Blood Flow

Kurachi et al. (1985) measured regional cerebral blood flow in schizophrenic and control subjects. They found decreased flow in the frontal regions and slightly increased flow in the left temporal regions of patients having auditory hallucinations.

More recently, Weinberger et al. (1988) reported that medication-free patients with chronic schizophrenia fail to activate the regional blood flow in the dorsolateral prefrontal cortex during performance of the Wisconsin Card-Sort Test relative to the regional blood flow present during a number-matching task. The patients' performance during the Wisconsin Card-Sort Test correlated with the degree of prefrontal activation. In our view, the fact that the regional blood flow in dorsolateral prefrontal cortex is normal during the number-matching task, but not during the Wisconsin Card-Sort Test, suggests that another area is involved in the activation of the frontal lobes during the performance of the Wisconsin Card-Sort Test (*see* Section 3.2.1.12.).

3.3.4. Dopamine Changes in Nucleus Accumbens and Basal Ganglia

The DA theory (Carlsson, 1978) proposes that the neurobiological mechanism underlying psychotic symptoms is a DA excess in the limbic regions. This hypothesis is partly based on the fact that amphetamine psychosis resembles acute paranoid schizophrenia and that this effect can be reversed by neuroleptics (*see* Carlton and Manowitz, 1984). Oades and Isaacson (1978) suggested that the effect of hippocampal lesions was similar to that produced by enhanced dopaminergic activity (*see* Lyon and Robbins, 1975). Interestingly, hippocampal lesions reproduce many of the behavioral effects of amphetamine administration: extinction is retarded, spontaneous alternation is abolished, shuttle-box avoidance is improved, active avoidance is improved, passive avoidance is impaired, tasks requiring lower rates of responding prove difficult, and superstitious behavior is present (Devenport et al., 1980). In addition, some of the effects of hippocampal lesions also can be reversed by neuroleptics (*see* Section 3.4.1.).

Although DA antagonists are effective in treating schizophrenia, the disorder does not seem to be associated with hyperactivity of dopaminergic neurons (Bowers, 1974). Rather, schizophrenia seems to be correlated with an increase in DA receptors in the nucleus accumbens and basal ganglia, and this increase is not entirely attributable to neuroleptic medication (Cross et al., 1983). As in the brains of schizophrenic patients, Reinstein (1980) found an increase in DA receptors in the caudate and nucleus accumbens in the brain of rats with hippocampal lesions. In agreement with Reinstein's result, Bar et al. (1980) found increased membrane phosphorylation in the nucleus accumbens and the caudate of rats with hippocampal lesions, a result that indicates altered synaptic activities. Therefore, it is possible that the increase in DA receptors in nucleus accumbens and basal ganglia found in schizophrenic patients is the result of hippocampal dysfunction.

Although not conclusive, the presented evidence suggests that hippocampal dysfunction is present in schizophrenia. This dysfunction might be the consequence of pyramidal cell disorientation in the CA1 region of the hippocampus, a neuronal loss in the entorhinal cortex, or cell disarrangement in cingulate cortex. These cytoarchitectonic abnormalities are reflected in the reduced volumes of these areas. Hippocampal dysfunction might result in the increased number of DA receptors found in the nucleus accumbens and basal ganglia of schizophrenic patients. It is not clear, however, if the reduced blood flow in the frontal cortex of the schizophrenic patients is the consequence of hippocampal dysfunction, the effect of neuroleptics, or is caused by additional abnormalities of the schizophrenic brain. In any case, because the hippocampus appears to be one of the neurobiological mechanisms affected in schizophrenia, the hippocampal-lesion model seems to comply with the third of McKinney and Bunney's (1969) criteria.

3.4. Reversal by Chemically Effective Treatments

For animals with hippocampal lesions to be considered animal models of schizophrenia, their deficits should be relieved by the treatments used to treat schizophrenic patients. Typically, schizophrenic patients are treated with DA blockers, such as phenothiazines (chlorpromazine, fluphenazine, thioridazine), butyrophenones (haloperidol, spiroperidol), or thioxanthenes (flupenthixol, chlorprothixene).

Oades and Isaacson (1978) found that animals with hippocampal lesions made more errors (visits to nonrewarded holes) when required to locate pellets of food located in four of 16 holes in an open field. Administration of haloperidol reduced the number of visits to nonrewarded holes in the lesioned animals, but not in normal animals. Devenport et al. (1980) found that animals with hippocampal lesions had an exaggerated reaction to reward, consisting of behavioral stereotypies and locomotor activity similar to those that follow D-amphetamine administration. Behavior returned to normal after haloperidol administration. Isaacson (1980) reported that the hyperactivity and hyperresponsiveness found after hippocampal damage can be reduced by acute administration of neuroleptics. As mentioned above, it has been reported that latent inhibition is absent in acute schizophrenics tested within the first week of the beginning of a schizophrenic episode (Baruch et al., 1988). Latent inhibition returns to normal in medicated patients (Lubow et al., 1987; Baruch et al., 1988). A similar experiment waits to be attempted in animals.

Because the effects of hippocampal lesions that have been experimentally explored are reversed by the chemical treatments that are effective in schizophrenics, the hippocampal-lesion model seems to comply also with the fourth criterion of McKinney and Bunney (1969).

In general, the hippocampal-lesion model seems to reproduce some aspects of the inducing conditions, behavioral changes, underlying neurobiological mechanisms, and reversibility by chemical treatments that are found in schizophrenia.

4. An Attentional Theory of Hippocampal Function

This section introduces an attentional model of the hippocampus that describes many effects of hippocampal lesions and, therefore, the behavior of schizophrenic patients in different behavioral paradigms.

4.1. The Model

Schmajuk (1989; Schmajuk and Moore, 1985, 1988) introduced a real-time version of Pearce-Hall's (1980) attentional model, designated the S-P-H model. A formal description of the model is presented in Schmajuk and Moore (1985,1988). When a CS is followed by a US, an association, V, is formed; this association can be regarded as the prediction of the US by the CS. Associations between CSs and the US are controlled by the attentional term α, which is proportional to the mismatch between the actual and the aggregate prediction of the intensity of environmental events, such as the US. α Is also proportional to the intensity of the orienting response. The "aggregate prediction" hypothesis (Schmajuk, 1984) proposes that the hippocampus computes the aggregate prediction of environmental events. The aggregate prediction is used to compute α, which modulates associative learning. The following paragraphs describe how variables in the S-P-H model can be mapped into the brain.

The nucleus accumbens receives a glutaminergic excitatory input from the hippocampus via the subiculum (Walaas and Fonnum, 1979) and a dopaminergic inhibitory input from the ventral tegmental area (Oades et al., 1986; Cools and Van Rossum,

1980). Yang and Mogenson (1986) showed that activation of dopaminergic neurons that originate from the ventral tegmental area attenuates the excitatory response of the nucleus accumbens to hippocampal stimulation. The GABAergic output of the nucleus accumbens inhibits ventral pallidal and subpallidal areas (Mogenson and Nielsen, 1983). In turn, ventral and subpallidal areas activate the mesencephalic locomotor regions, which presumably control exploratory and orienting movements. Therefore, as the animal builds predictive associations about the environment, the hippocampal output increases and activates the nucleus accumbens. Activation of the nucleus accumbens leads to the inhibition of the ventral pallidum and subpallidum, and to a decrease of the activation of the mesencephalic locomotor regions. Decreased activity of the mesencephalic locomotor region translates into decreased orienting responses and exploratory movements.

Figure 2 shows how variables in the S-P-H model can be mapped onto the brain. The association of CS_1 with the US (V_1) and the association of CS_2 with the US (V_2) may be stored in different brain regions, depending on the type of learning (e.g., the cerebellum, in the case of the classical conditioning of the nictitating membrane preparation [Thompson, 1986], or the thalamus, in the case of avoidance behavior [Gabriel et al., 1987]). Neurons in entorhinal cortex are assumed to receive information about the associations established by CS_1 and CS_2 with the US. The hippocampal output to the nucleus accumbens is proportional to the aggregate prediction of the US ($CS_1 V_1 + CS_2 V_2$). We assume that the dopaminergic input is proportional to the activating value of the US (Beninger, 1983). Therefore, the output of the subpallidal area increases with increasing values of the US and decreases with increasing values of the aggregate prediction, $CS_1 V_1 + CS_2 V_2$. The output of the subpallidal area is similar to the attentional term α in the S-P-H model: It decreases as the prediction of the US increases.

At the beginning of learning, the aggregate prediction, $CS_1 V_1 + CS_2 V_2$, is very small, and the hippocampus sends a weak excitatory signal to the nucleus accumbens, which in turn does not inhibit orienting responses (α is large). With learning, the

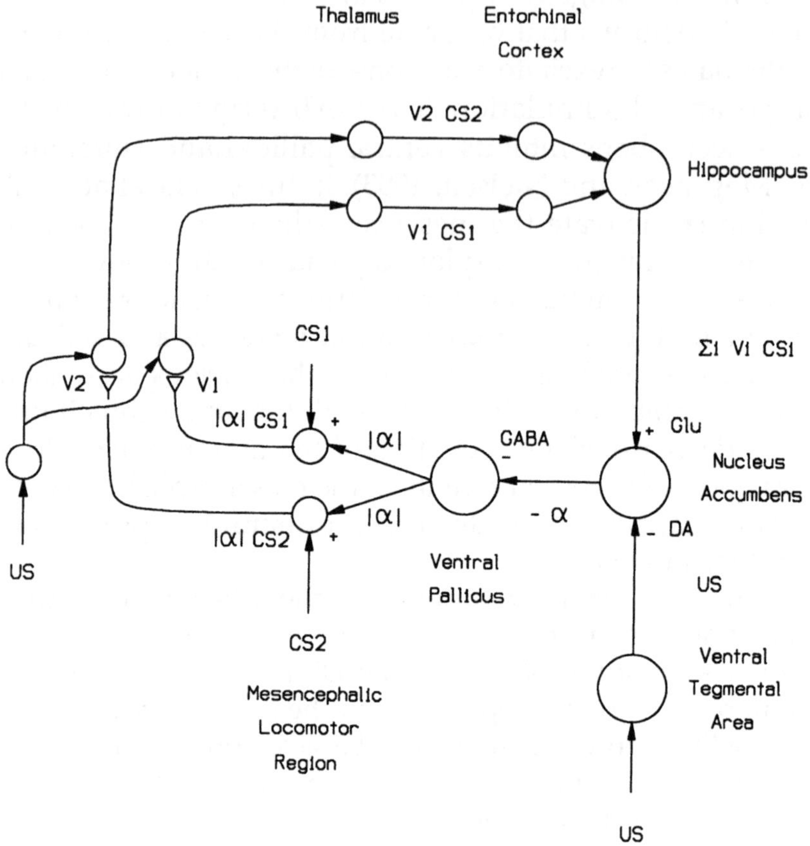

Fig. 2. An attentional model of hippocampal function. CS_1 and CS_2, conditioned stimuli; US, unconditioned stimulus; V_1 and V_2, CS–US associations; α, attentional term proportional to the mismatch between the actual and the aggregate prediction of the US intensity. Arrows represent nonmodifiable synapses. Triangles represent plastic synapses.

prediction of the US increases, and the hippocampus increases the excitation of the nucleus accumbens, which in turn inhibits orienting responses (α becomes small). In the case of blocking, when CS(A) predicts the US, orienting responses are inhibited and the animal does not learn the association between CS(B) and the US. In the case of latent inhibition, when CS(A) is predicted by the context, orienting responses to CS(A) are inhibited and the animal does not learn the association between CS(A) and the US.

4.2. Effects of Hippocampal Lesions

Because after hippocampal lesions aggregate predictions are no longer computed, α equals the value of the US and does not decrease as learning proceeds. Schmajuk (1989) and Schmajuk and Moore (1988) showed through computer simulations that the "aggregate-prediction" hypothesis correctly describes the effect of hippocampal lesions on a variety of learning paradigms. Attentional paradigms, such as blocking, are impaired because conditioning of CS(A) with the US does not decrease the value of α and, therefore, CS(B) can gain association with the US. Descriptions generated with the S-P-H attentional model under the aggregate-prediction hypothesis for hippocampal-lesioned animals performing in different behavioral tasks are shown in Table 1.

According to Fig. 2, hippocampal lesions and elevated DA activity have similar effects. Both procedures inhibit the inhibition of the nucleus accumbens on the motor regions controlling orienting responses, thereby increasing orienting responses and impairing selective attention to relevant stimuli. Figure 2 also shows that, because the hippocampal input to the nucleus accumbens is excitatory, hippocampal and nucleus accumbens lesions share many common effects (*see* Lorens et al., 1970).

4.3. Effects of Haloperidol

A possible site for haloperidol action is the nucleus accumbens. If the schizophrenic syndrome is related to a decreased or defective hippocampal input to the nucleus accumbens, blocking the activity of DA tends to reinstate the excitatory–inhibitory equilibrium and to decrease the generation of orienting responses and motor hyperactivity. This decrease in orienting responses results in a decrease in the formation of new associations, both relevant and irrelevant. Therefore, the effect of DA blockers is not functionally equivalent to the hippocampal function.

5. Discussion

Animals with hippocampal lesions seem to comply with the McKinney and Bunney (1969) criteria for animal models of schizophrenia: similarity of inducing conditions, similarity of behavioral states, similarity of underlying neurobiological mechanisms, and reversibility by chemical treatments. The good correlation between cognitive and psychophysiological processes in schizophrenic patients and hippocampal-lesioned animals (*see* Table 1) tends to support the belief that the hippocampus is impaired in schizophrenia (Mednick, 1974a) and the view that an attentional function is affected in this disorder (Oades, 1982).

The hippocampal-lesion model analyzed in this chapter supports the hippocampal view of schizophrenia. This view can be summarized in the following terms:

1. Schizophrenia is the result of disorientation of hippocampal pyramidal cells, loss of entorhinal cortex cells, or disarrangement in cingulate cortex cells.
2. These pathological alterations translate into an altered hippocampal function, as described by an attentional model of the hippocampus (Schmajuk and Moore, 1988).
3. This attentional impairment underlies the changes in cognitive processes found in schizophrenic patients and hippocampal lesioned animals.
4. Abnormal behavior is improved by neuroleptics because they reinstate a modulatory (inhibitory) function of the hippocampus on the formation of associations.

Albeit less complete than in the case of the hippocampus, some evidence suggests that the frontal cortex is compromised in schizophrenia. Some studies indicate that schizophrenic patients have significantly smaller frontal lobes (Andreasen et al., 1986; *but see* Suddath et al., 1989 and Andreasen et al., 1990), fail to increase regional blood flow in response to task demands (Weinberger et al., 1988), and show deficits in tests sensitive to frontal cortex dysfunction (Kolb and Wishaw, 1983). In addition, destruction of dopamine terminals in the medial prefrontal cortex of the rat with 6-hydroxydopamine produces presynaptic and postsynaptic hyperactivity in nucleus accumbens and striatum (Pycock et al.,1980), and depletion of DA in the frontal-lobe has effects similar to those of frontal cortex lesions (Brozoski et al., 1979). Unfortunately, the frontal-lobe-lesion animal model of schizophrenia is inconsistent with the finding of Brozoski et al. (1979) that the behavioral deficits resulting from the biochemical lesion of the frontal cortex can be reversed by peripheral L-dopa and apomorphine, drugs that exacerbate, rather than improve, schizophrenic symptoms.

In addition to the description of an animal model, the present chapter offers a model of the hippocampus that helps to understand attentional processes in normal and pathological behavior. Recently, Swerdlow and Koob (1987) proposed a unified model of cortico-striato-pallido-thalamic function in schizophrenia, mania, and depression. According to the model, the limbic system excites the nucleus accumbens, the nucleus accumbens inhibits the ventral pallidum, and the ventral pallidum inhibits the dorsomedial thalamic nucleus, which in turn sends back information to the limbic system. According to Swerdlow and Koob, increased DA activity inhibits the nucleus accumbens, and in consequence inhibits the positive-feedback loop between the limbic system and the thalamus. Inhibition of the nucleus accumbens results in impaired filtering of irrelevant information, and inhibition of corticothalamic interactions results in limited cognitive processes. It should be noted that, in Swerdlow and Koob's model, the effect of hippocampal (limbic system) lesions is equivalent to that of increased DA activity: They both result in decreased activity of the nucleus accumbens.

In conclusion, animals with bilateral lesions of the hippocampus seem to provide an adequate animal model for many of the cognitive and psychophysiological disorders associated with the positive symptoms of schizophrenia. A better comprehension of hippocampal function, through animal research and model building, may be fruitful for improved understanding and treatment of schizophrenic disorders.

Acknowledgments

The authors thank John Gabrieli and Daniel Luchins for their comments on this manuscript. This project was supported in part by BRSG SO7 RR07028-22, awarded by the Biomedical Research Support Grant Program, Division of Research Resources, National Institutes of Health.

References

Altshuler L. L., Conrad A., Kovelman J. A., and Scheibel A. (1987) Hippocampal pyramidal cell orientation in schizophrenia. *Arch. Gen. Psych.* **44**, 1094–1098.

Andreasen N. C., Ehrhardt J. C., Swayze V. W., Alliger R. J., Yuh W. T. C., Cohen G., and Ziebell S. (1990) Magnetic resonance imaging of the brain in schizophrenia. *Arch. Gen. Psych.* **47**, 35–44.

Andreasen N., Nasrallah H. A., Dunn V., Olson S. C., Grove W. M., Ehrhardt J. C., Coffman J. A., and Crossett J. H. W. (1986) Structural abnormalities in the frontal system in schizophrenia. *Arch. Gen. Psych.* **43**, 136–144.

Bagshaw M. H. and Kimble D. P. (1972) Bimodal EDR orienting response characteristic of limbic lesioned monkeys: Correlates with schizophrenic patients. Annual Meeting, Society for Psychophysiological Research.

Bagshaw M. H., Kimble D. P., and Pribram K. H. (1965) The GSR of monkeys during orientation and habituation and after ablation of amygdala, hippocampus, and inferotemporal cortex. *Neuropsychologia* **3**,111–119.

Bar P. R., Gispen W. H., and Isaacson R. L. (1980) Behavioral and regional sequelae of hippocampal destruction in the rat. *Pharmacol. Biochem. Behav.* **14**, 305–312.

Baribeau-Braun J., and Picton T. W., and Gosselin J. (1983) Schizophrenia: A neurophysiological evaluation of abnormal information processing. *Science* **219**, 874–876.

Baruch I., Hemsley D., and Gray J. A. (1988) Differential performance of acute and chronic schizophrenics in a latent inhibition task. *J. Nerv. Ment. Dis.* **176**, 598–606.

Bauman E. and Murray D. J. (1968) Recognition versus recall in schizophrenia. *Can. J. Psychol.* **22**,18–25.

Bender L. and Schilder P. (1930) Unconditioned reactions to pain in schizophrenics. *Amer. J. Psychiat.* **10**, 365–384.

Benes F. M and Bird E. D. (1987) An analysis of the arrangement of neurons in the cingulate cortex of schizophrenic patients. *Arch. Otolaryng.—Head Neck Surg.* **113**, 608–616.

Beninger R. J. (1983) The role of dopamine in locomotor activity and learning. *Brain Res. Rev.* **6**, 173–196.

Berger T. W. and Orr W. B. (1983) Hippocampectomy selectively disrupts discrimination reversal conditioning of the rabbit nictitating membrane response. *Behav. Brain Res.* **8**, 49–68.

Berstein A. (1987) Orienting response in schizophrenia, Where we have come and where we might go. *Schizophr. Bull.* **13**, 623–641.

Bleuler E. (1911/1950) *Dementia Praecox or the Group of Schizophrenias* (Translated by Zinkin E.) International Universities Press, New York.

Bogerts B., Meertz E., and Schonfeld-Bausch R. (1985) Basal ganglia and limbic system pathology in schizophrenia. *Arch. Gen. Psychiatry* **42**, 784–791.

Bowers M. B. (1974) Central dopamine turnover in schizophrenic syndromes. *Arch. Gen. Psychiatry* **31,** 50–54.

Brown R., Colter N., Corsellis J. A. N., Crow T. J., Frith C. D., Jagoe R., Johnstone E. C., and Marsh L. (1986) Postmortem evidence of structural brain changes in schizophrenia. *Arch. Gen. Psychiatry* **43,** 36–42.

Brozoski T. J., Brown R. M., Roswold H. E., and Goldman P. S. (1979) Cognitive deficit caused by regional depletion of dopamine in prefrontal cortex of Rhesus monkey. *Science* **205,** 929–932.

Buchanan S. L. and Powell D. A. (1980) Divergences in Pavlovian conditioned heart rate and eyeblink responses produced by hippocampectomy in the rabbit *(Oryctolagus Cuniculus). Behav. Neural Biol.* **30,** 20–38.

Bures J., Buresova O., and Huston J. (1976) *Techniques and Basic Experiments for the Study of Brain and Behavior.* Elsevier/North Holland Biomedical Press, Amsterdam.

Cameron, N. (1951) Perceptual organization and behavior pathology, in *Perception: An Approach to Personality* (Blake R. R. and Ramsey G. V., eds.) Ronald, New York, pp. 283–306.

Carlsson A. (1978) Does dopamine have a role in schizophrenia? *Biol. Psychiatry* **13,** 3–21.

Carlton P. L. and Manowitz P. (1984) Dopamine and schizophrenia: An analysis of the theory. *Neurosci. Biobehav. Rev.* **8,** 137–151.

Chapman L. J. (1958) Intrusion of associative responses into schizophrenic conceptual performance. *J. Abnorm. Soc. Psychol.* **56,** 374–379.

Conrad A. J. and Scheibel A. B. (1987) Schizophrenia and the hippocampus: The embryological hypothesis extended. *Schizophr. Bull.* **13,** 567–577.

Cools A. R. and Van Rossum J. M. (1980) Multiple receptors for brain dopamine in behavior regulation: Concept of dopamine–E and dopamine-I receptors. *Life Sci.* **27,** 1237–1253.

Cross A. J., Crow T. J., Ferrier I. N., Johnstone E. C., McCreadie R. M., Owen F., Owens D. G. C., and Poulter M. (1983) Dopamine receptor changes in schizophrenia in relation to the disease process and movement disorder. *J. Neural Trans. (Suppl.)* **18,** 265–272.

Crowne D. P. and Riddell W. I. (1969) Hippocampal lesions and the cardiac component o the orienting response in the rat. *J. Comp. Physiol. Psychol.* **69,** 748–755.

Davison K. and Bagley C. R. (1969) Schizophrenia-like psychosis associated with organic disorders of the central nervous system: A review of the literature. *Brit. J. Psychiatry* **4,** 113–187.

Devenport L. D. (1978) Schedule-induced polydipsia in rats: Adrenocortical and hippocampal modulation. *J. Comp. Physiol. Psychol.* **92,** 651–660.

Devenport L. D. (1979) Superstitious bar pressings in hippocampal and septal rats. *Science* **205,** 721–723.

Devenport L. D. (1980) Response-reinforcer relationships and the hippocampus. *Behav. Neural Biol.* **29,** 105–110.

Devenport L. D., Devenport J. A., and Holloway F. A. (1980) Reward-induced stereotypy: Modulation by the hippocampus. *Science* **212,** 1288,1289.

Dewan M. J., Pandurangi A. K., Lee S. H., Ramachandran T., Levy B., Boucher M., Yozawitz A., and Major L. (1983) Central brain morphology in chronic schizophrenic patients: A controlled study. *Biol. Psychiatry* **18,** 1133–1139.

DeWied D. and Bohus B. (1966) Long term and short term effects on retention of a conditioned avoidance in rats treated with long acting pitressin and MSH. *Nature* **212,** 1484–1486.

Dunn J. D. and Orr S. E. (1984) Differential plasma corticosterone responses to hippocampal stimulation. *Exp. Brain Res.* **54,** 1–6.

Falkai P., Bogerts B., and Rozumek M. (1988) Limbic pathology in schizophrenia: The entorhinal region—a morphometric study. *Biol. Psychiatry* **24,** 515–521.

Fowles D. C., Watt N. F., Maher B. A., and Grinspoon L. (1970) Autonomic arousal in good and premorbid schiophrenics. *Brit. J. Soc. Clin. Psychol.* **9,** 135–147.

Frith C. D and Done D. J. (1988) Towards a neuropsychology of schizophrenia. *Brit. J. Psychiat.* **153,** 437–443.

Gabriel M., Sparenborg S., and Stolar N. (1987) Hippocampal control of cingulate cortical and anterior thalamic information processing during learning in rabbits. *Exp. Brain Res.* **67,** 131–152.

Gaffan D. (1972) Loss of recognition memory in rat lesions of the fornix. *Neuropsychologia* **10,** 327–341.

Goldman M. B. and Luchins D. J. (1987) Prevention of episodic water intoxication with target weight procedure. *Amer. J. Psychiat.* **144,** 3.

Goldstein M. J. and Acker C. W. (1967) Psychophysiological reactions to films by chronic schizophrenics: II. Individual differences in resting levels and reactivity. *J. Abnorm. Psychol.* **72,** 23–29.

Gray A. L. (1975) Autonomic correlates of schizophrenia: A reaction time paradigm. *J. Abnorm. Psychol.* **84,** 189–196.

Gray J. A. (1982) *The Neuropsychology of Anxiety: An Inquiry into the Functions of the Septo-Hippocampal System.* Oxford University Press, New York.

Handford H. A. (1975) Brain hypoxia, minimal brain dysfunction and schizophrenia. *Amer. J. Psychiat.* **132,** 192–194.

Harvey P. D. and Neale J. M. (1983) The specificity of thought disorder to schizophrenia: Research methods in their historical perspective, in *Progress in Experimental Personality Research* vol. 12 (Maher B. A., ed.) Academic, New York, pp. 153–180.

Hemsley D. R. and Zawada S. L. (1976) "Filtering" and the cognitive deficits in schizophrenia. *Brit. J. Psychiat.* **128,** 456–461.

Hendrickson C. W., Kimble R. J., and Kimble D. P. (1969) Hippocampal lesions and the orienting response. *J. Comp. Physiol. Psychol.* **67,** 220–227.

Hirsh R. (1974) The hippocampus and contextual retrieval of information from memory: A theory. *Behav. Biol.* **12,** 421–444.

Hirsh R., Holt L., and Mosseri A. (1978) Hippocampal mossy fibers, motivational states, and contextual retrieval. *Exp. Neurol.* **62,** 68–79.

Huag O. (1982) Pneumoencephalographic evidence of brain atrophy in acute and chronic schizophrenia. *Acta Psychiatr. Scand.* **66**, 374.

Hunt J. McV. (1936) Psychological experiments with disordered persons. *Psychol. Bull.* **33**, 1–58.

Hunt J. Mcv., and Cofer C. N. (1944) Psychological deficit, in *Personality and the Behavior Disorder* (Hunt J. McV., ed.), Ronald Press, New York, pp. 971–1032.

Isaacson R. L. (1980) A perspective for the interpretation of limbic system function. *Physiol. Psychol.* **8**, 183–188.

Isaacson R. L. (1982) *The Limbic System* (Plenum, New York).

Isaacson R. L. Douglas R. J., and Moore R. Y. (1961) The effect of radical hippocampal ablation on acquisition of an avoidance response. *J. Comp. Physiol. Psychol.* **54**, 625–628.

Isaacson R. L. and Woodruff M. (1975) Spontaneous alternation and passive avoidance behavior in rats after hippocampal lesions, in *Experimental Psychology* (Hart B., ed.) Freeman Press, San Francisco.

Jacob H. and Beckman H. (1986) Prenatal developmental disturbances in the limbic allocortex in schizophrenics. *J. Neural Trans.* **65**, 303–326.

Kesner R. P. and Novak J. M. (1982) Serial position curve in rats: Role of the dorsal hippocampus. *Science* **218**, 173–175.

Kimble D. P. (1963) The effects of bilateral hippocampal lesions in rats. *J. Comp. Physiol. Psychol.* **56**, 337–340.

Kimble D. P. (1968) Hippocampus and internal inhibition. *Psychol. Bull.* **70**, 285–295.

Knigge K. M. (1966) Feedback mechanisms in neural control of adrenohypophyseal function: Effect of steroids implanted in amygdala and hippocampus. Second International Congress on Hormonal Steroids, Milan, Italy.

Kolb B. and Wishaw I. Q. (1983) Performance of schizophrenic patients on tests sensitive to left or right frontal, temporal, or parietal function in neurological patients. *J. Nerv. Ment. Dis.* **171**, 435–443.

Kovelman J. A. and Scheibel A. B. (1984) A neurohistological correlate of schizophrenia. *Biol. Psychiat.* **19**, 1601–1621.

Kurachi M., Kobayashi K., Matsubara R., Hiramatsu H., Yamaguchi N., Matsuda H., Maeda T., and Hisada K. (1985) Regional cerebral blood flow in schizophrenic disorders. *Eur. Neurol.* **24**, 176–181.

Lang P. J. and Buss A. H. (1965) Psychological deficit in schizophrenia: II. Interference and activation. *J. Abnorm. Psychol.* **70**, 77–106.

Lawson J. S., McGhie A., and Chapman J. (1967) Distractibility and organic cerebral disease. *Brit. J. Psychiat.* **113**, 527–535.

Leaton R. N. (1981) Habituation of startle response, lick suppression, and exploratory behavior in rats with hippocampal lesions. *J. Comp. Physiol. Psychol.* **95**, 813–826.

Lesch A. and Bogerts B. (1984) The diencephalon in schizophrenia: Evidence for reduced thickness of the periventricular grey matter. *Eur. Arch. Psychiatr. Neurol. Sci.* **234**, 212–219.

Lorens S. A., Sorenson J. P., and Harvey J. A. (1970) Lesions of the nuclei accumbens septi of the rat: Behavioral and neurochemical effects. *J. Comp. Physiol. Psychol.* **73,** 284–290.

Lubow R. E., Weiner I., Schlossberg A., and Baruch I. (1987) Latent inhibition and schizophrenia. *Bull. Psychon. Soc.* **25,** 464–467.

Luchins D., J. Pollin W., and Wyatt R. J. (1980) Laterality in monozygotic schizophrenic twins: An alternative hypothesis. *Biol. Psychiat.* **15,** 87–93.

Lyon L. and Robbins R. (1975) The action of central nervous system stimulant drugs: A general theory concerning amphetamine effects. *Curr. Dev. Psychopharmacol.* **2,** 81–162.

Magaro P. A. (1973) Skin conductance basal level and reactivity in schizophrenia as a function of chronicity, premorbid adjustment, diagnosis and medication. *J. Abnorm. Psychol.* **76,** 242–248.

Mastroiani P. P. (1979) Hippocampal lesions and the generalization of auditory stimuli. *Neuropsychologia* **17,** 401–412.

Matthysse S. (1981) Nucleus accumbens and schizophrenia, in *Neurobioiogy of the Nucleus Accumbens* (Chronister R. B. and DeFrance J. F., eds.) Haer Institute, Brunswick, ME, pp. 351–359.

Mays L. L. (1934) Studies on catatonia: V. Investigation of the perseverational tendency. *Psychiatr. Quart.* **8,** 728.

McKinney W. T. and Bunney W. E., Jr. (1969) Animal models for depression. A review of evidence: Evidence for Research. *Arch. Gen. Psychiat.* **21,** 240–248.

Mednick S. A. (1974a) Breakdown in individuals at high risk for schizophrenia: Possible predispositional perinatal factors, in *Genetics, Environment, and Psychopathology* (Mednick A., Schulsinger F., Higgins J., and Bell B., eds.) North Holland, Amsterdam, pp. 249–262.

Mednick, S. A. (1974b) Electrodermal recovery and psychopathology. in *Genetics, Environment, and Psychopathology* (Mednick A., Schulsinger F., Higgins J., and Bell B., eds.) North Holland, Amsterdam, pp. 135–148.

Mednick S. A., Parnes J., and Schulsinger F. (1987) The Copenhagen high-risk project, 1962–1986. *Schizophr. Bull.* **5,** 460–479.

Meltzer H. Y. (1987) Biological studies in schizophrenia. *Schizophr. Bull.* **13,** 77–111.

Mogenson G. J. and Nielsen M. A. (1983) Evidence that an accumbens to subpallidal GABAergic projection contributes to locomotor activity. *Brain. Res. Bull.* **11,** 309–314.

Morris R. G. M., Garrud P., Rawlins J. N. P., and O'Keefe J. (1982) Place navigation impaired in rats with hippocampal lesions. *Nature* **297,** 681–683.

Nachmani G. and Cohen B. D. (1969) Recall and recognition free learning in schizophrenics. *J. Abnorm. Psychol.* **74,** 511–516.

Naficy A. and Willerman L. (1980) Excessive yielding to normal biases is not a distinctive sign of schizophrenia. *J. Abnorm. Psychol.* **89,** 697–703.

Neale J. M. and Cromwell R. L. (1968) Size estimation in schizophrenics as a function of stimulus presentation time. *J. Abnorm. Psychol.* **73,** 44–49.

Neale J. M. and Oltmanns T. F. (1980) *Schizophrenia.* Wiley, New York.

Oades R. D. (1982) *Attention and Schizophrenia.* Pitman, London.

Oades R. D. and Isaacson R. L. (1978) The development of food search behavior by rats: The effects of hippocampal damage and haloperidol treatment. *Behav. Biol.* **24,** 327–337.

Oades R. D., Taghzouti K., Rivet J. M., Simon H., and LeMoal M. (1986) Locomotor activity in relation to dopamine and noradrenaline in the nucleus accumbens, septal and frontal areas: A 6–hydroxydopamine study. *Neuropsychobiology* **16,** 37–42.

O'Keefe J. and Nadel L. (1978) *The Hippocampus as a Cognitive Map.* Clarendon, Oxford, UK.

Oltmanns T. F. (1978) Selective attention in schizophrenia and manic psychoses: The effect of distraction on information processing. *J. Abnorm. Psychol.* **87,** 212–225.

Osborne B. and Seggie J. (1980) Behavioral, corticosterone, and prolactin responses to novel environment in rats with fornix transections. *J. Comp. Physiol. Psychol.* **94,** 536–546.

Paller K. A., Zola-Morgan S., Squire L. R., and Hillyard S. A. (1988) P3-like brain waves in normal monkeys and in monkeys with medial temporal lesions. *Behav. Neurosci.* **102,** 714–725.

Payne R. W., Matussek P., and George E. I. (1959) An experimental study of schizophrenic thought disorder. *J. Ment. Sci.* **105,** 627–652.

Pearce J. M. and Hall G. (1980) A model for Pavlovian learning: Variations in the effectiveness of unconditioned but not unconditioned stimuli. *Psychol. Rev.* **87,** 532–552.

Pffafman C. and Schlosberg H. (1930) The conditioned knee jerk in psychotic and normal individuals. *J. Psychol.* **1,** 201–206.

Port R. L., Romano A. G., Paterson M. M. (1986) Stimulus duration discrimination in the rabbit: Effects of hippocampectomy on discrimination and reversal learning. *Physiol. Psychol.* **14,** 124–129.

Powell D. A. and Buchanan S. (1980) Autonomic-somatic relationships in the rabbit *(Oryctologus cuniculus)*: Effects of hippocampal lesions. *Physiol. Psychol.* **8,** 455–462.

Pritchard W. S. (1986) Cognitive event-related potential in schizophrenics. *Psychol. Bull.* **100,** 43–66.

Pycock C. J., Kerwin R. W., and Carter C. J. (1980) Effect of lesion of cortical dopamine terminals on subcortical dopamine receptors in rats. *Nature* **286,** 74–76.

Redding F. K. (1967) Modification of sensory cortical evoked potentials by hippocampal stimulation. *Electroencephlogr. Clin. Neurophysiol.* **22,** 74–83.

Reinstein D. K. (1980) Behavioral and Biochemical Changes after Hippocampal Damage. Unpublished PhD Dissertation, SUNY-Binghamton, New York.

Rickert E. J. and Bennett T. L., and French J. (1978) Hippocampectomy and the attenuation of blocking. *Behav. Biol.* **22,** 597–609.

Rickert E. J., Lorden J. F., Dawson R., Smyly E., and Callahan M. F. (1979) Stimulus processing and stimulus selection in rats with hippocampal lesions. *Behav. Neural Biol.* **29,** 454–465.

Robbins T. W. and Koob G. F. (1980) Selective disruption of displacement behaviour by lesions of the mesolimbic dopamine system. *Nature* **285,** 409–412.

Saitoh O., Niwa S. I., Hiramatsu K. I., Kameyama T., Rymar K., and Itoh K. (1984) Abnormalities in late positive components of event-related potentials may reflect a genetic predisposition to schizophrenia. *Biol. Psychiat.* **19,** 293–303.

Scheibel A. B. and Kovelman J. A. (1981) Disorientation of the hippocampal pyramidal cell and its processes in the schizophrenic patient. *Biol. Psychiat.* **16,** 101–102.

Schmajuk N. A. (1984) A model for the effects of hippocampal lesions on Pavlovian conditioning. *Abstracts 14th Ann. Meeting Society for Neuroscience* **10,** 124.

Schmajuk N. A. (1987) Animal models for schizophrenia: The hippocampally lesioned animal. *Schiz. Bull.* **13,** 317–327.

Schmajuk N. A. (1989) The hippocampus and the control of information storage in the brain, in *Dynamic Interactions in Neural Networks: Models and Data* (Arbib M. and Amari S. I., eds.), Springer-Verlag, New York.

Schmajuk N. A. and Isaacson R. L. (1984) Classical contingencies in rats with hippocampal lesions. *Physiol. Behav.* **33,** 889–893.

Schmajuk N. A. and Moore J. W. (1988) The hippocampus and the classically conditioned nictitating membrane response: A real-time attentional-associative model. *Psychobiology* **46,** 20–35.

Schmajuk N. A., Spear N. E., and Isaacson R. L. (1983) Absence of overshadowing in rats with hippocampal lesions. *Physiol. Psychol.* **11,** 59–62.

Schmajuk N. A. and Moore J. W. (1985) Real-time attentional models for classical conditioning and the hippocampus. *Physiol. Psychol.* **11,** 278–290.

Schmaltz L. W. and Theios J. (1972) Acquisition and extinction of a classically conditioned response in hippocampectomized rabbits *(Oryctolagus cuniculus). J. Comp. Physiol. Psychol.* **79,** 328–333.

Seidman L. J. (1983) Schizophrenia and brain dysfunction: An integration of recent neurodiagnostic findings. *Psychol. Bull.* **94,** 195–238.

Shagass C., Roemer R. A., Straumanis J. J., and Amadeo M. (1978) Evoked potentials of schizophrenics in several sensory modalities. *Biol. Psychiat.* **13,** 163–184.

Shipley W. C. (1934) Studies of catatonia: VI. Further investigation of preseverative tendency. *Psychiat. Quart.* **8,** 736–744.

Solomon P. R. (1977) Role of the hippocampus in blocking and conditioned inhibition of the rabbit nictitating membrane response. *J. Comp. Physiol. Psychol.* **91,** 407–417.

Solomon P. R. and Moore J. W. (1975) Latent inhibition and stimulus generalization of the classically conditioned response in rabbits following dorsal hippocampal ablations. *J. Comp. Physiol. Psychol.* **89,** 1192–1203.

Suddath R. L., Casanova M. F., Goldberg T. E., Daniel D. G., Kelsoe J. R., and Weinberger D. R. (1989) Temporal lobe pathology in schizophrenia. *Amer. J. Psychiatry* **146,** 464–472.

Swerdlow N. R. and Koob G. F. (1987) Dopamine, schizophrenia, mania, and depression: Toward a unified hypothesis of cortico-striato-pallido-thalamic function. *Behav. Brain Sci.* **10,** 197–245.

Thayer J. and Silber D. E. (1971) Relationships between levels of arousal and responsiveness among schizophrenics and normal subjects. *J. Abnorm. Psychol.* **77,** 162–173.

Thompson R. F. (1986) The neurobiology of learning and memory. *Science* **233,** 941–947.

Torrey E. F. and Peterson M. R. (1974) Schizophrenia and the limbic system. *Lancet* **II,** 942–946.

Venables P. H. (1966) Psychophysiological aspects of schizophrenia. *Br. J. Med. Psychol.* **39,** 289–297.

Venables P. H. (1973) Input regulation and psychopathology, in *Psychopathology* (Hammer M., Salzinger K., and Sutton S., eds.), Wiley, New York, pp. 261–284.

Venables P. H. (1974) The recovery limb of the skin conductance response in "high-risk" research, in *Genetics, Environment, and Psychopathology* (Mednick A., Schulsinger F., Higgins J., and Bell B., eds.) North Holland, Amsterdam, pp. 117–134.

Venables P. H. (1977) The electrodermal physiology of schizophrenics and children at risk for schizophrenia: Controversies and development. *Schiz. Bull.* **3,** 28–48.

Vinogradova N. V. (1962) Protective and stagnant inhibition in schizophrenics. *Zh. Vyssh. Nerv. Deiat.* **12,** 426–431.

Walaas I. and Fonnum F. (1979) The effects of surgical and chemical lesions on neurotransmitter candidates in the nucleus accumbens of the rat. *Neuroscience* **4,** 209–216.

Weckowicz T. E. (1957) Size constancy in schizophrenic patients. *J. Ment. Sci.* **103,** 475–486.

Weckowicz T. E., and Blewett D. B. (1959) Size constancy and abstract thinking in schizophrenic patients. *J. Ment. Sci.* **105,** 909–934.

Weikart C. and Berger T. W. (1986) Hippocampal lesions disrupt classical conditioning of cross-modality reversal learning of the rabbit nictitating membrane response. *Behav. Brain Res.* **22,** 85–90.

Weinberger D. R., Berman K. F., and Illowsky B. P. (1988) Physiological dysfunction of dorsolateral prefrontal cortex in schizophrenia. *Arch. Gen. Psychiatry* **45**, 609–615.

Weinberger D. R., Wagner R. L., and Wyatt R. J. (1983) Neuropathological studies of schizophrenia: A selective review. *Schizophr. Bull.* **9**, 193–212.

Wickelgren W. O. and Isaacson R. L. (1963) Effect of the introduction of an irrelevant stimulus in runway performance of the hippocampectomized rat. *Nature* **200**, 48–50.

Yang C. R. and Mogenson G. J. (1986) Dopamine enhances the terminal excitability of hippocampal-nucleus accumbens neurones by D2 receptor. *J. Neurosci.* **6**, 2470–2478.

Zahn T. P. (1964) Autonomic reactivity and behavior in schizophrenia. *Psychiat. Res. Rep.* **19**, 156–172.

Zahn T. P., Rosenthal D., and Lawlor W. G. (1968) Electrodermal and heart rate orienting in chronic schizophrenia. *J. Psychiat. Res.* **6**, 117–134.

An Animal Model of Stimulant Psychoses

Mathew T. Martin-Iverson

1. Issues Facing Animal Models of Stimulant Psychosis

1.1. *Issues of Validity*

The rationale for animal models of psychomotor stimulant-induced psychosis is both simple and complex. The simple rationale is that, since stimulants induce psychosis, an understanding of the neurochemical effects of stimulants will provide a similar knowledge of the neurochemical basis of psychosis. The most thorough way of determining the neurochemical effects of stimulants is by studying nonhuman animals. Close inspection of each element in this rationale reveals its true complexity.

First, the statement that "stimulants induce psychosis" is incomplete. Initial evidence that psychosis develops with sustained stimulant use was based on a selected sample, primarily stimulant abusers (Connell, 1958). Further evidence for stimulant-induced psychosis was that stimulants given to schizophrenic patients can induce a florid psychotic episode that is virtually indistinguishable from the patients' "normal" psychotic episodes (Janowsky and Davis, 1974). Interestingly, methylphenidate appears to be a more potent psychotomimetic than amphetamine (Janowsky and Javis, 1974; Lieberman et al., 1987). These data suggest that the stimulant-induced psychosis may be closely related to schizophrenic psychosis, especially of the paranoid type. However, the evidence also raises the possibility

From: *Neuromethods, Vol. 18: Animal Models in Psychiatry I*
Eds.: A. Boulton, G. Baker, and M. Martin-Iverson

that stimulants exacerbate an underlying personality disorder that predisposes those individuals to psychosis. Stimulant use may therefore be causally related to psychosis only indirectly, as one of many possible general contributing factors (e.g., as a "stressor"). Subsequent evidence has suggested that the causal relationship between stimulants and psychosis is somewhat stronger than that. Normal volunteers given stimulants, or Parkinson's disease patients treated with any of a variety of indirect or direct dopamine agonists, without other signs of predisposition to psychosis, can also develop stimulant psychoses (Angrist and Gershon, 1977; Griffith et al., 1968). It is interesting that, even with continuous intravenous infusion of high doses, psychosis takes a number of hours to occur (sometimes <24, but apparently not within a few hours of infusions). It should be pointed out that there are very few studies of stimulant psychosis in normal volunteers, because of ethical issues. By far the majority of studies have been carried out using subjects with a history of stimulant abuse, making difficult the resolution of the question of the importance of chronic exposure to stimulants as a determining factor in psychoses.

Taken altogether, the evidence suggests that the stimulant induction of psychosis is clearly not a simple relationship. There does seem to be a subpopulation of humans that are predisposed toward developing stimulant psychosis. Even among schizophrenics, only about 40% exhibit a psychotic reaction to a dose of stimulant that is subpsychotomimetic in normal volunteers (Lieberman et al., 1987). Whether this is a result of a subtype of schizophrenia, a predisposition exclusively to stimulant-induced psychoses, individual differences in the pharmacokinetics or pharmacodynamics of stimulants, or a difference in some exogenous factor (such as treatment regimen, past drug history, and the like) is not known. This is unfortunate, because without a clear understanding of the factors in humans that result in stimulant use having the ability to induce psychosis, how can we assume that nonhuman animals will be like those humans who develop psychosis, and not be like those who do not? One strategy for resolving this problem is to look at stimulant effects in animals exhibiting similar individual differences. Of course, there are many reasons for individual differences in responses to stimulants. Identifying individual differences in animals that are homologous to those in humans can hardly be as easy as identifying reasons for variations in response in humans, itself a formidable task.

The second element of the rationale for animal models of stimulant psychosis also poses difficulties. That understanding the neurochemical effects of stimulants will generalize to an understanding of the neurochemical substrates of psychosis is not an assumption that is easy to defend, even given, for the sake of argument, a clear causal relationship between stimulant use and psychosis. Psychomotor stimulants produce many and diverse neurochemical effects, even if such factors as dose, treatment regimen, time of treatment, and so on, are kept constant. Determining the subset of effects that are related to the induction of psychosis is not an easy task. The availability of a range of drugs that can induce psychosis in humans may make this task easier, for we can look for common effects across psychotomimetics. On the other hand, different drugs may well produce psychosis via actions on different systems. Another strategy to resolving this issue is to look for functional homologies between stimulant-induced behaviors in animals and stimulant-induced psychosis. This approach will narrow the selection of neurochemical effects of stimulants to those that correlate with such homologies. However, such a strategy replaces one set of problems with another: identifying functional homologies.

As will be discussed later, there is a large number of factors that greatly alter the effects of stimulants, including temporal pattern of stimulant administration, duration of treatment, dose, route of administration, and timing of administration in relation to circadian rhythms. The relationship of these factors to the induction of psychosis is not known, but is critical to the development of appropriate animal models.

The final element of what seemed a simple rationale for animal models of stimulant psychosis is the relationship between stimulant effects in humans and those in nonhumans (primarily rodents). Even if there is close similarity between the brains of nonhuman animals and those of humans, it is conceivable that the brain processes that produce stimulant psychosis would be just those processes that are unique to humans. This possibility is strengthened by the nature of psychosis, a uniquely human disorder of cognition. If this is the case, as it may well be, then functional homologies between animal behavior and psychosis will not exist, and animal models will not, perforce, model psychosis.

On the other hand, there does seem to be a remarkable degree of similarity between the organizations of rodent and of human brains, and functional homologies relevant to psychosis may well exist. Identifying such homologies is problematic. Two major deficits in our knowledge contribute to this difficulty. First, "psychosis" itself is not well characterized from a cognitive or behavioral viewpoint. Second, most of the measurement of stimulant-induced behavior in animals has been rather crude and beset with interpretational difficulties (Fray et al., 1980; Rebec and Bashore, 1984; Szechtman et al., 1988). Although a number of general theories of stimulant effects on behavior have been proposed (e.g., Lyon and Robbins, 1975), this remains a murky area (Martin-Iverson et al., 1989). In short, our understanding of both stimulant psychosis and stimulant-induced behavior is rather poor, making it difficult to characterize the validity of animal models.

Animal models of stimulant psychosis have invariably used chronic treatment regimens. This is a result of the assumption that psychosis results from repeated use of stimulants, as observed with stimulant abusers. This assumption may not be warranted. Segal and Schuckitt (1983) have argued that psychosis can be induced with a single administration of amphetamine. Studies have shown that psychosis in normal volunteers can be induced with intravenous infusions of a few hours (*see* Segal and Schuckitt, 1983). On the other hand, psychosis is relatively rare after initial use of stimulants, and those cases that do occur may be attributed to underlying schizophrenic disorders. Thus, the duration of treatment is an issue of stimulant psychosis that needs to be resolved, especially since chronic stimulant treatment produces behavioral and neurochemical effects different from those produced by acute treatments.

All the complexities just related are problems in determining the validity of animal models of stimulant psychosis. However, attempts to develop such models are not spurious; no matter the actual validity as a model, animal models have and will shed considerable light on the nature of the behavioral and neurochemical effects of stimulants. Such information can only aid in the understanding of stimulant psychosis. Furthermore, there are certain strategies that can be followed that will maximize the probability of developing valid models, given these limitations. The next section describes some of these approaches.

1.2. Issues of Current Animal Models

As mentioned above, the major models of stimulant psychosis have focused on relatively long-term treatment regimens. This strategy has been adopted as the most likely to elucidate homologies to stimulant psychosis. Four different animal models, which differ primarily in treatment regimen, have been used. These are:

1. Repeated, once-a-day injections of relatively low doses, usually of amphetamine;
2. Continuous infusions of relatively low doses of (typically) amphetamine;
3. Repeated injections, sometimes two or three times a day, of high doses of methamphetamine;
4. Repeated injections of methamphetamine, beginning with a relatively low dose, but with progressive increases in dose.

1.2.1. Repeated Intermittent Injections

Repeated intermittent injections with low to medium doses, primarily with (+)-amphetamine, but also with methylphenidate, cocaine, and apomorphine (a direct agonist) has been a common animal model of stimulant psychosis. This regimen results in behavioral sensitization, a gradual augmentation of locomotor activity or "stereotypies" (depending on dose) that is apparent during treatment and also upon a subsequent injection days, weeks or months after cessation of treatment. Repeated intraperitoneal (ip) or subcutaneous (sc) injections of (+)-amphetamine in the dose range of 1.5–3.0 mg/kg results in an augmentation of sniffing, whereas doses of 0.5–1.0 mg/kg will cause sensitization of locomotion. Interestingly, it has been reported that licking and gnawing induced by daily injections (2–12 mg/kg) exhibits tolerance (or no effect with the two lower doses) for the first 17 days, and sensitization only after 25–35 days and only with the two lower doses (Eichler et al., 1980). Sensitization of locomotion is particularly prominent with direct agonists selective for the dopamine (DA) D2 receptor subtype. It has been argued that the enhanced stereotypies are similar to compulsive, stereotyped behaviors that occur in humans experiencing stimulant psychosis (Segal and Schuckitt, 1983). The intermittent-treatment regimen results in an augmentation of DA release when stimulated by amphetamine, cocaine, KCl, or stress (Robinson and Becker, 1986). This augmentation of stimulated DA release has been attributed to subsensitivity of inhibitory DA autoreceptors (Antelman and Chiodo, 1981; Baudry et al., 1977; Martres et al., 1977; Muller and Seeman, 1979), but not all have found such an effect (Conway and Uretsky, 1982; Kuczenski et al., 1983; Riffee and Wilcox, 1985). A similar behavioral and neurochemical sensitization occurs with stressors, and there is cross-sensitization between the effects of amphetamine and stress (For review, *see* Robinson and Becker, 1986). The observation of both sex and individual differences (in males) in the degree of sensitization (e.g., Camp and Robinson, 1988) has been taken by some (Robinson, 1988; Robinson and Becker, 1986) to be an important area of study for understanding individual differences in the susceptibility of humans to stimulant psychosis.

1.2.1.1. Role of Dopamine Receptor Subtypes in Sensitization

Consistent alterations in DA D2 receptors have not been found following repeated intermittent injections of amphetamine (*see* Robinson and Becker, 1986, for review). Interestingly, chronic treatment with a selective DA D1-receptor antagonist, SCH 23390, leads to an augmented behavioral response to either a mixed D1/D2-receptor agonist (Gandolfi et al., 1988; Vaccheri et al., 1987) or to a selective D2-receptor agonist (Hess et al., 1986; Martin-Iverson et al., 1988a). This suggests the possibility that sensitization to repeated injections with amphetamine could result from D1-receptor supersensitivity. Receptor supersensitivity after chronic treatment with an agonist appears incompatible with the usual concept of receptor regulation. However, it has been reported that chronic treatment with bromocriptine, a selective D2 agonist, produces D1 receptor supersensitivity as well as D2-receptor subsensitivity (Wei-Dong et al., 1988). It has also been observed that L-dopa treatment of either patients with Parkinson's disease (Rinne et al., 1985) or rats with 6-OHDA-induced DA depletions (Parenti et al., 1986) increases the density of D1 receptors and decreases the density of D2 receptors. Furthermore, we observed that continuous infusions of a D1 antagonist, although initially blocking the nocturnal locomotor stimulant effects of a coadministered D2 agonist, eventually produces an augmented behavioral effect (Martin-Iverson et al., 1988a). This led us to propose initially that the behavioral sensitization produced by repeated stimulants may result from D1-receptor supersensitivity (Martin-Iverson et al., 1988a). It was recently reported that a D1 antagonist infused into the ventral tegmental area blocked the sensitization that is normally produced by repeated injections of amphetamine (Stewart and Vezina, 1989). In addition, persistent augmentation of amphetamine-induced dopamine release by pretreatment with cocaine can be blocked by either D1- or D2-receptor antagonists (Peris and Zahniser, 1989), indicating that the facilitated amphetamine-induced release is a secondary effect of receptor activation. It therefore seems likely that behavioral sensitization to repeated treatments with stimulants results from an enhanced neurotransmission at the D1-receptor site.

1.2.1.2. Problems with a Sensitization Model of Psychosis

There are two serious problems with this model that have not as yet been addressed. One is that sensitization to the acute effects of the stimulants can be mimicked by simply increasing the dose of drug. If sensitization to stimulants produces psychosis, then psychosis should be produced just as readily after acute injections, if the dose is sufficient. Whether or not such is the case is controversial. Related to this problem is one arising from the observations that sensitization to amphetamine can be produced by chronic administration of a variety of drugs, such as antidepressants and anticholinergics (Martin-Iverson et al., 1983).

In addition, repeated treatments with a direct agonist, such as apomorphine, can produce behavioral sensitization (Mattingly and Rowlett, 1989), but does not augment amphetamine-induced DA release.

A second issue is one of exclusion; to my knowledge no investigator has yet shown whether or not chronic continuous treatment with stimulants produce a similar augmentation of stimulated DA release. Stereotyped behaviors exhibit tolerance, not sensitization, during daytime continuous infusions (data shown in subsequent section). It must be demonstrated that stimulant treatment following a regimen that does not produce behavioral sensitization does not produce augmentation of stimulated DA release, if we are to conclude that such augmented release is causally related to behavioral sensitization.

A final problem is the relationship of compulsive, purposeless repetitive motor acts to amphetamine delusional disorder (classification 292.11 in DSM-III-R [American Psychiatric Association, 1987]), which consists primarily of a "highly organized paranoid delusional state indistinguishable from the active phase of Schizophrenia" (p. 109), with associated features of distortion of body image, misperception of people's faces, and formication ("the hallucination of bugs or vermin crawling in or under the skin [which] can lead to scratching or skin excoriations," p. 137). Stereotyped orofacial behaviors and simple motor acts are not intuitively similar to amphetamine psychosis. The concept that paranoia is a stereotyped idea seems to be stretching a point.

1.2.1.3. Contribution of Pavlovian Conditioning to Sensitization

Segal (1975) has argued that sensitization to amphetamine effects cannot be attributed solely to conditioning effects. This holds true for sensitization to a direct DA D2 agonist, in that sensitization occurs with repeated 12-hour daily infusions of (+)-4-propyl-9-hydroxynaphthoxazine to rats in their home cages (Martin-Iverson et al., 1988a). These infusions were made with modified Alzet® osmotic pumps, containing saline, but connected to a long length of polyethylene tubing glued into a tight spiral. This tubing contained the drug and a dye in short sections, interspersed with a substance immiscible with drug, to allow 12-hour infusions of drug and 12-hour infusions of the inert substance (perfluorodecalin) throughout a 14 day infusion period. With this procedure, there are no external environmental cues that are uniquely associated with the presence of the drug, but sensitization still occurs.

Although sensitization is unlikely to be a function of conditioning, conditioning processes undoubtedly contribute to the sensitization normally observed with daily intermittent injections. This can be an important confounding factor. We have shown that the conditioning of stimulant effects with either amphetamine or a direct D2 agonist is independent of effects on DA systems (Martin-Iverson and McManus, 1990). Neither blockade of both D1 and D2 receptors nor inhibition of DA synthesis (unpublished data)—both treatments that result in blockade of the unconditioned locomotor stimulant effects of these drugs—block conditioning of locomotion. Very few studies control for the possible contribution of these non-DA-mediated actions when assessing the neurochemical effects of repeated stimulant treatments.

1.2.2. Continuous Infusions of Stimulants

1.2.2.1. Dependence of Effects of Continuous Infusions on Day/Night Cycle

The effects of continuous infusions of psychomotor stimulants are reviewed in the chapter by Ellison in this volume, so they will not be covered in detail here. Continuous infusions of amphetamine result in the development of tolerance to focused oral stereotypies, sniffing, head movements, and locomotion, rather than sensitization. The tolerance to licking and gnawing occurs irrespective of the phase of the light–dark cycle, but tolerance to sniffing, head movements, and locomotor stimulation occurs only during the light phase of the light–dark cycle (Martin- Iverson and Iversen, 1989; data presented in this chapter). Beha-vioral sensitization is observed during the night for locomotion, rearing, repetitive mouth movements, and grooming. Tolerance to the licking and gnawing behaviors appears to be related to a long–term depletion of DA in the striatum, but not in the nucleus accumbens (Eison et al., 1983; Ellison, 1983; Gately et al., 1987; data shown later in this chapter). However, daytime tolerance to similar effects of direct agonists and cocaine also occurs, but these treatments do not produce long-term depletion of striatal DA levels (e.g., Ryan et al., 1988).

Continuous infusions of amphetamine, therefore, can produce both sensitization, as seen with intermittent injections, and tolerance, depending on the phase of the light–dark cycle and the behavior measured. Since the rate of drug delivery and, presumably, pharmacokinetic factors remain fairly constant across different phases of the light–dark cycle, this procedure allows investigation of the neurochemical basis for differential development of sensitization or tolerance while maintaining most drug-treatment conditions constant as well as circumventing the confound produced by Pavlovian conditioning processes. Thus, the continuous-infusion paradigm may be superior to the intermittent-injection procedure for elucidating the neurological correlates of amphetamine's behavioral effects.

1.2.2.2. Stress and D1/D2 Receptor Interactions

In addition, daytime tolerance to low, but not to higher, doses of amphetamine can be reversed by presentation of mild stressors or arousing stimuli (Martin-Iverson and Iversen, 1989). This stress–stimulant interaction appears even more strongly with continuous infusions of rats with a selective DA D2 receptor agonist, (+)-4-propyl-9-hydroxynaphthoxazine (Martin-Iverson et al., 1987,1988a,1988b). The reversal of daytime tolerance to the locomotor-stimulant effects of a D2 agonist by arousing stimuli can be blocked by a D1-receptor antagonist, SCH 23390, and can be mimicked by injections of a D1-receptor agonist, SKF 38393. This evidence suggests that stress-stimulant interactions may depend on a D1–D2 receptor interaction. It also suggests that the continuous-infusion paradigm shares the property of stress interaction with the intermittent-injection regimen.

1.2.2.3. Emergent "Hallucination-Like" Behaviors

Some investigators (*see* Ellison's chapter on animal models of hallucinations in this book) have found "emergent hallucination-like" behaviors in a variety of species after continuous infusions of amphetamine or apomorphine (Davis et al., 1985) over four to seven days. The label of "hallucination-like" was applied because of similarities to behaviors induced by acute injections of hallucinogens, such as *d*-lysergic acid diethylamide or mescaline. In rats, these behaviors include "abortive grooming," limb flicks, body shakes, sudden orienting behaviors, and abnormal mouth movements. Hallucination-like behaviors are more likely to occur after a challenge dose of amphetamine after drug withdrawal, as long as the withdrawal period is not much longer than 12 h (Ellison's chapter; personal observations). We have observed continuous grooming of certain spots on an animal's forelimb in some, but not all, rats given 14 days of administration of a relatively high dose of amphetamine (8 mg/d), without withdrawal or subsequent challenge. Such grooming extends for much longer periods than is normal, and sometimes results in loss of hair and abrasion of the skin. About one in six animals exhibit such behavior to the point of skin abrasion. Similar behavior has been observed in primates (Nielsen et al., 1983). It has been likened to a relatively common delusion among stimulant-induced psychotics of "bugs" crawling under the skin (formication, *see* Section 1.2.1.2. for the DSM-III-R description). This delusion sometimes results in the drug abusers trying to dig the "bugs" out with sharp implements. We have not observed increases in "wet-dog shakes," body twitches, or limb flicks during continuous treatment. On the other hand, these are relatively rare behaviors, which may not be susceptible to accurate frequency assessments by our time-sampling methods, described subsequently. Ellison discusses additional abnormal social behaviors that may also reflect paranoid psychosis in his chapter of this volume. The abnormal aggressive behaviors described by Ellison and the formication-like abnormal grooming behavior induced by chronic, continuous administration of psychomotor stimulants are the most intuitively appealing potential homologs of stimulant psychosis.

1.2.2.4. Problems with a Continuous-Infusion Model

A major issue in this protocol is whether or not the emergent hallucination-like behaviors are related to stimulant-induced psychosis. It is intuitively attractive to interpret the focused grooming of forelimbs as a homolog to formication, especially as it occurs only after chronic treatment. Indeed, the behavioral parallels are striking. Of course, the "delusional" nature of this behavior is not open to study. Alternate interpretations of these hallucination- or delusion-like behaviors are available. One possibility is that continuous treatments result in drug accumulations in brain tissue in such a way that these behaviors are a function of gradually increasing doses. However, we have found that brain levels of amphetamines remain constant during this treatment regimen (*see* Section 2.4.1.). In the case of formication-like behavior, increasing dose is not a feasible explanation, as this behavior has not been reported to occur in acute treatment studies. It is still possible that the accumulation of some active metabolite of amphetamine, such as para-hydroxyamphetamine or para-hydroxynorephedrine (Matsuda et al., 1989), or the increase in some trace amine, such as 2-phenylethylamine, as a function of the monoamine oxidase activity of higher doses of amphetamine is responsible for these effects. It is also possible that the hallucination-like behavioral effects of chronic, continuous amphetamines are a function of actions on serotonergic systems (Trulson and Jacobs, 1979), but some evidence against this view has been presented (Ellison and Ratan, 1982).

1.2.3. Intermittent High-Dose Methamphetamine Regimens

Intermittent high doses (6–20 mg/kg) of methamphetamine, given 1–3 times/day, leads to degeneration of dopaminergic and serotonergic terminals, especially in the striatum (for review, *see* Seiden and Ricaurte, 1987). This procedure also produces an increase in Substance P accumulation in terminals in the ventral tegmental area, which is thought to result in a loss of DA-induced release (Hanson et al., 1986). Neurotoxicity after high doses of methamphetamine may occur as a result of metabolism of methamphetamine to 6-hydroxydopamine (Seiden and Vosmer, 1984). Some investigators have rejected methamphetamine-induced neurotoxicity as a model of stimulant psychoses because a decrease in DA is incompatible with the DA hypothesis of schizophrenia, which posits an increase in DA neurotransmission.

The lack of behavioral consequences of methamphetamine neurotoxicity similar to those observed after 6-hydroxydopamine-induced depletions (cf, Seiden and Ricaurte, 1987) has been attributed to insufficient depletions, which usually are no greater than 70% reductions relative to controls (Seiden and Ricaurte, 1987). It is well known that behavioral deficits require DA depletions of 80% or more before they become obvious. This view is strengthened by the recent findings, using in vivo microdialysis techniques, that extracellular dopamine levels are *not* reduced in animals with methamphetamine-induced neurotoxicity, and that extracellular DA levels can be altered appropriately by subsequent amphetamine challenge (Robinson et al., 1990). Very little in the way of comprehensive behavioral analyses has been carried out with this regimen, although tolerance to focused oral stereotypies occurs.

1.2.4. Progressively Higher Doses of Methamphetamine

If repeated injections of methamphetamine begin with a relatively low dose, progressively increasing over days, then many of the neurotoxic effects of a repeated-high-dose regimen are circumvented (Schmidt et al., 1985). The effects that are avoided or reduced in magnitude include inhibition of DA and serotonin synthesis, DA and serotonin depletion, reduction in D2-receptor binding, and the increase in Substance P-like reactivity in neuronal terminals in the ventral tegmental area (Schmidt et al., 1985). This "inoculation-like" effect may be similar to effects observed in stimulant abusers, who often begin by using lower doses and increase the dose over a long period of time. On the other hand, the behavioral consequences of this treatment regimen have not been well characterized. Escalating doses of (+)-amphetamine can result in behavioral sensitization, and, although basal DA release as determined by in vivo microdialysis is unaffected, amphetamine-induced release is increased (Robinson et al., 1988).

2. Continuous Infusions with Day and Night Monitoring

2.1. *Rationale for Choosing a Continuous-Infusion Model*

I have chosen to work with the continuous-infusion treatment regimen as a potential animal model of stimulant psychosis for a variety of reasons:

1. By monitoring the behavior of animals given continuous infusions of psychomotor stimulants during both day and night, one can simultaneously study both the sensitization model typical of intermittent-injection regimens (night) and the stereotypy tolerance with "emergent hallucination-like behavior" model of continuous-infusion regimens (day). Since the behavioral response is largely a function of the duration of treatment and the day/night cycle, all other factors can be kept constant by using the same animal, the same dose, the same treatment regimen, and so forth. Thus, it is unlikely that differences observed between nocturnal and diurnal phases of the light–dark cycle can be attributed to protocol artifacts.
2. At least intuitively, the formication-like behavior observed in animals after chronic and continuous infusions of amphetamine appears to be the clearest example of a behavioral homology to behaviors observed during stimulant psychosis.
3. Drug–stress or drug–arousal interactions can be assessed with this procedure.
4. The confounding factor of Pavlovian conditioning of drug effects can be avoided.
5. The occurrence of formication-like behavior is not constant across animals; therefore, the factor of individual differences is present in this model, and neurochemical changes that are induced by the drugs and unique to this subpopulation can be assessed.

2.2. Behavioral Measures

The most difficult part of observational analysis of behavior is determining what method of behavioral categorization to use and the criteria to adopt within a particular method. Some investigators choose definitions of specific behavioral acts that are mutually exclusive. Others use rating scales that focus on only certain components of the behavioral repertoire, and that make possibly unwarranted assumptions concerning the nature of the drug effects on behavior (*see* Fray et al., 1980; and Rebec and Bashore, 1984, for discussions of the limitations of behavior rating scales). A morphogenetic approach utilizing the Eshkol-Wachmann Movement Notation has been used by some investigators (e.g., Szechtman et al., 1988). A different approach is to analyze certain spatial properties of an animal's movement using computerized tracking methods.

Important factors in deciding on a behavioral scoring system include the purposes of the experiment, the kind of environment in which the animals will be tested, and convenience. The failure to achieve a widespread adoption of any one method, or even specific definitions within any one mood, is partially a function of differences in these factors.

Our purpose is to characterize behavioral effects of chronic stimulant treatments that may be homologous to stimulant-induced psychoses. Evidence has implicated the dopaminergic systems in psychoses, and these systems are integral to the behavioral effects of psychomotor stimulants. Although the dopaminergic systems of the brain appear to mediate many of the behavioral effects of stimulants, such as increases in locomotion, sniffing, stereotyped oral behaviors, and so on, it is just as clear that neither the nigrostriatal nor the mesolimbic DA systems provide the basic organization of motor outputs that subserve these behaviors. For example, although the mesolimbic DA system appears responsible for much of the locomotion-inducing effect of stimulants, the brain area that underlies locomotion is the "mesencephalic locomotor region." The functions of the DA systems have been variously described as arousal, motivation, regulating goal-directed behavior, underlying response initiation and/or maintenance, or responsibility for "switching" between behaviors. Thus, it appears that the DA system is involved in the coordination and selection of behavior over time; that is, organizing behavior in such a way that the most appropriate response is emitted in the appropriate context. It also seems reasonable that psychosis is not a deficit in performance of behaviors, but in the organization of percepts, thoughts, and behaviors.

These considerations suggest that a behavioral assessment of chronic stimulant treatment should include an analysis of *sequences* of behaviors, as well as of frequencies and durations of behaviors. Such an approach was advocated by Norton in 1968. It was found that amphetamine had significant effects on the sequencing of behaviors (Norton, 1973). I have applied such an analysis of amphetamine effects on sequences of behaviors (Martin-Iverson, et al., 1989), with provocative results. Alteration of the normal sequence in which behaviors occur appears after acute injections of amphetamine that are too low to have significant effects on frequencies or durations of specific behavioral acts. Furthermore, the sequences of behaviors that are most sensitive to the effects of amphetamine are transitions between behaviors that are thought to be mediated by different dopaminergic systems (nigrostriatal and mesolimbic), whereas transitions between behaviors subserved by the same system are relatively resistant. This suggests that the coordination of different dopaminergic projection systems is most sensitive to disorganization after stimulant treatment. In addition, increasing doses of amphetamines decrease the frequency of bouts of behaviors, but increase the duration of each bout of striatum-mediated behaviors, and increase the frequency, but not the bout duration, of nucleus accumbens-mediated behaviors. These findings suggest the possibility that psychoses may result from loss of coordination of function of different dopaminergic systems.

In order to assess the sequential nature of behavior, a categorization of ongoing behavior into discrete behavioral acts is necessary. Having decided on this approach to behavioral scoring, one is left with the more difficult problem of defining such acts. Grooming behavior provides an excellent illustration of the difficulty in making definitions of a behavior. One could make one category, labeled "grooming," and this category could include face washing, penile grooming, flank stroking, flank licking, and hindlimb scratching. However, penile grooming can be dissociated pharmacologically from other forms of grooming. It is also questionable whether or not hindlimb scratching of the body should be included as a separate category. In addition, hindlimb scratching consists of repeated strokes of the body, neck, or head, followed by licking of the hindlimb, and then more strokes. Should the licking of the hindlimb be defined as a category separate from hindlimb scratching? Should a bout of hindlimb scratching be defined as terminated by licking of the hindlimb, with another bout initiated after the paw licking, or should the whole sequence of scratching and paw licking be considered one behavior? Should mouth movements or licking that are emitted while an animal is grooming be considered as part of grooming, or as separate coincident behaviors? Is the posture of an animal during grooming a separate aspect of behavior; for example, grooming while sitting may be included in one category, and grooming while standing or lying could be included in different categories. For that matter, grooming, itself a functionally defined behavior, may be too global a definition of behavior, and behavioral acts consisting of movement of specific muscle groups could be used. The resolution of the problem of behavioral definition rests ultimately on the purposes of the experiment.

Our definitions of specific behavioral acts are constrained by a number of factors. First, it is necessary to have a comprehensive number of definitions, so that sequences of behaviors are adequately described. That is, there should not be periods of time in which uncategorized behaviors occur. Second, the frequency of the defined acts has to be high enough for statistical analysis. This often necessitates "lumping" of behaviors (as opposed to "splitting"). Third, some rare behaviors must be included, as infrequent "emergent" hallucination-like behaviors have been reported to occur with chronic stimulant treatments. Finally, the maximum number of behaviors that an observer can comfortably score is about 30.

The decision of what kind of behavioral categories to define should be made only after some hours of observation of the animals in the specific testing environment to be used, since the behaviors that will occur depend to some extent on the environment in which they are tested. Furthermore, animals that have been treated with the drugs to be used should also be observed in preliminary experiments, to ensure that behaviors that are specific to the drug treatments are adequately defined in advance of the experiment.

2.3. Experimental Protocol

2.3.1. Equipment

In this model, male Sprague-Dawley rats (250–400 g) are individually housed in each of 48 stainless steel cages (24 [width] × 16 [height] × 26 [length] cm), with one Plexiglas™ side and a wire-mesh floor, arranged in 4 rows of 12 boxes on a cage rack. In front of the cages is a set of four tracks, on each of which runs a truck. Each track is placed at a height that enables an infrared-sensitive CCD video camera (Canon Ci-20R) mounted on the truck to view one row of cages, one cage at a time. The four cameras are equipped with a 12mm auto-iris lens (Canon C61205) and an infrared light source (IR-20W) that does not provide light in the visible spectrum. The movement of the camera trucks is completely programmable with a (very inexpensive) Commodore 64 computer in an adjacent room. An electromagnetic sensor within each track signals the position of a truck with respect to each cage. Each cage is equipped with two photocell assemblies placed 3 cm from the floor, spaced 13 cm apart, and equidistant from the front and back walls, in such a way that the length of the cage is bisected by two infrared beams. Interruptions of either beam are counted by a single 4-bit memory chip and recorded via an interface with the computer that is controlling the movement of the camera trucks. Since the maximum number of counts of each chip is 255, the computer accesses the counter memories every 60 seconds, reading the count and clearing it. The video signal from each camera is forwarded to a video monitor (Hitachi VM-900), a time–date generator (Pelco TDG200DT), and each of three extended-play video cassette recorders (NEC

PV-1200A), all of which are kept in a room adjacent to the testing room. Each VCR records a consecutive eight hour period, so that all 24 hours of each day are recorded, with the time of recording programmed into each VCR. Thus, a total of four monitors and time-date generators, and 12 VCRs, are used in the data acquisition. The cages contain removable food dishes and water bottles. The cages, cage racks, photobeam assemblies, camera trucks and tracks, and computer interfaces were made to order by Acadia Instruments Ltd., Saskatoon, Saskatchewan, Canada, a security systems and behavioral-testing-apparatus design and assembly company.

The program for moving the camera trucks specifies that each camera remains in front of a particular cage for 295 seconds, and then moves to the next cage (2.5 seconds). It therefore takes a camera truck 59.5 minutes to view all the cages in its row. It takes a further 20 seconds for the truck to move from the last cage to the first cage, where it waits the appropriate interval (about 10 seconds) before starting the next 295-second timing interval. The camera therefore records each animal for 295 second/hour under both day and night conditions. Every hour, while the camera trucks move back to the starting position, the photobeam counts are saved onto a disk and printed out.

2.3.2. *Drug Treatments*

The rats live in the test cages for 10–14 days before the initiation of any drug treatments, to acclimatize them to the cages and the light cycle. Food and water is available *ad libitum* throughout the experiment. Changes of food and water, and cleaning of the litter from the pans underneath the cages, are performed at specific times, since these procedures strongly influence the behavioral effects of the drugs. The lights are on from 7:00AM to 7:00PM.

Drugs are administered by Alzet® osmotic minipumps (Model 2ML2, ALZA Corporation, Palo Alto, CA), which infuse the solution at a rate of approx 5 μL/h for 14–15 d. (+)-Amphetamine sulfate is dissolved in propylene glycol at concentrations ranging from 16.67 to 66.67 mg/mL (2–8 mg/d). The dopamine agonists, (+)-4-propyl-9-hydroxynaphthoxazine (PHNO, D2-receptor selective) and SKF 38393 (D1-receptor selective) were dissolved in double-distilled water in doses of 0.5–2 mg/mL (PHNO, 66–264 μg/d) or 5–20 mg/mL (SKF 38393, 0.66–2.64 mg/d). There is some question as to whether dosages of drugs given by Alzet® osmotic minipumps should be calculated on the basis of the animals' body weights or simply given as a constant dose over time. The latter is most convenient, but a computer program for weight-adjusted filling concentrations has been published (Greenshaw, 1986). As will be shown in a subsequent section, the brain levels of amphetamine in rats given a constant dose regardless of body weight does not significantly vary with body weight, within a rather restricted size range (400–700*g*). In these experiments drug dosages have not been adjusted for body weight. Amphetamine sulfate was supplied courtesy of Bruce Lodge, Chemistry Section, Bureau of Dangerous Drugs, Health and Welfare, and Ron Coutts, Faculty of Pharmacy and Pharmaceutical Sciences, University of Alberta. PHNO was supplied courtesy of Merck Sharp and Dohme Ltd, and SKF 38393 was purchased from Research Biochemicals Inc. (Natick, MA).

Implantation of the pumps is a simple procedure. The rats are anesthetized with ether or methoxyflurane. A patch of hair is shaved off in the midscapular region of the back, and the skin is cleaned and disinfected with an alcohol solution. A small incision is made with a scalpel, and a subcutaneous pocket is made with hemostats. The pump is inserted, the wound is closed with wound clips, and a powdered antibiotic is sprinkled over the wound. The process requires about 3 minutes. Pump implants are made in the afternoon, between 1:00 and 3:00PM, and behavior is recorded beginning at 7:00PM. The pumps are primed by keeping them in a physiological saline solution at 37°C for about 4 hours prior to implantation. The order of drug treatments of the 48 rats is randomized within a counterbalanced format, so that each row of 12 boxes contains a similar number of rats from each group. This is important because the height of a cage from the floor will determine the intensity of illumination from the overhead fluorescent lights, which affects the amplitude of the circadian rhythm of motor activity.

2.3.3. Behavior Rating

Video records from the first (from 7:00PM on the day of implant) and last 24 h (night and day 14) of drug treatments are analyzed by an observer blind to the drug conditions of the animals. I have developed a computer program in Turbo Pascal, called BEBOP (Basic Experimental Behavioral Observation Program), that can record, by keypunches on an IBM or IBM-compatible computer, the frequencies, durations, and sequences of up to 30 behaviors. These behaviors can occur simultaneously and independently of each other. In addition, an analysis of which behaviors occur simultaneously, and the degree of overlap, can be conducted. The acts that have been defined include

- Postural behaviors (standing, sitting, lying down, rearing, adjustments of body position),
- Whole-body-movement behaviors (locomoting, backwards walking, jumping),
- Some functionally defined behaviors (eating, drinking, grooming, hindlimb scratching, retrieving food from the food dish, gnawing, licking, and sleeping),
- Some behaviors referring to movements of certain parts of the body ("wet-dog" shakes, limb flicks, body twitches, mouth movements, sniffing, head movements, stretching, yawning and "scrabbling"),
- One behavior that refers to a relationship between the animal and its environment (snout contact with cage surface), and
- One behavior defined as the momentary cessation of all other behaviors, other than postural behaviors (pause).

A final "behavior" is defined as "can't tell," to include those instances when the observers cannot confidently ascribe the animal's behavior to one (or more) of the other categories. With BEBOP, a key is pressed when a behavior is initiated, and then pressed again when the behavior ceases. Data is collected as a listing of the times of day (to the hundredth of a second) of the initiation and terminations of each behavior, and this listing is saved to a computer file. From this record, the frequencies, durations, and sequences of initiations of behaviors can be abstracted with additional Pascal analysis programs.

The behavioral definitions that I have used have been determined after many hours of watching animals with no treatment, injections of a wide range of doses of amphetamines (both (+)-amphetamine and an amphetamine analog), and chronic treatment with (+)-amphetamine. They are essentially definitions based, to a large part, on what intuitively appeared to be organized behaviors, some functionally defined and some not. These behaviors generally consist of a number of apparently associated movements, although certain less-molar definitions have been used. Many different behaviors are sometimes grouped under one definition. For example, "head movements" include both head bobbing and head weaving, as well as subcategories of these movements consisting of a variety of different intensities or rates. Likewise, no attempt was made to account for the intensity dimension of other behaviors, such as locomotion (i.e., no distinction between walking and running). The intensity dimension has been avoided because of the extremely subjective nature of the demarcation lines between different intensities. Although the subjective component cannot be removed from the more simple definitions of behavioral acts without an intensity valuation, it is much reduced.

Statistical analysis of the frequencies, durations, and mean bout durations (duration/frequency) of the behavioral measures is done using the appropriate analysis of variance. Behavioral sequences can be easily analyzed by the method of adjusted residuals (Haberman, 1973), comparing the actual behavioral transitions with the random transition, given the frequency of the initiations of behavior. Readers are referred to an excellent review of a variety of methods of analyzing sequences of behavior by Van Hoof (1982). Firstly, an $R \times C$ cross-contingency table is made, with the first behavior in a pair along the row (R), the second behavior in a pair in the column (C), and the frequency of transitions from R_i to C_j in the cell, thus making a table with each cell designated by o_{ij} (observed transitions), where subscript i refers to a row, and subscript j refers to a column. The frequency of transitions expected (e_{ij}) assuming a random relationship between behaviors can then be calculated as the product of the respective row and column totals in the observed transition table, divided by the total of all transitions:

$$e_{ij} = (o_i^* \times o_j^*) / o^{**} \qquad (1)$$

To determine if specific transitions are significantly different from the random pattern, by calculating the adjusted residuals, a standardized residual (q_{ij}) and an estimate of its variance (v_{ij}) are calculated, and the adjusted residual (d_{ij}) is the standardized residual divided by the square root of the variance:

$$q_{ij} = (o_{ij} - e_{ij}) / (e_{ij})^{1/2} \qquad (2)$$

$$v_{ij} = (1 - o_i^* / o^{**})(1 - o^* j / o^{**}) \qquad (3)$$

$$d_{ij} = q_{ij} / (v_{ij})^{1/2} \qquad (4)$$

The adjusted residual is a normal deviation, with the significance level found on a table of the normal distribution.

2.3.4. Neurochemical Measures

After drug treatment on day 1, 3, 7 or 14, the rats were killed by guillotine decapitation, during the light phase of the day–night cycle. The brains were removed, and the striatum, nucleus accumbens, and olfactory tubercle were dissected out, and the tissue (including the "rest of brain") was rapidly frozen in isopentane on dry ice. The tissue was stored at –80°C until neurochemical analyses at a later date. The dissected brain regions were subjected to high pressure liquid chromatography with electrochemical detection (HPLC–EC) for determination of the concentrations of monoamines and acidic metabolites. The "rest of brain" was used for amphetamine-level determinations. Ongoing experiments are determining the effects of amphetamines during both phase of the light–dark cycle.

Amphetamine concentrations in brain tissue are measured using a modification of the electron-capture gas chromatographic procedure of Rao et al. (1986). This method involves extractive derivatization of amphetamine with pentafluorobenzenesulfonyl

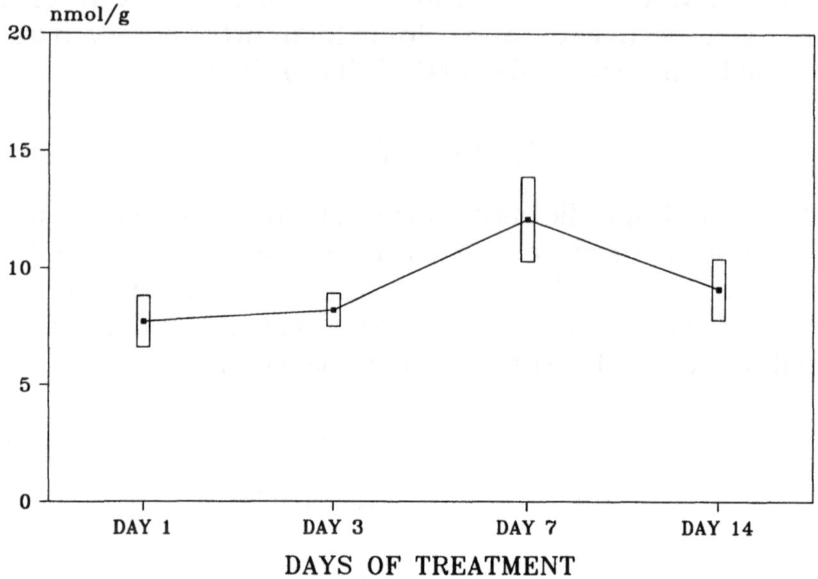

Fig. 1. Effects of continuous subcutaneous infusions of (+)-amphetamine (40 µmol/d) by minipump for 1–14 d on whole-brain (minus striatum, accumbens, and olfactory tubercle) levels of (+)-amphetamine (nmol/g tissue). Points represent the mean value of 8 animals ± SEM (boxes around points). Analysis of variance revealed no significant effects of duration of treatment, indicating that brain amphetamine levels reach a steady-state by 24 h.

chloride prior to injection of the sample into a gas chromatograph equipped with a fused silica capillary column. Concentrations of the monoamines, DA, noradrenaline, and 5-hydroxytryptamine, and their major acidic metabolites (homovanillic acid [HVA], 3,4-dihydroxyphenylacetic acid [DOPAC], and 5-hydroxyindoleacetic acid), are determined with HPLC–EC (Baker et al., 1987).

2.4. Results of a Continuous-Infusion Model

2.4.1. Brain Levels of Amphetamine

The effects of continuous infusions of 40 µmol/d (+)-amphetamine for 1, 3, 7, or 14 days on brain levels of (+)-amphetamine in rats ($N = 8$ per group) are depicted in Fig. 1. There was a slight trend toward an increase in amphetamine levels with

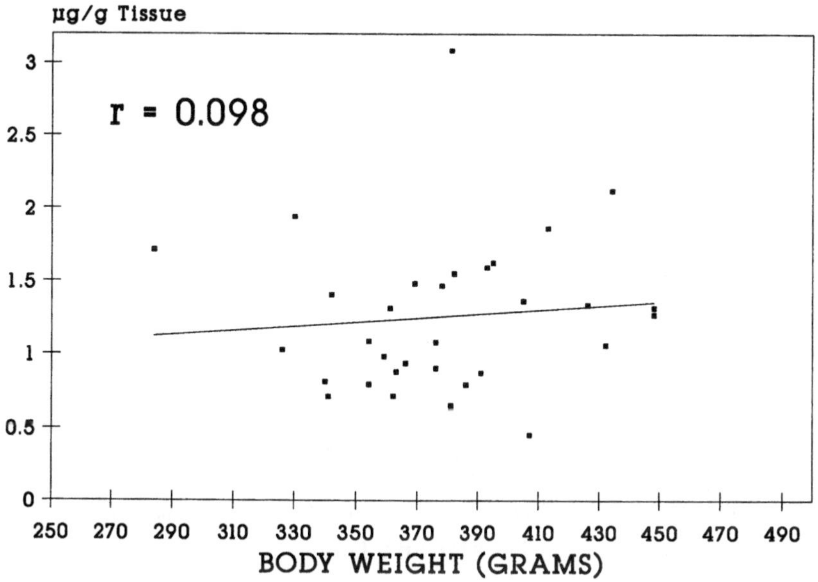

Fig. 2. Scattergram indicating the relationship of brain amphetamine concentrations to body weights on the last day of treatment of 32 rats treated for 1, 3, 7, or 14 d (8 rats/group). All rats were given a constant dose of 40 µmol/d, regardless of body weight. There was no significant correlation between body weights and amphetamine concentrations (r, Pearson correlation coefficient, equals 0.098).

duration of treatment, especially after 7 days, but there was no significant effect of duration of treatment observed with analysis of variance (time effect: $F(3,28) = 2.31$, $p > 0.05$). It can therefore bc concluded that brain levels of amphetamine reach steady state after the first 24 hours, without further significant accumulation of the parent compound.

The dosages of amphetamine were given on a *per diem* basis, without regard to body weight. Figure 2 indicates that the brain levels of amphetamine, across all durations of treatment, were not significantly correlated with body weight ($R = 0.098$, $p > 0.05$). This suggests that, at least for amphetamine and animals with body weights between 280 and 450 g, body weight does not significantly affect the brain levels achieved with a constant dose per day.

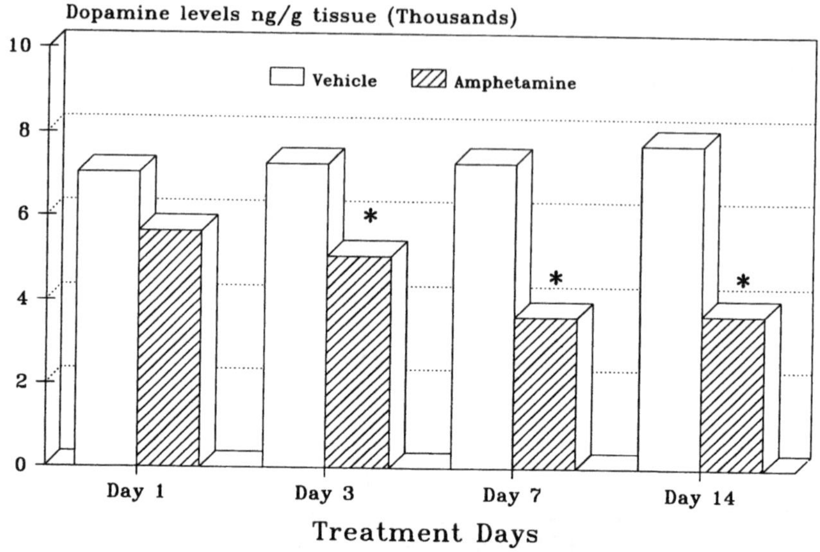

Fig. 3. Effects of vehicle or (+)-amphetamine (40 μmol/d), on DA levels of the stratium. A persistent depletion, significant by d 3 and reaching a maximum after 7 d, was observed (□, vehicle; ⊟, amphetamine).

2.4.2. Brain Regional Concentrations of Monoamines and Metabolites

Consistent with previous reports of a neurotoxic effect of continuous infusions of amphetamine specific for the striatum, we found depletions of DA in the striatum, but not the nucleus accumbens, beginning after 3 days of infusions, and reaching a maximum depletion after 7 days (Fig. 3). No DA depletions were found in the nucleus accumbens (Fig. 4). Interestingly, DOPAC levels were decreased substantially after 24 hours of treatment, and remained at the same levels throughout the treatment period (Fig. 5). This is consistent with the view that DOPAC reflects intraterminal metabolism of DA, and that amphetamine preferentially releases DA from newly synthesized stores that are susceptible to degradation by intraterminal monoamine oxidase. On the other hand, HVA levels initially decrease (Fig. 6), but then return to normal, even when DA concentrations are at their lowest. This suggests that DA turnover increases, and is consistent with an in vivo microdialysis study showing that extra-

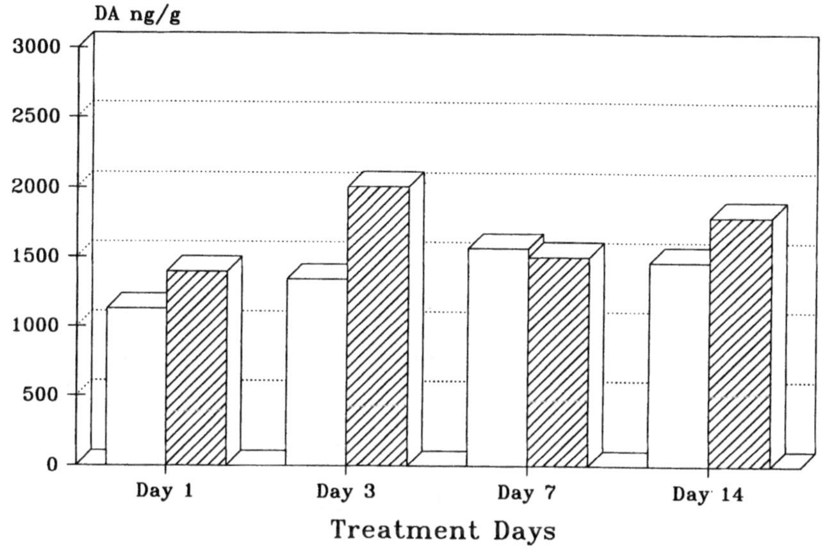

Fig. 4. Effects of vehicle or (+)-amphetamine (40 µmol/d), on DA levels of the nucleus accumbens. No statistically significant effects were observed, in spite of a trend to increased DA levels on day 3. Although striatal DA levels were depleted by amphetamine, no such effect was observed in the nucleus accumbens (□, vehicle; ▤, amphetamine).

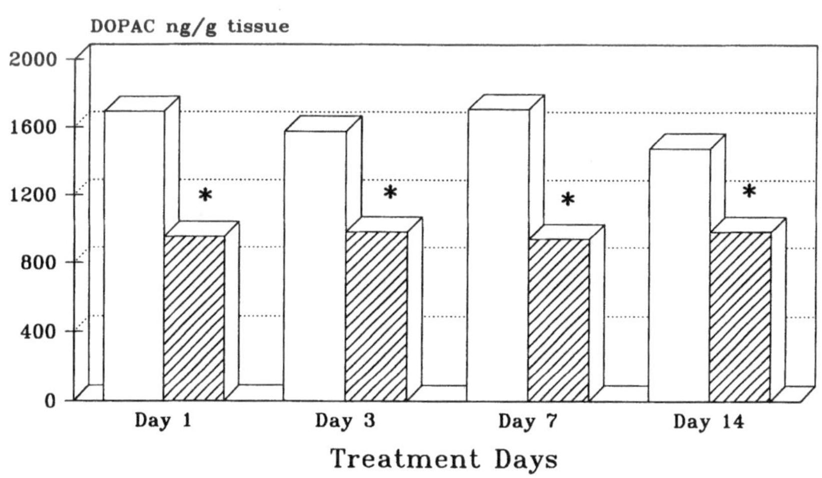

Fig. 5. Infusions of (+)-amphetamine (40 µmol/d) produced a consistent reduction in the acid metabolite of DA, DOPAC, thought to represent the intraterminal metabolism of DA. A similar effect (not shown) was found in the nucleus accumbens (□, vehicle; ▤, amphetamine).

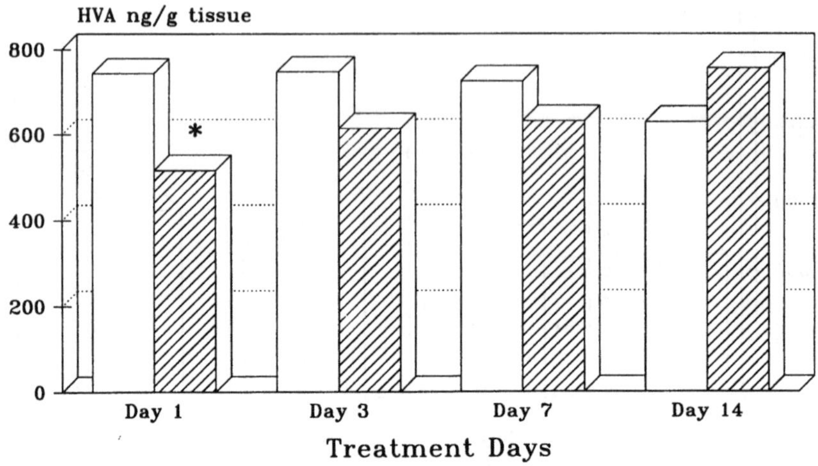

Fig. 6. Infusions of (+)-amphetamine (40 μmol/d) produced an initial decrease, followed by recovery to normal levels, of homovanillic acid, an acidic metabolite of DA that is thought to represent extraneuronal metabolism. A similar effect (not shown) was found in the nucleus accumbens (*, sig. from vehicle; #, sig. from amphetamine; □, vehicle; ▤, amphetamine).

neuronal DA is maintained at normal levels after neurotoxic doses of methamphetamine (Robinson et al., 1990).

2.4.3. Behavioral Effects of Continuous Infusions of Stimulants

2.4.3.1. Vehicle Effects

The next series of figures illustrate the effects of one dose of amphetamine (8 mg/d); one dose of a D2-receptor-selective agonist, PHNO (0.125 mg/d); one dose of a D1-selective agonist, SKF 38393 (1.25 mg/d), and a combination of the two direct agonists on behavioral measures with 3–8 rats/group. These results are from preliminary experiments and are depicted with *kinematic* graphs, in which cylinders represent frequencies (diameter of cylinder) and durations (length of cylinders), and arrows connecting behaviors represent transitions significantly above the expected random frequency (width of arrows represent frequencies of transitions). Random transitions and transitions that occur less frequently than expected are not shown. In these figures, some of the behavioral categories have been combined, and not all of the behaviors measured are shown, for the sake of clarity. "Groom" includes both grooming and hindlimb scratching. Locomotion and rearing have been combined, since they vary together, and sniffing and head movements have been combined for the same reason. Oral behaviors have been combined, so that category includes mouth movements, gnawing, and licking.

Some of the behaviors observed in vehicle-infused animals, averaged across the first and fourteenth days and nights, are shown in Fig. 7. All behaviors, except sleeping and lying down, show increased frequencies and durations during the night. Concomitant with a decreased frequency, there are fewer significant transitions during the day. However, all the daytime transitions, except that from "stand" to "lie," also occur at night.

2.4.3.2. Amphetamine Effects

Figure 8 depicts the effects of amphetamine on nocturnal behaviors, contrasting the effects of the first and last day of drug treatment. Behaviors that increased in frequency on the first night of treatment included "stand," "oral," "contact," "sniff/head," "groom," and "sit." Locomotion and rearing did not increase in either frequency or duration, relative to vehicle-infused animals. All of the behaviors that increased in frequency also increased in duration, except for "oral" and "sit." All the significant transitions between behaviors that occurred on the first night of amphetamine also occurred in the vehicle-infused rats, with one exception. After standing, animals significantly switched to snout contact with a cage surface. Therefore, amphetamine results in only one occurrence of a behavioral transition out of the realm of normal sequences. However, some transitions that occur in vehicle-infused animals did not occur after amphetamine. It can therefore be concluded that, during the first 12 hours of amphetamine infusions, the number of pairs of behaviors that exhibit significant transitions between each other decreases. This is in spite of increases in the frequencies of these behaviors.

After 14 days of amphetamine infusion, two general trends in the behavioral effects can be observed. First, some behaviors exhibit tolerance, and other behavioral effects are increased.

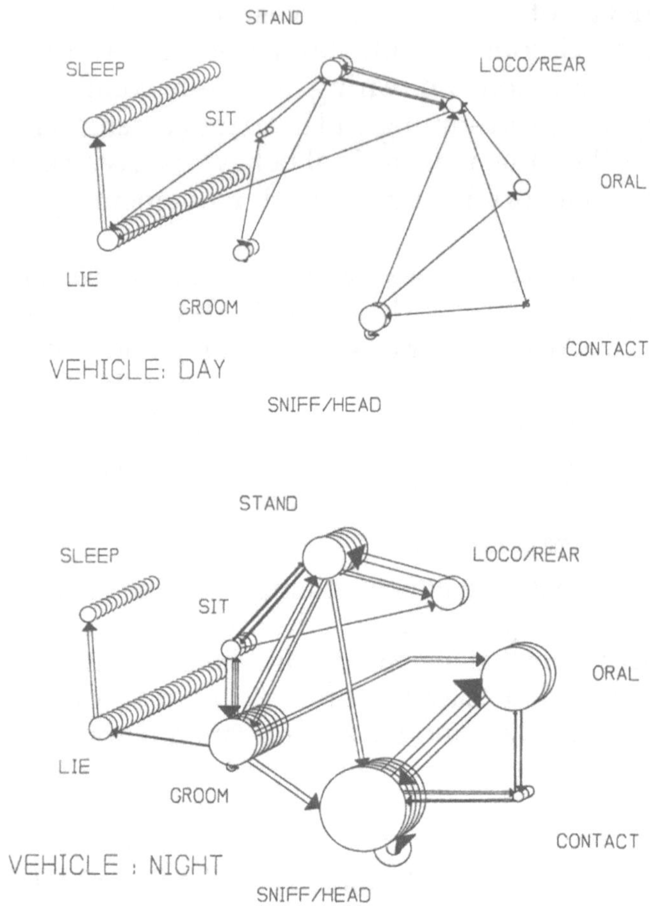

Fig. 7. Kinematic graph indicating the relative frequencies (diameter of cylinder), durations (length of cylinder, each arc representing 2 min) and frequencies of transitions (width of arrow) from one behavior initiation to another (direction of transition indicated by arrow) for a variety of behaviors for rats given infusions of vehicle. Only transitions that occur significantly more frequently than expected on a random basis, given the frequencies of the behaviors, are shown. Lack of transitions to or from a behavior reflect a random number of transitions or fewer transitions than expected. More complete definitions of behaviors are in the text. The top graph represents the means from twelve 5-min observations made each hour of the day, averaged across the first and 14th day of treatment, and the bottom shows the effects at night. The pattern of transitions was found to be very reliable both across different rats and across the same rat over different days (or nights).

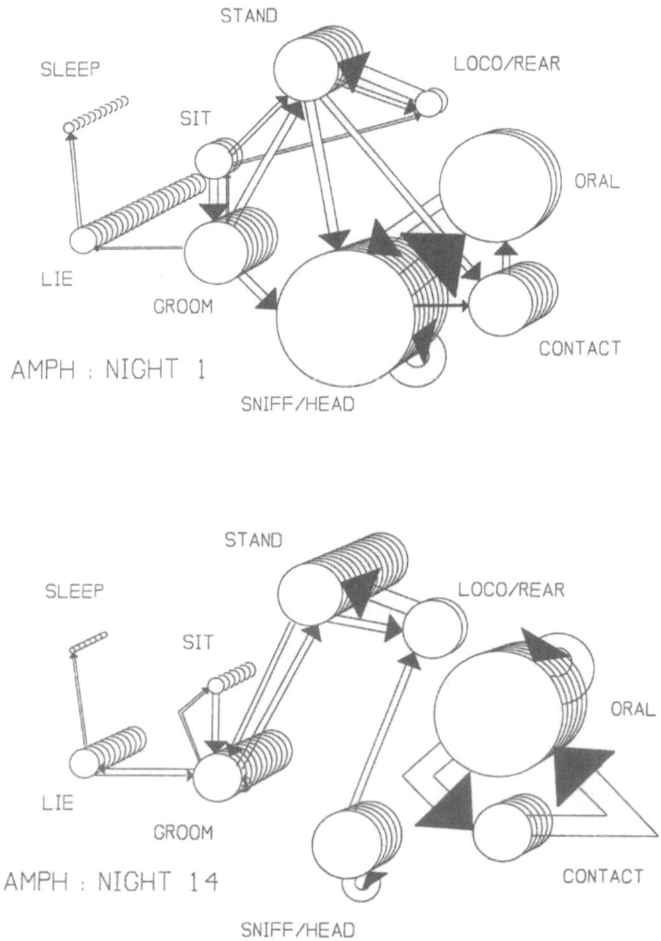

Fig. 8. Kinematic graph of the effect of (+)-amphetamine (AMPH, 8 mg/24-h period) given by osmotic minipump on the first and 14th *night* of treatment. Note that tolerance occurs to sniffing and head movements, but sensitization occurs to oral behaviors, locomotion and rearing, the duration of standing, grooming, and sitting; and in the reduction in sleeping and lying down. Note also that the number of significant transitions is reduced with chronic treatment, so that a number of behaviors occur randomly.

Sniffing and head movements decrease in both frequency and duration. Snout contact with a cage surface remains relatively constant during the night. "Groom" and "sit" decrease in frequency but show large increases in duration, and this phenomenon reflects the occurrence of "formication-like" behavior, which involves the continued licking of a certain area of a forelimb while sitting. Locomotion and rearing increase in frequency, but not duration, indicating a decrease in the mean bout duration of these behaviors, an effect not observed after acute injections. Oral behaviors increase predominantly in duration, and this largely reflects increases in mouth movements and licking, but not gnawing. Both sleeping and lying down are much reduced in duration. The second general trend is that the number of significant transitions between behaviors decreases even further from the first night of treatment. Thus, sniffing and head movements occur in a random relationship to the other behaviors, rather than after standing, grooming, snout contact or oral behaviors. In addition, oral and contact behaviors, although leading to each other in an orderly fashion, do not significantly follow the occurrences of "sniff/head," or "groom," as in vehicle-infused rats. Again, although there was only one occurrence of an unusual behavioral transition, this time there were significant transitions between different components of oral behaviors.

The daytime effects of amphetamine differ greatly from the nocturnal effects, as can be observed in Fig. 9. During the first day (following the first night), locomotion and rearing were increased in both frequency and duration to an extent much greater than that of the observed increase in oral behaviors (during the first night, locomotion and rearing were not affected, but large increases in oral behaviors were noted). In addition, neither grooming nor sitting were much affected during the first day. Generally speaking, but excluding locomotion and rearing, durations of behaviors were increased more than frequencies during the day, as compared to the night. In spite of the increases in frequencies of behaviors, beyond that normally observed even at night in vehicle-infused animals, the number of significant transitions did not increase substantially relative to control daytime, and were less than those observed in controls during the night. Both abnormal transitions observed at night ("stand" → "contact" [night 1], and "oral" → "oral" [night 14]) occurred on the first day of treatment.

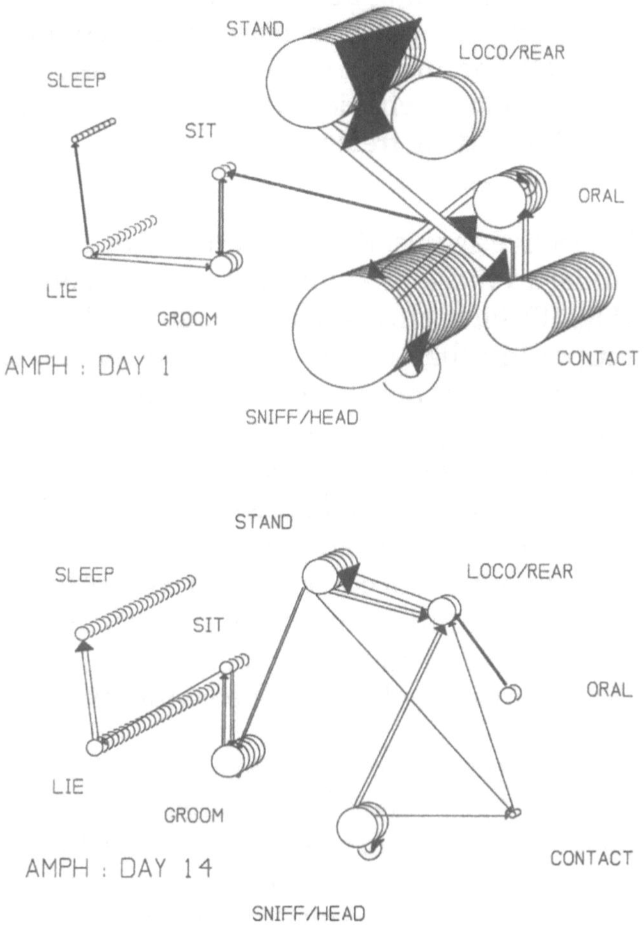

Fig. 9. Kinematic graph of the behavioral effects of (+)-amphetamine (AMPH, 8 mg/24-h period) given by osmotic mimpump on the first and 14th *day* of treatment. Note that tolerance occurs to all of the behavioral effects except grooming and sitting. The number of significant transitions are reduced at both times of treatment, so that a number of behaviors occur randomly. These data are from 12–24+ h of amphetamine treatment.

The greatest difference in the effects of amphetamine on behavior depending on the light cycle was observed on the last 24 hours of drug treatment. Whereas many of the behaviors increased from night 1 to night 14, most of the behaviors exhibited complete or near complete tolerance to amphetamine effects by day 14. Sitting and grooming behaviors are the notable exceptions to this rule. Grooming was increased from day 1 to day 14, and this reflected the occurrence of "formication-like" behavior. With this one exception, behavior was otherwise quite normal during the last day of drug treatment.

2.4.3.3. D2 Agonist Effects

The D2-selective agonist PHNO produced nocturnal effects generally similar to those of amphetamine with some notable exceptions (Fig. 10). First, effects on locomotion and rearing were greater in proportion than effects on oral behaviors, an effect similar to the first day of amphetamine, but different from the first night effects of amphetamine. Effects on "stand," "sniff/head," and "contact" were greater with PHNO than with amphetamine. PHNO effects on oral behaviors, "groom," and "sit" were less than the effects of amphetamine. The duration of grooming was actually reduced, relative to control levels. There was a slightly greater reduction in the number of significant transitions after PHNO than after amphetamine. Furthermore, three abnormal transitions between behaviors were observed after PHNO.

On night 14 of treatment, the effects of PHNO on frequencies of "stand," and "loco/rear" were augmented. "Groom" and "sit" were increased relative to controls. Although the frequency of grooming was greater than that observed after amphetamine, the duration was less. No obvious "formication-like" behavior was noted. Effects on "sniff/head" and oral behaviors remained relatively constant. A much smaller reduction in the number of transitions was observed after chronic treatment with PHNO, and three abnormal transitions were evident, only one of which was identical to an abnormal transition observed after amphetamine.

The effects of PHNO on the first day were nearly identical to the effects of amphetamine (Fig. 11). Increases in duration of "sniff/head" were greater after PHNO than after amphetamine, with corresponding decreases in oral behaviors and "contact." The pattern of transitions was somewhat, but not remarkably, different. As with amphetamine, chronic treatment produced

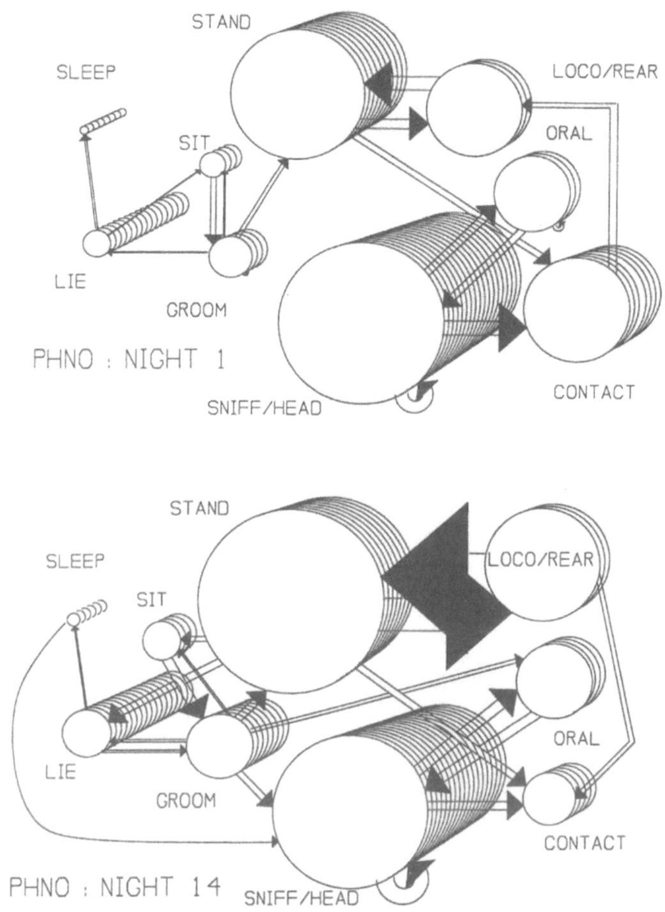

Fig. 10. Kinematic graph of the behavioral effects of a selective DA D2 agonist, PHNO (0.125 mg/d), on the first and 14th *night* of treatment. This drug produced greater effects than amphetamine on all behaviors except grooming and oral behavior. In contrast to amphetamine, tolerance developed only to snout contact with a cage surface (CONTACT), and sensitization or maintenance of initial effects occurred with all other behaviors. On the other hand, there was a smaller reduction in the number of significant transitions with PHNO than with amphetamine.

tolerance to most of the effects of PHNO during the day, except for grooming. Thus, the pattern of daytime tolerance, and nocturnal sensitization observed with amphetamine also is evident with a D2 agonist.

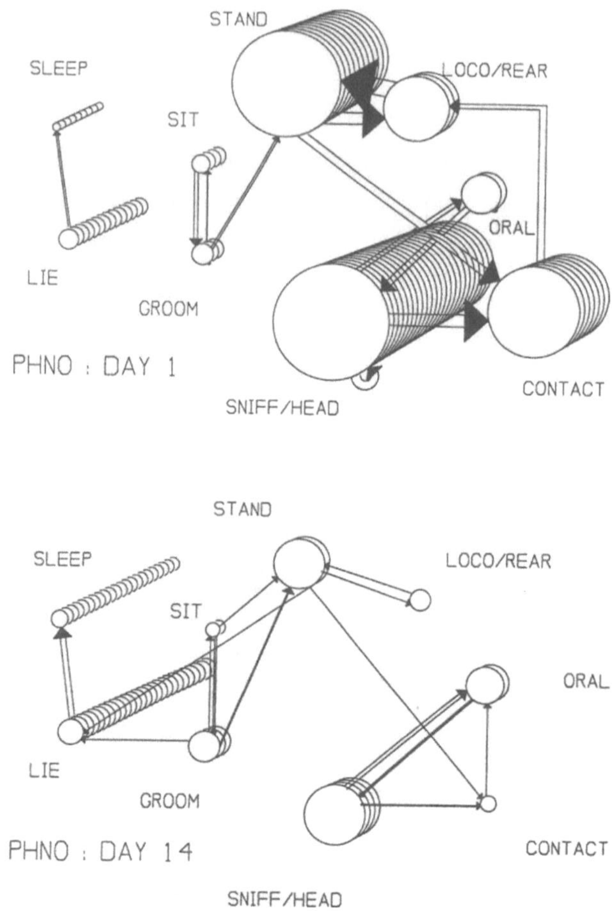

Fig. 11. Kinematic graph of the behavioral effects of a selective DA D2 agonist, PHNO (0.125 mg/d), on the first and 14th *day* of treatment. Daytime effects of this drug was very similar to those of amphetamine.

2.4.3.4. D1-Agonist Effects

Figure 12 shows that the primary nocturnal effect of SKF 38393, a D1 agonist, was to increase the frequency (but not duration) of grooming, and the duration (but not frequency) of sniffing/head movements. In addition, there was a loss of some behavioral transitions, with the addition of three novel transitions. One of these "novel" transitions was observed in control animals during the day, but not during the night. The effects observed on night 14 did not significantly differ from the effects

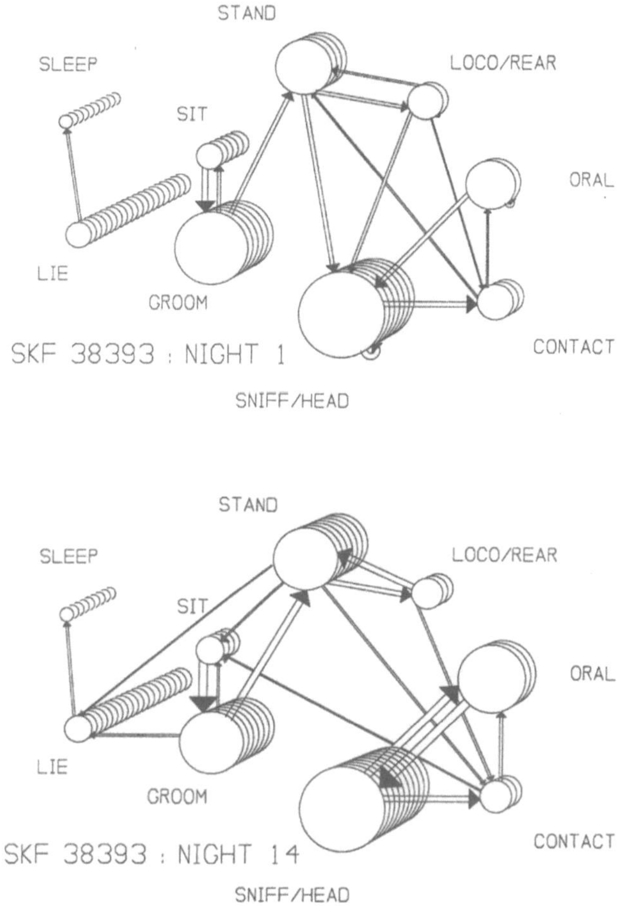

Fig. 12. Kinematic graph of the behavioral effects of a selective DA D1 agonist, SKF 38393 (1.25 mg/d), on the first and 14th *night* of treatment. Only grooming was significantly increased in frequency. There was some reduction in the number of significant transitions.

on night 1. There is a great deal of similarity between the nocturnal effects of SKF 38393 and the nocturnal effects of vehicle.

On the other hand, SKF 38393 tended to increase most of the behaviors during the day (Fig. 13). Although these effects are much less profound than the effects of amphetamine or PHNO, it is clear that SKF 38393 enhances daytime "active" behaviors, and reduces duration of sleeping. The D1 agonist therefore can be concluded to have a mild "arousal" action. As with

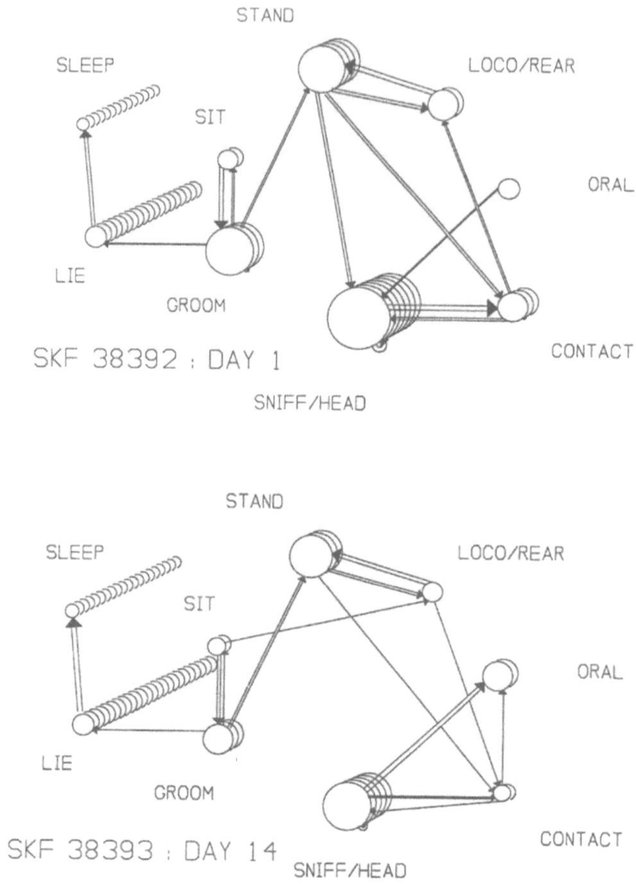

Fig. 13. Kinematic graph of the behavioral effects of a selective DA D1 agonist, SKF 38393 (1.25 mg/d), on the first and 14th *day* of treatment. Most active behaviors were increased, although not nearly to the same extent as with amphetamine or PHNO. Very little evidence of either tolerance or sensitization is apparent.

the nocturnal effects of this drug, there was little evidence of either tolerance or sensitization, although sleeping duration did increase after 14 days of treatment, with some slight decrease in duration of "sniff/head" and "stand." Behavioral transitions are largely unaffected by this drug at night.

2.4.3.5. D2 + D1 Agonist Effects

The addition of the D 1 agonist to the D2 agonist influenced nocturnal behavior in an interesting fashion (Fig. 14). The effects

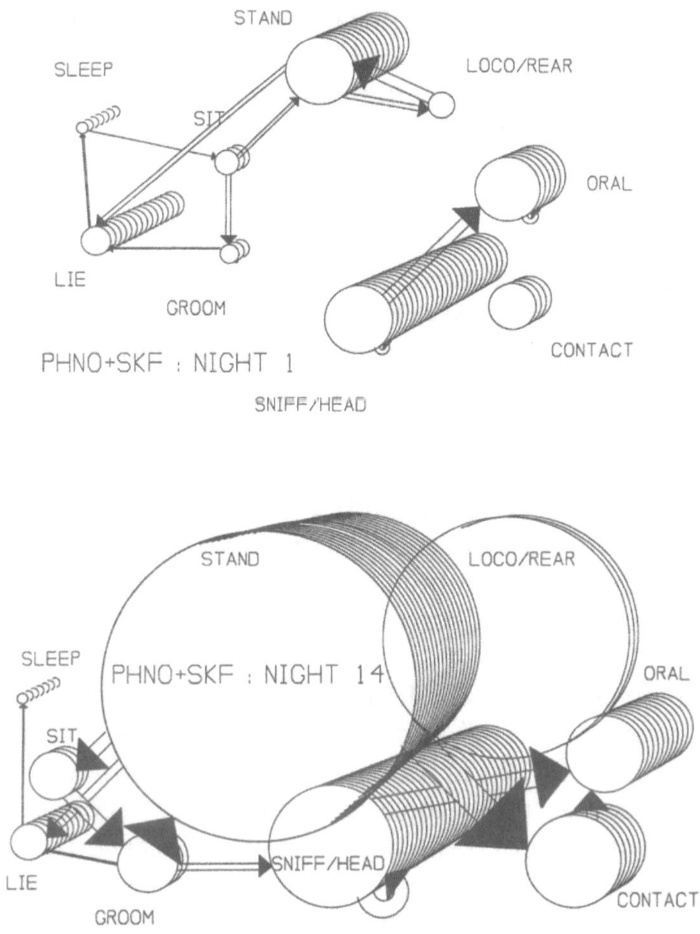

Fig. 14. Kinematic graph of the behavioral effects of combined treatment with a D2 agonist, PHNO (0.125 mg/d), and a D1 agonist, SKF 38393 (1.25 mg/d), on the first and 14th *night* of treatment. On the first night, SKF 38393 tended to decrease behavioral frequency effects of PHNO, but increased the behavioral duration effects. This suggests that D2 receptors function to facilitate the initiation of behaviors, and that D1 receptors inhibit the termination of bouts of behaviors. The combination of treatments reduces the number of significant transitions between behaviors to a greater extent than PHNO alone or amphetamine, with a marked effect on the first night. Chronic combined treatment results in a massive sensitization of the behavioral effects, far in excess of that observed with the D2 agonist or amphetamine. Note that significant transitions occurred between "stand" and "loco/rear" and vice versa, but are not shown, since the large size of the arrows would obscure the cylinders.

of the D2 agonist on frequencies of behaviors were dramatically reduced. The increases in the durations of "stand," "sniff/head," and "oral" behaviors produced by PHNO were enhanced by SKF 38393. Grooming was significantly reduced on the first night by combined treatment. There were fewer significant transitions after combined treatments with PHNO and SKF 38393 than after any of the other treatments.

Chronic treatment with both PHNO and SKF 38393 produced the greatest increases in the frequencies of the most behaviors of any treatments during the night. The sensitization observed after chronic coadministration of these two drugs at night is truly remarkable. In particular, the animals would exhibit short bursts of locomotion, interrupted by a partial rear, interrupted by a brief stand, and then another brief bout of locomotion, accompanied by virtually continuous sniffing and head movements. Grooming occurred in normal frequencies, but with reduced durations. The number of significant transitions between behaviors were much decreased, with no abnormal transitions apparent.

The effects of PHNO during the first day on all behaviors except "oral," "groom," and "sit" were reduced by coadministration of SKF 38393 (Fig. 15). Increases in these three behaviors were much increased by the addition of the D1 agonist. Reductions in the number of transitions that were observed during the night were also found during the day. The effects on oral behaviors and "sniff/head" developed daytime tolerance with coadministration over 14 days, but the effects on "loco/rear" and frequency (but not duration) of "stand" increased. Chronic treatment with either PHNO or amphetamine resulted in daytime tolerance to the locomotor- and rearing-stimulant effects of these drugs, but coadministration of SKF 38393 and PHNO produces daytime sensitization of these behaviors. There were fewer significant transitions on day 14 with cotreatment of these drugs than in any other condition.

Summary

Work with both day and night monitoring of continuous infusions of stimulants is preliminary, but has already produced

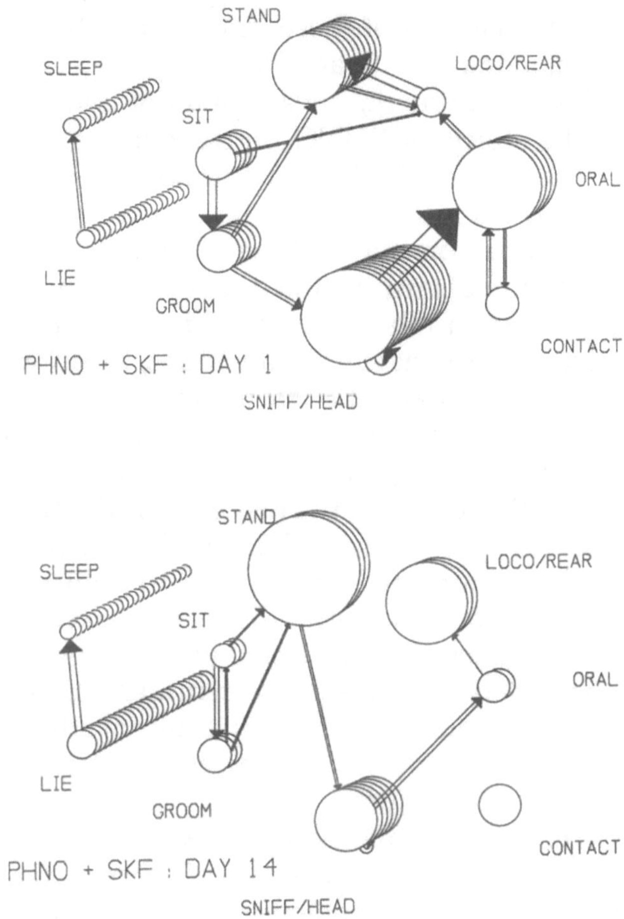

Fig. 15. Kinematic graph of the behavioral effects of combined treatment with a D2 agonist, PHNO (0.125 mg/d), and a D1 agonist, SKF38393 (1.25 mg/d), on the first and 14th *day* of treatment. Little evidence of either tolerance or sensitization is apparent. However, there is a great reduction in the number of significant transitions between behaviors.

some intriguing results. First, the tolerance to the motor-stimulant effects that has been reported after chronic continuous infusions is found to be specific to daytime. During the night, many behaviors display sensitization or are at least maintained at the initial high levels. This pattern is similar to that observed with selective D2-receptor agonists, but not with D1 agonists. Combinations of D1- and D2-receptor agonists result in daytime sensitization to some behaviors and in degrees of sensitization at night that are far in excess of those observed with either a D2-receptor agonist alone or with amphetamine alone. On the first night of combined treatment, it was apparent that the D1 agonist reduced the effects of PHNO on frequencies of behavior, but increased the effects of PHNO on durations of behavior. This suggests that DA D2 receptors may be intimately involved in the initiation of behaviors and that the D1 receptors may play a role in the inhibition of termination of bouts of behaviors, once initiated. In addition, amphetamine produces a reduction in the number of significant behavioral transitions, as previously reported by Norton (1973), and this effect is increased with chronic administration. A similar, though weaker, effect is observed with a D2 agonist, and a stronger effect is observed with combinations of D1 and D2 agonists. If the loss of logical transitions between behaviors is related to stimulant psychoses, then it would be predicted that treatment of humans with both agonists, as has been suggested for Parkinson's disease, would have greater psychotomimetic effects than treatment with amphetamine. The decrease in the variety of transitions seen in animals treated with stimulants raises the question of whether or not stimulant psychoses are a cognitive parallel to a reduced variety of sequences of thoughts and/or behaviors.

The occurrence of "formication-like" behavior was restricted to amphetamine. Whether higher doses of the receptor agonists could produce a similar phenomenon is yet to be determined. At present, it cannot be concluded that this behavioral syndrome is a function of amphetamine-induced activity at DA receptors. On the other hand, it is safe to conclude that the stereotyped oral behaviors (licking, gnawing, and vacuous mouth movements) require both D1- and D2-receptor activation; either agonist alone is not sufficient to produce these behaviors.

It is clear that brain levels of amphetamine reach a steady-state within a few days of initiation of continuous infusions. The effects observed after 14 days of treatment, therefore, cannot be attributed to accumulation of the drug, although accumulation of an active metabolite, such as para-hydroxyamphetamine or para-methoxyamphetamine (a potent hallucinogen), cannot be ruled out.

Further studies determining the neurochemical and receptor-binding effects of continuous stimulant treatments are presently underway. Of particular interest are the neural substrates for the day/night differences. Although these differences are the most striking behavioral effects of chronic stimulant treatments, little research has previously been directed to this question. It will also be of interest to specify the regions and neurochemical correlates of the changes in the sequences of behaviors.

Finally, since this model of stimulant psychoses incorporates the major features of the two most studied and conflicting models of stimulant psychoses used previously, it may serve as an ideal paradigm for resolving the differences of the sensitization and continuous-infusion models.

References

American Psychiatric Association (1987) *Diagnostic and Statistical Manual of Mental Disorders* (3rd ed., revised), Washington, DC.

Angrist B. (1983) Psychoses induced by central nervous system stimulants and related drugs, in *Stimulants: Neurochemical, Behavioral, and Clinical Perspectives* (Creese, I., ed.), Raven, New York, pp. 1–30.

Angrist B. and Gershon S. (1977) Clinical response to several dopamine agonists in schizophrenics and nonschizophrenic subjects (Costa E. and Gessa G. L., eds.), Raven Press, New York, pp. 677–680.

Antelman S. M. and Chiodo L. A. (1981) Dopamine autoreceptor subsensitivity: A mechanism common to the treatment of depression and the induction of amphetamine psychosis. *Biol. Psychiat.* **16,** 717–727.

Baker G. B., Coutts R. T., and Rao T. S. (1987) Neuropharmacological and neurochemical properties of *N*-(2-cyanoethyl)-2-phenylethylamine, a prodrug of 2-phenylethylamine. *Br. J. Pharmacol.* **92,** 243–255.

Baudry M., Costentin J., Marcais H., Martres M. P., Protais P., and Schwartz J. C. (1977) Decreased responsiveness to low doses of apomorphine after dopamine agonists and the possible involvement of hyposensitivity of dopamine "autoreceptors." *Neurosci. Lett.* **4,** 203–207.

Camp D. M. and Robinson T. E. (1988) Susceptibility to sensitization. I. Sex differences in the enduring effects of chronic D-amphetamine treatment on locomotion, stereotyped behavior and brain monoamines. *Behav. Brain Res.* **30,** 55–68.

Connell P. H. (1958) Amphetamine psychosis. *Maudsley Monograph No. 5.* Chapman Hall, London.

Conway P. G. and Uretsky N. J. (1982) Role of striatal dopaminergic receptors in amphetamine-induced behavioral facilitation. *J. Pharmacol. Exp. Ther.* **221,** 650–655.

Davis R. E., Sant W. W., and Ellison G. (1985) Continuous low-level apomorphine administration induces motor abnormalities and hallucinogen-like behaviors. *Psychopharmacology* **85,** 1–7.

Eichler A. J., Antelman S. M., and Black C. J. (1980) Amphetamine stereotypy is not a homogeneous phenomenon: sniffing and licking show distinct profiles of sensitization and tolerance. *Psychopharmacology* **68,** 287–290.

Eison M. S., Eison A. S., and Iversen S. D. (1983) Two routes of continuous amphetamine administration induce different behavioral and neurochemical effects in the rat. *Neurosci. Lett.* **39,** 313–319.

Ellison G. (1983) Phasic alterations in dopamine metabolites following continuous administration of amphetamine. *Brain Res.* **268,** 387–389.

Ellison G. and Ratan R. (1982) The late stage following continuous amphetamine administration to rats is correlated with altered dopamine but not serotonin metabolism. *Life Sci.* **31,** 771–777 .

Fray P. J., Sahakian B. J., Robbins T. W., Koob G. F., and Iversen S. D. (1980) An observational method for quantifying the behavioral effects of dopamine agonists: Contrasting effects of *d*-amphetamine and apomorphine. *Psychopharmacology* **69,** 253–259.

Gandolfi O., Dall'Olio R., Vaccheri A., Roncada P., and Montanaro N. (1988) Responses to selective D-1 and D-2 agonists after repeated treatment with selective D-1 and D-2 antagonists. *Pharmacol. Biochem. Behav.* **30,** 463–469.

Gately P. F., Segal D. S., and Geyer M. A. (1987) Sequential changes in behavior induced by continuous infusions of amphetamine in rats. *Psychopharmacology* **91,** 217–220.

Greenshaw A. J. (1986) Osmotic Mini-pumps: A convenient program for weight-adjusted filling concentrations. *Brain Res. Bull.* **16,** 759–761.

Griffith J. D., Oates J., and Cavanaugh J. (1968) Paranoid episodes induced by a drug. *J. Am. Med. Assoc.* **205,** 39.

Haberman S. J. (1973) The analysis of residuals in cross-classified tables. *Biometrics* **29,** 205–220.

Hanson G. R., Ritter J. K., Schmidt C. J., and Gibb J. W. (1986) Response of mesolimbic substance P to methamphetamine treatment. *Eur. J. Pharmacol.* **128,** 265–268.

Hess E. J., Albers L. J., Le H., and Creese I. (1986) Effects of chronic SCH23390 treatment on the biochemical and behavioral properties of D_1 and D_2 dopamine receptors: Potentiated behavioral responses to a D_2 dopamine agonist after selective D_1 dopamine receptor upregulation. *J. Pharmacol. Exp. Ther.* **238,** 846–854.

Janowsky D. S. and Davis J. M. (1974) Dopamine, psychomotor stimulants, and schizophrenia: Effects of methylphenidate and the stereoisomers of amphetamine in schizophrenics, in *Neuropsychopharmacoloy of Monoamines Regulatory Enzymes* (Usdin E., ed.), Raven, New York, pp. 317–323.

Kuczenski R., Leith N. J., and Applegate C. D. (1983) Striatal dopamine metabolism in response to apomorphine: The effects of repeated amphetamine pretreatment. *Brain Res.* **258,** 333–337.

Lieberman J. A., Kane J. M., and Alvir J. (1987) Provocative tests with psychostimulant drugs in schizophrenia. *Psychopharmacology* **91,** 415–433.

Lyon M., and Robbins T. W. (1975) The action of central nervous system stimulant drugs: A general theory concerning amphetamine effects, in *Current Developments in Psycholpharmacology,* vol. 2 (Essman W. and Valzelli L., eds.), Spectrum, New York, pp. 79-163.

Martin-Iverson M. T. and Iversen S. D. (1989) Day and night locomotor activity effects during administration of (+)-amphetamine. *Pharmacol. Biochem Behav.* **34,** 465–471.

Martin-Iverson M. T. and McManus D. J. Stimulant-conditioned locomotion is not affected by blockade of D1 and/or D2 dopamine receptors during conditioning. *Brain Res.* **521,** 175–184.

Martin-Iverson M. T., Coutts R. T., and Baker G. B. (1989) Behavioral effects of (+)-amphetamine and a side-chain fluorinated (+)-amphetamine. *J. Neurochem.* **52 (Suppl),** S200D.

Martin-Iverson M. T., Iversen S. D., and Stahl S. M. (1988a) Long-term motor stimulant effects of (+)-4-propyl-9-hydroxynaphthoxazine (PHNO), a dopamine D-2 receptor agonist: Interactions with a doparnine D- 1 receptor antagonist and agonist. *Eur. J. Pharmacol.* **149,** 25–31.

Martin-Iverson M. T., Leclere J. F., and Fibiger H. C. (1983) Cholinergic-dopaminergic interactions and the mechanisms of action of antidepressants. *Eur. J. Pharmacol.* **94,** 193–201.

Martin-Iverson M. T., Stahl S. M., and Iversen S. D. (1987) Factors determining the behavioural consequences of continuous treatment with 4-propyl-9-hydroxynaphthoxazine, a selective dopamine D2 agonist, in *Parkinson's Disease: Clinical and Experimental Advances* (Rose F. C., ed.), John Libbey, London, pp. 169–177.

Martin-Iverson M. T., Stahl S. M., and Iversen S. D. (1988b) Chronic administration of a selective dopamine D-2 agonist: Factors determining behavioral tolerance and sensitization. *Psychopharmacology* **95,** 534–539.

Martres M. P., Costentin J., Baudry M., Marcais H., Protais P., and Schwartz J. C. (1977) Long-term changes in the sensitivity of pre- and postsynaptic dopamine receptors in mouse striatum evidenced by behavioral and biochemical studies. *Brain Res.* **136,** 319–337.

Matsuda L. A., Hanson G. R., and Gibb J. W. (1989) Neurochemical effects of amphetamine metabolites on central dopaminergic and serotonergic systems. *J. Pharmacol. Exp. Ther.* **251,** 901–908.

Mattingly B. A. and Rowlett J. K. (1989) Effect of repeated apomorphine and haloperidol treatments on subsequent behavioral sensitivity to apomorphine. *Pharmacol. Biochem. Behav.* **34,** 345–347.

Muller P. and Seeman P. (1979) Presynaptic subsensitivity as a possible basis for sensitization by long-term dopamine mimetics. *Eur. J. Pharmacol.* **55,** 149–157.

Nielsen E. B., Lyon M., and Ellison G. (1983) Apparent hallucinations in monkeys during around-the-clock amphetamine for seven to fourteen days: Possible relevance to amphetamine psychosis. *J. Nerv. Ment. Dis.* **171,** 222–233.

Norton S. (1968) On the discontinuous nature of behavior. *J. Theor. Biol.* **21,** 229–243.

Norton S. (1973) Amphetamine as a model for hyperactivity in the rat. *Physiol. Behav.* **11,** 181–186.

Parenti M., Flauto C., Parati E., Vescovi A., and Goppetti A. (1986) Differential effects of repeated treatment with L-dopa on doparnine D-1 or D-2 receptors. *Neuropharmacology* **25,** 331–334.

Peris J. and Zahniser N. R. (1989) Persistent augmented doparnine release after acute cocaine requires dopamine receptor activation. *Pharmacol. Biochem. Behav.* **32,** 71–76.

Rao T. S., Baker G. B., and Coutts R. T. (1986) Pentafluorobenzenesulfonyl chloride as a sensitive reagent for the rapid gas chromatographic analysis of tranylcypromine in tissues and body fluids. *Biochem. Pharmacol.* **35,** 1925–1928.

Rebec G. V. and Bashore T. R. (1984) Critical issues in assessing the behavioral effects of amphetamine. *Neurosci. Biobehav. Rev.* **8,** 153–159.

Riffee W. H. and Wilcox R. E. (1985) Effects of multiple pretreatment with apomorphine and amphetamine on amphetamine-induced locomotor activity and its inhibition by apomorphine. *Psychopharmacology* **85,** 97–101.

Rinne J. O., Rinne J. K., Laakso K., Lonnberg P., and Rinne U. K. (1985) Dopamine D-1 receptors in the Parkinsonian brain. *Brain Res.* **359,** 306–310.

Robinson T. E. (1988) Stimulant drugs and stress: Factors influencing individual differences in the susceptibility to sensitization, in *Sensitization of the Nervous System* (Kalivas P. W. and Barnes C., eds.), Telford, Caldwell, NJ, pp. 145–173.

Robinson T. E., and Becker J. B. (1986) Enduring changes in brain and behavior produced by chronic amphetamine administration: A review and evaluation of animal models of amphetamine psychosis. *Brain Res. Rev.* **11,** 157–198.

Robinson T. E., Jurson P. A., Bennett J. A., and Bentgen K. M. (1988) Persistent sensitization of dopamine neurotransmission in ventral striatum (nucleus accumbens) produced by prior experience with (+)-amphetamine: A microdialysis study in freely moving rats. *Brain Res.* **462,** 211–222.

Robinson T. E., Yew J., Paulson P. E., and Camp D. M. (1990) The long-term effects of neurotoxic doses of methamphetamine on the extracellular concentration of dopamine measured with microdialysis in striatum. *Neurosci. Lett.* **110,** 193–198.

Ryan L. J., Martone M. E., Linder J. C., and Groves P. M. (1988) Cocaine, in contrast to D-amphetamine, does not cause axonal terminal degeneration in neostriatum and agranular frontal cortex of Long-Evans rats. *Life Sci.* **43,** 1403–1409.

Schmidt C. J., Gehlert D. R., Peat M. A., Sonsalla P. K., Hanson G. R., Wamsley J. K., and Gibb J. W. (1985) Studies on the mechanism of tolerance to methamphetamine. *Brain Res.* **343,** 305–313.

Segal D. S. (1975) Behavioral and neurochemical correlates of repeated D-amphetamine administration. *Adv. Biochem. Psycopharmacol.* **15,** 247–262.

Segal D. S. and Schuckitt M. A. (1983) Animal models of stimulant-induced psychosis, in *Stimulants: Neurochemical, Behavioral, and Clinical Perspectives* (Creese I., ed.), Raven, New York, pp. 131–167.

Seiden L. S. and Ricaurte G. A. (1987) Neurotoxicity of methamphetamine and related drugs, in *Psychopharmacology: The Third Generation of Progress* (Meltzer H. Y., ed.), Raven Press, New York, pp. 359–366.

Seiden L. S. and Vosmer G. (1984) Formation of 6-hydroxydopamine in caudate nucleus of the rat brain after a single large dose of methyl-amphetamine. *Pharmacal. Biochem. Behav.* **21,** 29–31.

Stewart J. and Vezina P. (1989) Microinjections of SCH-23390 into the ventral tegmental area and substantia nigra pars reticulata attenuate the development of sensitization to the locomotor activating effects of systemic amphetamine. *Brain Res.* **495,** 401–406.

Szechtman H., Eilam D., Teitelbaum P., and Golani I. (1988) A different look at measurement and interpretation of drug-induced stereotyped behavior. *Psychobiology* **16,** 164–173.

Trulson M. E. and Jacobs B. L. (1979) Long-term amphetamine treatment decreases brain serotonin metabolism: Implications for theories of schizophrenia. *Science* **205,** 1295–1297.

Vaccheri A., Dall'Olio R., Gandolfi O., Roncada P., and Montanaro N. (1987) Enhanced stereotyped response to apomorphine after chronic D-1 blockade with SCH 23390. *Psychopharmacology* **91,** 394–396.

Van Hoof J. A. R. A. M. (1982) Categories and sequences of behavior: Methods of description and analysis, in *Handbook of Methods in Nonverbal Behavior Research* (Scherer K. R. and Ekman P., eds.), Cambridge University Press, Cambrdge, UK, pp. 362–439.

Wei-Dong L., Xiao-Da Z., and Guo-Zhang J. (1988) Enhanced stereotypic behavior by chronic treatment with bromocripdne accompanies increase of D-1 receptor binding. *Life Sci.* **42,** 1841–1845.

Animal Models of Hallucinations

Continuous Stimulants

Gaylord D. Ellison

1. Introduction

When amphetamine or cocaine is administered to humans every hour or so for several days, either during the "speed runs" of addicts or in controlled laboratory settings, there reliably results a psychosis that is similar to paranoid schizophrenia in a number of important aspects. This unique regimen of amphetamine intake, involving the continuous presence of stimulants over a prolonged period of time, can be simulated in animals using subcutaneously implanted slow-release silicone pellets containing *d*-amphetamine or cocaine base. Monkeys and rats implanted with amphetamine pellets develop stages of behavioral alterations that are somewhat similar in sequence to those observed in humans who self-administer frequent doses of amphetamine. An initial period of hyperactivity and exploratory behavior is followed by the gradual development of motor stereotypies, then a period of relative inactivity, and finally, at three to five days after pellet implantation, by a late-stage. This final stage is characterized by "wet-dog" shakes, parasitotic-like grooming episodes, and a variety of other forms of hallucinatory-like behavior. Because continuous amphetamine administration also induces distinctive, neurotoxic alterations in dopaminergic innervations of the caudate nucleus, but not in mesolibic dopamine innervation of the nucleus accumbens or in several other neurotransmitte systems, we initially hypothesized that the hallucinatory behaviors were a reflection, in part, of the

From: *Neuromethods, Vol. 18: Animal Models in Psychiatry I*
Eds: A. Boulton, G. Baker, and M. Martin-Iverson

damaged caudate dopamine terminals. However, we have recently discovered that continuous cocaine is even more potent at inducing the distinctive late-stage behaviors, yet clearly does not produce the distinctive neurotoxic effect in the caudate. A theory is presented, emphasizing that continuous stimulants may reproduce some aspects of the prolonged excitation that accompanies an acute psychotic episode and be a fruitful model for the clarification of schizophrenia.

1.1. Problems with Animal Models of Hallucinations

Animal models are essential for the development of more advanced treatment strategies for human psychopathologies, just as they are in all fields of medicine. The problem is how to know when one has a valid model, or merely a superficial analogy, because some types of animal models are much easier to accept as valid than other types. For example, when a clear behavioral deficit, such as the motor deficits in Parkinson's disease, are accompanied by a reproducible brain lesion, such as damage to dopamine cell bodies, the goals of animal models are clear. In such cases, the degree of accuracy of the model can be clearly measured.

However, this problem is considerably more difficult when one deals with animal models of what are often purely subjective or psychological concepts; a primary example is "hallucinations." One simple *tour de force* is to define a "hallucinatory behavior" in an animal as any behavior that is distinctively induced by any drug that humans call a "hallucinogen," cf. Jacobs et al. (1976). This strategy led to the discovery that such behaviors as "limb flicks" and "wet-dog shakes" in cats were distinctively induced by hallucinogens, such as LSD or mescaline and that S2 serotoninergic receptors were especially involved in the elicitation of these behaviors. This is clearly a heuristic model of hallucinations, but several problems soon became apparent.

1. A "hallucination" is a subjective construct, for it is defined as a sensory experience not based in reality. The problem is that it may or may not lead to a behavioral reaction, and, without verbal questioning, how can one objectively dis-

cern a hallucination in an animal? One can never assess the presence of sensory experiences that do not result in behavior. It is possible that sometimes a pressure to emit a behavior can build up to point at which the behavior can be emitted in the absence of any external stimuli (as with the so-called vacuum behaviors of ethologists). Are such behaviors, emitted out of context, to be classified as "hallucin-atory"? Or might they not sometimes be mere motor discharges, albeit highly organized ones?

2. The problem of low-frequency behaviors: Incessantly performed behaviors, such as motor stereotypies, are a joy to the behavioral observer, for they can be rated so conveniently. Because they do not change dramatically over time, infrequent time-sampling is sufficient to accurately describe their form and frequency. The opposite is true for "hallucinatory behaviors:" They are rare in occurrence and sometimes highly unique in form. Furthermore, they exhibit extreme tolerance to repeated elicitation, for one of the major problems with the study of the effects of hallucinogens is that they often show rapid habituation over time when the drug is given on a repeated basis. It is consequently difficult to obtain statistical significance in studies relying on these behaviors, which have a very low rate of occurrence that can be quantified only by human observers.
3. Lack of agreement on subtypes: In humans, there are clearly distinctive types of hallucinations. Some are described as pleasant, dreamy experiences; some are frightening, paranoia-inducing; still others are brief and less organized. Humans also clearly attribute hallucinations to either internal or external causes, and it is probably a critical point in a psychotic episode when the hallucination is first attributed to external causes. Yet these fundamental distinctions have never been raised in animal models of hallucinations.

1.2. Stimulant Drug- and Paranoid Schizophrenia-Based Models

Compared to the classic LSD- or mescaline-induced syndrome, paranoid hallucinations are particularly amenable and important for modeling for three reasons:

1. There are a limited number of clearly-defined drug states that induce paranoid behaviors in humans, most notably amphetamine and cocaine psychoses. As many authors have noted, the degree to which these drugs can reproduce the paranoid state is remarkable. Before this was widely known, many humans were misdiagnosed as paranoid schizophrenics when they were really just drug abusers. Indeed, in *DSM*-III, the symptoms of paranoid schizophrenia and those of amphetamine psychosis are listed as identical (with the exception of needle marks on the arms and the presence of amphetamine in the urine).
2. The behavior in these syndromes is quite clear and focused, for the paranoid acts out in a quite direct manner. Pure paranoids characteristically behave quite openly, rather than becoming catatonic, engaging in constant motor stereotypies, or becoming autistically withdrawn. Consequently, there are certain very clear and distinctive behavioral aberrations that accompany paranoid behavior, including aberrant social behaviors, which have certain distinctive features, such as fixations on a certain enemy, chase and flight behaviors, and also certain distinctive and bizarre thought patterns.
3. This kind of drug-induced paranoia is now an especially important phenomenon, because it has become a severe societal problem. Amphetamine and cocaine psychoses are now widespread and are highly correlated with a substantial crime problem in several modern societies.

1.3. Amphetamine Psychosis: Addict Reports

In schizophrenic patients, amphetamine activates or worsens the preexisting psychotic symptoms, implying an action of the drug on fundamental symptoms of schizophrenia (Angrist et al., 1974b). This was one of the origins of the dopamine theory of schizophrenia, and it led to conclusions that the closest drug model to schizophrenia is amphetamine psychosis (Snyder et al., 1974). Amphetamine psychosis was first described by Young and Scoville (1938), and its classical description is contained in Connell's monograph (1958) reporting on 42 amphetamine addicts who developed psychoses. According to Connell and others (Bell, 1965), amphetamine psychosis can have features indistinguishable from paranoid schizophrenia, with profound psychological alterations, including aberrant behavior and ideation, in the absence of physical signs.

Yet, others have disputed this conclusion, arguing that there are clear differences between the symptomatology seen in schizophrenia and that seen in amphetamine psychosis. Visual hallucinations are more prominent in amphetamine psychosis, auditory hallucinations are more prevalent in schizophrenia. The answer to this question—how closely amphetamine psychosis mimics schizophrenia—seems to depend in large part on the type of drug regimen used to precipitate the psychosis, since several different types of amphetamine-induced psychoses have been described.

At one extreme are the rare but well-documented psychotic states that develop in certain drug-naive individuals shortly after they first ingest relatively low doses of amphetamine (Gold and Bowers, 1978). Such states sometimes persist for considerable periods of time in the absence of further drug intake, and the symptomatology can closely parallel that of schizophrenia; however, these acute, low-dose cases are usually ascribed to a latent psychosis in the individual, which has been triggered by the ingestion of the drug. Although amphetamine seems particularly able to act as "the key which unlocks the psychotic episode" in such cases (Gold and Bowers, 1978), presumably other drugs or events could also precipitate a developing psychosis in the specially vulnerable, borderline individual.

Most humans do not become psychotic on initial exposure to *d*-amphetamine, but only after a chronic pattern of abuse has developed, and this presents a problem in documenting the actual drug-intake pattern as well as the quantity and purity of the drug that precipitated the psychosis. Such information can come only from the unreliable reports of the addicts themselves. Two less-biased sources for information on amphetamine psychosis are reports of the drug culture from medical workers in drug rehabilitation centers, and direct observations made during controlled administration studies on human volunteers.

A variety of quite detailed descriptions of amphetamine addicts and the subculture that can develop around its use have been published. Ellinwood (1967) described hospitalized addicts in Kentucky, who reported well-formed delusions of perse-cution that grew out of an initial fear, suspiciousness, and an awareness of being watched. The hallucinations that occurred in over half the patients were visual (developing from fleeting glimpses of just-recognizable images into formed images of God, tormentors, animals, and so on), auditory (developing from simple voices that merely whispered the patient's name into voices with which the patient conversed), and tactile (reports of infestations of microanimals and the presence of vermiform and encysted skin lesions, which they felt as well as saw). Three patients incurred punctuate scars when they attempted to dig out the imaginary encysted parasites. Smith (1969) described parasitotic hallucinations in amphetamine addicts, adding that it was common to find open sores or scabs on their faces or arms resulting from picking or cutting out hallucinated bugs. It is noteworthy that cocaine induces this identical syndrome ("crank bugs").

A detailed description of the events that lead to amphetamine psychosis was compiled by the staff of a California institution for the treatment of civilly committed drug abusers (Kramer et al., 1967). They reported that, after initial phases of experimentation with the drug, a pattern of usage emerged in which the abuser injected the drug about every two hours around the clock for a period of three to six days. The addict remained awake continuously, using the amphetamines to ward off sleep, knowing full well that discontinuation of the drug would result in a rebound depression. During these prolonged bouts of continuous drug intoxication, which came to be known as "speed runs," behavior gradually became less organized and two distinctive behavioral alterations were reported to develop in the abusers. During a "run," the user pushed activities that progressively became less varied and more compulsive, eventually resulting in the development of a perseverative hyperactivity that consisted of repeatedly performing simple motor tasks. These "motor stereotypies" were particularly evident in mechanically-inclined addicts.

The other alteration was a gradually developing paranoia. Its eventual appearance was accepted by the addicts as inevitable. Relatively early in the "speed run" the addict would experience suspicious thoughts. Although these were initially recognized as drug-induced, as the "run" progressed, the paranoid thoughts began to be accepted as real, and paranoid delusions began to develop. This initially appeared on the second or third day of the "run" and then gradually intensified as the "run" progressed (although experienced users sometimes became paranoid with a single dose of amphetamine even after prolonged abstinence).

1.4. Amphetamine Psychosis: Controlled Studies

In an attempt to reproduce this phenomenon in a controlled hospital setting, Griffith et al. (1972) administered small (10-mg) doses of *d*-amphetamine to volunteers hourly over a prolonged period. Stages of behavioral alterations were observed, with eight of nine subjects developing a paranoid psychosis within one to five days after the start of amphetamine administration. During the course of the experiment, the subjects gradually became hypochondriacal and dysphoric, spending most of their time in bed and were increasingly irritable. A distinct prepsychotic phase then appeared, during which they confined themselves to their rooms and shut their doors. The onset of paranoia was usually abrupt. Griffith's subjects spoke freely about their paranoid delusions and ideas of reference. Although visual and auditory hallucinations were not present, all subjects reported pins-and-needles paresthesias. In a subsequent study, Angrist and Gershon (1970) were able to elicit a variety of hallucinations, including auditory, cutaneous, olfactory, and visual, using higher doses than Griffith, but again administering the drug every few hours for several days. In this study, the subjects again evidenced progressive flattening of affect and, prior to the onset of the paranoid state, they were also reported to have thought disorders.

We have previously argued (Ellison and Eison, 1983) that the symptomatology observed during these nearly continuous but relatively low-dose amphetamine regimens is more similar to that of schizophrenia than that following other schedules of drug administration, inasmuch as it combines progressive flattening of affect with well-formed delusions that are not predominantly visual; and the subjects talk freely about their delusions. There is a possible exception to this conclusion, for one other controlled experiment has reported the elicitation of amphetamine psychosis without prolonged drug intoxication. Bell (1973) was able to elicit a psychosis within an hour after injection of a single, very large dose of methamphetamine administered intravenously to former addicts. Although this dose had an effect on blood pressure that was much greater than that in Griffith's subjects, Bell reported that his subjects for the most part concealed their psychosis rather well, with most of them admitting to having had a hallucination only upon later questioning. Although this type of immediately elicited, high-dose psychosis resembles paranoid schizophrenia in several ways, there are important differences. The hallucinations reported by his subjects were primarily visual. Also, emotional responsivity in such rapid, high-dose psychoses is usually brisk, often involving considerable anxiety (Bell, 1965; Slater, 1959), whereas schizophrenics (and addicts toward the end of "speed runs") often exhibited a blunted affect. The interpretation of this study is also complicated by the fact that in his study, Bell administered methamphetamine, a drug that is more potent and longlasting than *d*-amphetamine and that has some quite different actions not shared by *d*-amphetamine, such as strong neurotoxic effects on serotonin neurons (Hotchkiss and Gibb, 1980; Ricaurte et al., 1980; Steranka and Sanders-Bush, 1980).

Even more important is the fact that, when Bell administered the methamphetamine, his former abusers, many of whom had been drug-free for only a few days before testing, said that they experienced the immediate reoccurrence of the *identical paranoid delusions* that they had previously experienced while taking amphetamines. Indeed, whereas paranoia was elicited in 12 of 14 patients who had previously experienced psychoses while taking amphetamines, no psychosis was elicited in two patients who were eventually found not to have been previous regular users of amphetamines and neither of whom had had a previous episode of amphetamine psychosis. In striking contrast, Griffith's study included five addicts who had no previous psychotic experiences with the drug, but who developed paranoid reactions when given the drug in frequent, low doses. This led to the conclusion that the relatively covert kinds of paranoid delusions studied by Bell may, in part, represent a conditioning phenomenon—i.e., the reinstatement of behavior learned under prior drug experiences (Ellinwood and Kilbey, 1977). Kramer (1969) has observed that, among addicts, paranoia rarely develops on initial exposure to amphetamine, that it progressively develops using a "speed run," but that, once experienced, it can readily return upon reexposure to the drug, even after a prolonged period of abstinence .

1.5. Continuous Stimulation and the Acute Psychotic Episode

We conclude that the rapid-onset psychosis elicited by Bell had a number of unusual features, and that the most reliable way to induce a model paranoid psychosis in psychologically healthy individuals is by exposing them to frequent, low doses of *d*-amphetamine for several days. There is a theoretical reason why this particular drug regimen, involving the continuous but low-level presence of psychoactive stimulants over an extended period of time, might be especially relevant to the study of schizophrenia: It results in a prolonged period of excitation similar to that which has been reported to occur during an acute psychotic episode. Bowers (1968) interviewed individuals who had recently undergone such a "psychotic break" and studied the diaries they kept during this period. Prominent in his description of the setting in which the acute psychotic episode developed was a prolonged and irresolvable conflict. The individual began to dwell on this problem and entered a lengthy period characterized by heightened arousal, sleeplessness, and anorexia, and by other stress-like alterations in emotionality, such as experiences of heightened awareness and increased perceptual acuity. Each of these alterations in behavior could also be induced by stimulants, such as amphetamine. During the further progressive stages described by Bowers, ideas of reference and, finally, hallucinatory episodes or delusional fixations developed. This parallel between the prolonged period of heightened arousal and insomnia produced by continuous amphetamines and that which often occurs during acute psychotic episodes has led to the hypothesis (Wyatt, 1978) that in schizophrenia an endogenous amphetamine-like compound is produced in a constant, low-level manner. These considerations suggest that the continuous administration of amphetamines to animals might uniquely mimic some of the alterations in emotionality and brain chemistry that occur during the psychotic onset and the early stages of schizophrenia, although they may not be at all relevant to the behavioral and brain states that are present in the chronic, long-term patient.

2. Continuous Amphetamines and Behavior in Animals

Although there have been many animal studies on the effects of chronic amphetamine administration (e.g., Ellinwood and Sudilovsky, 1973; Segal and Mandell, 1974), most of the earlier studies involved one or two ip injections daily, whereas the human studies demonstrating amphetamine psychosis have used small, frequent oral doses that achieved a continuous plasma level of the drug for up to six days (Griffith et al., 1972; Angrist et al., 1974a). The previous research stimulated our interest in observing rats in colonies during continuous amphetamine administration, for one of the most notable characteristics of paranoid schizophrenia is the type of social interactions that develop, and these could be observed only in rats in social colonies, such as ours.

In order to study this continuous-drug regimen, we developed relatively inexpensive silicone pellets that continuously release amphetamine base when implanted subcutaneously. In our initial studies (Huberman et al., 1977), we found that, by employing a drug reservoir and a silastic window of variable sizes, through which the drug can diffuse, it is possible to precisely control diffusion rate and duration of drug action. Rats were implanted subcutaneously with these capsules and their brains assayed for amphetamine 2, 5, 7, or 10 days after pellet implantation. It was found that this pellet produced an initial brain level of amphetamine nearly equivalent to that produced by a 2.0 mg/kg *d*-amphetamine sulfate injection (ip) and that the release rate very gradually declined following implantation. Rats implanted with these pellets and housed in cages were very hyperactive for three days, during which they engaged in intense and nearly continuous motor stereotypies. There was then a striking decrease in motor activity, even though appreciable amphetamine was still present in the brain. During the period from 4–10 days they showed increased startle responses and were irritable to handling. We later modified this dosage considerably (Nielsen and Ellison, 1980), using a "30-day" pellet that induces only very mild, or no, stereotypies.

Because the role of neurotoxicity in the production of the unique effects of continuous amphetamine has been questioned (e.g., Robinson and Becker, 1986), it is important to note that we have found that continuous, low-level amphetamine can be administered with this second, longer-lasting pellet in behaviorally significant doses without producing neurotoxic effects. By altering the silicone window through which the drug can diffuse and changing the concentration of drug within the pellet, one can produce either neurotoxic or nondamaging doses.

2.1. Effects of Continuous Amphetamine in Social Rat Colonies

The study of drug effects on social behavior ("sociopharmacology") has gradually evolved over the last four decades into a highly refined science (McGuire et al., 1982), and the close observation of social behaviors can sometimes be an extremely sensitive way to record drug effects and can provide data obtainable in no other manner. Often another member of the same nonhuman species can recognize behaviors no human could, and can also serve as a stimulant object to elicit highly unique and meaningful reactions from a cohort.

We conducted a colony study comparing the stages of behavior of rats implanted with slow-release amphetamine pellets with the behavior of controls. Male rats were raised in two different colonies for several months, marked, and then half the animals from each colony were implanted with amphetamine pellets and half with control pellets (Fig. 1). We conducted the entire experiment twice in order to ensure that the results obtained were not the result of some idiosyncratic selection of animals. In one colony, the largest and most dominant rat ("king rat") was placed in the amphetamine group; in the replication, in the control group.

During the first few hours after pellet implantation, the amphetamine-implanted rats were hyperactive and exploratory, locomoting about the enclosure, approaching the human observers, and peering off ledges, sniffing the air (Fig. 2). Gradually the amount of space traversed by these animals decreased, and their behavior evolved over the next 24 hours into motor stereotypies of an increasingly more circumscribed nature. During the second and third days after amphetamine-pellet implantation, they engaged in intense stereotypies, constantly self-grooming, sifting the straw with their forepaws, or sniffing surfaces. Compared to the results from caged, isolated animals treated identically, in this complex, enriched environment, exceedingly complicated motor stereotypies develop.

During these initial three days, the amphetamine-implanted animals had been out in the arena more than control animals throughout the day and night. The motor stereotypies then gradually disappeared, and the animals entered a second phase of constant amphetamine intoxication. During this period the anorexic effects of amphetamine waned, as did the arousal properties, and the implanted animals began to eat and briefly retreated to the burrows after three days of hyperactivity. Concurrently, the amphetamine-implanted animals began to show exaggerated startle responses, frequent vocalizations, and, finally, on day five, increased social behaviors. Whereas the amphetamine-implanted animals had not engaged in social interactions during the initial stages of constant amphetamine intoxication, during days 5–7 after implantation, their behavior was characterized by extraordinarily heightened social behaviors, principally involving heightened fight and flight behaviors. The particular type of social behavior observed during this phase was correlated with dominance standing prior to pellet implantation: The largest amphetamine-implanted animals showed more aggres-

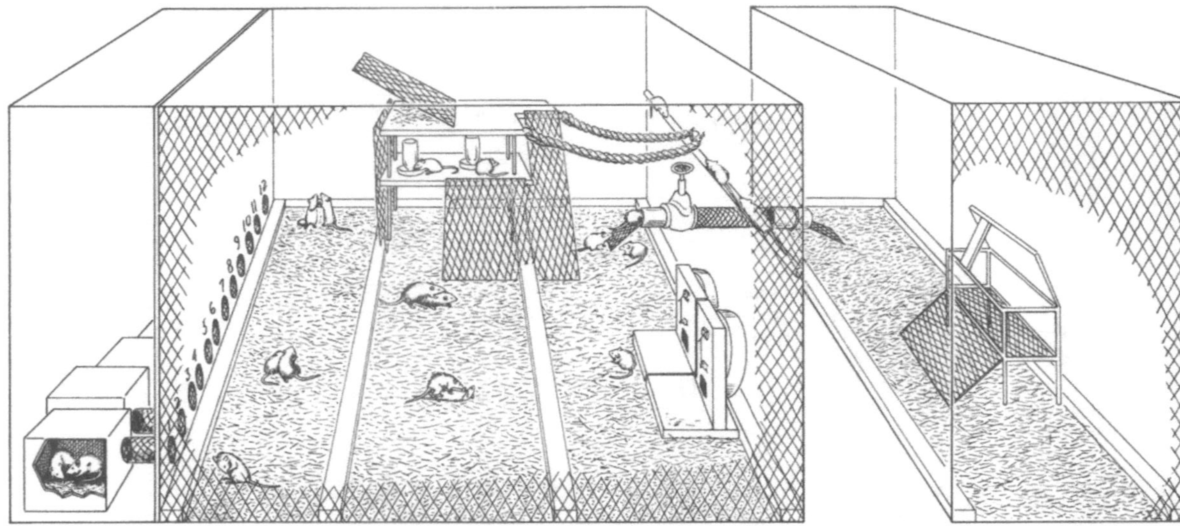

Fig 1. Drawing of one of the two spacious rat colony enclosures we have constructed at UCLA. On the left is the burrows area, kept in constant dim illumination. In the center is the behavioral arena, with saw-covered floor and numerous play objects. On the right is the feeding enclosure, to which the rats are allowed access one hour daily. A tube and valve connects the feeding arena with the behavioral arena.

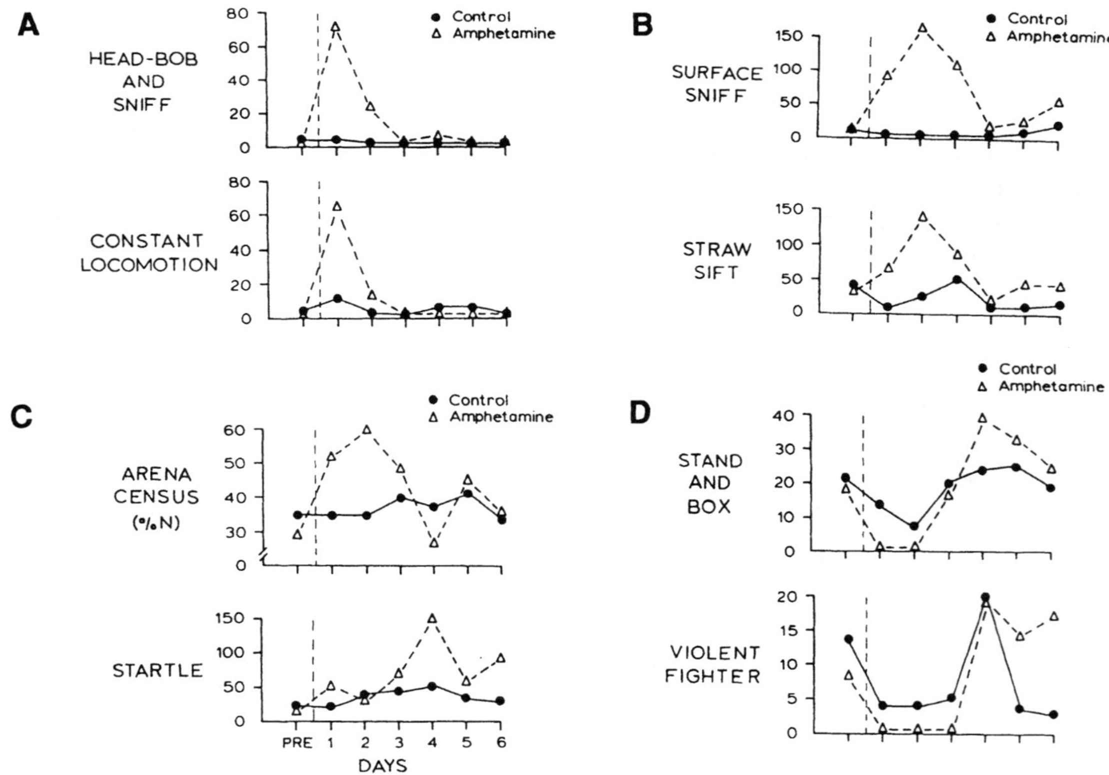

Fig 2. Clear stages of behavioral alterations during continuous amphetamine. "A": During the first day, pellet animals engage in exploratory behaviors and locomotion. Then, "B", motor stereotypies predominate throughout days two and three. On day four, ("C"), the pellet animals remain in the burrows but, when out, emit startle responses. Finally, there is a "late-stage", "D", of heightened fight and flight responses.

sive fighting, violent episodes, and chasing, whereas the smaller amphetamine animals were more frequently chased. Several interesting details of these social interactions were unexpected, in that they had never been previously observed in our colony studies. Although the smaller amphetamine-implanted animals were frequently chased by larger implanted animals, they actively solicited these chases, approaching larger animals and goading them over and over again. More interestingly, stable pairs of implanted animals formed social bonds of aggressor–aggressee. We painted numbers on the flanks of the amphetamine animals so we could identify individuals, and we found that during this late phase of constant amphetamine intoxication, the same pairs of amphetamine-implanted animals would seek each other out and fight. That is, each implanted rat had as its major source of social interactions one other implanted rat, and the aggressive encounters between these two individuals were repeated over and over again in form of social stereotypy.

This late and social phase of constant amphetamine intoxication, which appears after four to seven days, represents an entirely different phenomenon from the earlier forms, and it has a number of similarities to amphetamine psychosis in humans. The stages of behavior reported by Griffith et al. (1972), and others, in humans given continuous amphetamines: (initial activation, gradual appearance of irritability and depression after several days, then withdrawal to their rooms and self-isolation, and finally the paranoid stage) is remarkably similar to the stages we observed in our implanted rats; an initial excited motor-stereotypy phase, a transient period when the implanted rats withdrew to the burrows, and then a period of increased social interactions, including exaggerated fight or flight responses, fear-like startle responses and vocalizations, and a habit of focusing attacks on only one other animal. This latter effect seems similar to the fixations characteristic of a paranoid delusion. We concluded that, after four to five days of continuous amphetamine stimulation, colony rats entered a phase of behavior that is not observable in caged rats and that has a number of similarities to the behavior of humans exhibiting amphetamine psychosis (Table 1).

Table 1
Stages of Continuous Amphetamine

	Human	Monkey	Rat
Initial stage, acute effects	Euphoria and exhilaration, except in low doses to addicts	Increased activity Fixated staring Motor sterotypy	Increased activity Motor stereotypy
Crash stage	Increased irritability Anhedonia, intellectual flattening Depression and self-isolation	General inactivity hunched posture Motor dyskinesias, including tongue protrusions, tremors	Retreat to burrows Heightened startle
Late stage	Paranoid delusions Parasitosis Hallucinations	Hallucinatory-like behaviors Parasitosis Sudden orient and fleeing "wet-dog" shakes	Parasitosis Hallucinogen-like behavior (limb flick "Wet-dog" shakes) Sudden startle Aberrant social behavior: fixations Fight or flight

Another colony was implanted with amphetamine pellets, but after three days the pellets were removed. These animals did not enter a late phase of heightened social interactions, indicating that the late phase was not simply a rebound from amphetamine. We also tested other animals by reimplanting amphetamine pellets seven days after initial pellet implantation and found the motor stereotypies induced by the implantation were less than those observed following initial pellet implantation, implying that a tolerance had developed to the motor-stereotypy-inducing effects of amphetamine.

It should be noted that the types of spontaneous startle responses shown by these rats are completely different from those studied in the typical "startle" paradigm in rats. Other researchers have reported that rats implanted with continuous-amphetamine pellets and tested with very loud, sudden probe noises actually show a decreased elicited startle amplitude (Gately et al., 1987). This decreased responsivity is presumably a result of the caudate dopamine depletions that are developing, for this experiment found, as we have, that the later stages of continuous amphetamines lead to an increase in dopamine in the nucleus accumbens, but to decreases in the caudate. However, this is not attributable to a lack of replication of our findings in the colony. We recently tested amphetamine-pellet-implanted animals in a highly computerized activity cage that continuously monitored activity around the clock, and recorded integrated activity levels every second. We found that after three to four days of continuous amphetamines, the animals began to show infrequent, but highly distinctive, sudden "startle" responses. These were extremely large in amplitude, and of a short duration, and no such sudden spontaneous startles were ever observed in the control animals housed next to the amphetamine-implanted animals, nor in the amphetamine-implanted animals during the stereotypy phase. It is clear that these "spontaneous startle" responses represent an entirely different phenomenon from the "elicited startle" responses typically studied in many laboratories, which are elicited by very loud, sudden noises during relatively constant and intense white noise. These unusual, spontaneous startle behaviors may be a fruitful way to measure "late-stage" behaviors in an automated fashion.

2.2. *Effects of Continuous Amphetamines on Monkeys*

In similar studies using adult monkeys, we found that the stages of alterations in behavior following continuous amphetamine administration are similar to those obtained in rats, but that the forms of the behavior, particularly the "late-stage" behaviors, are much more complex and vivid (Ellison et al., 1981a). Following a brief period of heightened activity, each monkey either developed highly idiosyncratic episodes of extremely complex motor stereotypies or began to engage in prolonged bouts of staring at a distant, circumscribed spot on the wall. This suggests that, for higher organisms, such as monkeys, staring is apparently an equivalent of motor stereotypies in rats. This phase was followed by a period of less-focused activity, when only fragments of the initial stereotypies appeared. During this phase, the monkeys were often inactive and seemed lethargic, although they did not sleep. On the fourth to sixth day after capsule implantation, several new behaviors developed. There were increases in sudden orienting responses and the appearance of motor dyskinesias, such as limb tremors and oral protrusions. At about the same time, discrete episodes of hallucinatory behaviors began to appear, including, at different times, attacks, sudden fleeing responses, capturing movements by the hands in midair, and excited grooming episodes suggestive of parasitosis. These results indicate that monkeys show stages of continuous amphetamine intoxication similar to those seen in rats, but that the behaviors are much more varied and the late-stage behaviors much more vivid. It is noteworthy that the onset of the hallucinatory behaviors in the monkeys was almost exactly at the same point (four to five days) at which rats entered the "late-stage."

2.3. Reactivation of "Late-Stage" Behaviors

The problem with studying these late-stage behaviors is that they have a very low rate of occurrence. Hence, we attempted to develop a procedure in which rats primed with the amphetamine pellet were given a brief drug-free period, then "reactivated" with a low dose of amphetamine. Using this paradigm, it was demonstrated that following several days of continuous administration, *d*-amphetamine comes to have properties that are more like those of a hallucinogen than like those of acute *d*-amphetamine.

In these experiments (Nielsen et al., 1980a) we first automatically videotaped rats housed in isolation cages every half hour for seven days following pellet implantation. The behaviors we observed during the initial stages were those with which we were familiar, such as incessant motor stereotypies, but several behaviors of which we had not previously been aware were observed during the late-stage. The rats showed increased "wet-dog shakes" and episodes of parasitotic-like frantic grooming, as had the monkeys, and they also showed limb-flicks, a behavior normally observed only after the administration of hallucinogens (Jacobs and Trulson, 1979). However, all these behaviors were infrequent, and many hours of videotape had to be observed to obtain significant results. We then developed a method for reliably inducing these behaviors at a high frequency upon demand. This method (the "reactivation" paradigm) capitalizes on the fact that, just before rats enter the "late-stage" of continuous amphetamine administration, they enter a crash phase, during which they are inactive. We removed the amphetamine pellets $4^1/_2$ days after implantation and gave the rats a 12 hour drug-free period so that residual amphetamine could be cleared and the animals could rest. The rats were then given either a saline or an amphetamine injection and observed. The results were quite striking. Pellet-pretreated animals injected with saline were extremely inactive, but, when injected with a low dose of *d*-amphetamine, these animals showed increased limb flicks, wet-dog shakes, and frantic grooming with the teeth. This is extremely important theoretically, for these behaviors are decreased in frequency if the *d*-amphetamine is administered to drug-naive animals, yet most of these behaviors are increased in drug-naive animals by injections of the hallucinogens LSD and mescaline.

One of the most convincing findings that this altered responsivity to amphetamine has some close relationship to amphetamine psychosis is that this altered response to *d*-amphetamine (i.e., the induction of a hallucinogen-like behavioral syndrome) does not appear until the animal has been on the drug for at least four days, and that it very rapidly disappears after removal of the pellet (the syndrome is virtually nonexistent if the animals are tested 36 hours, rather than 12 hours, after pellet removal). These results indicated that we could reliably elicit in amphetamine-pellet-pretreated animals behaviors that are unique measures of the late-stage of continuous amphetamine. The behaviors observed were not species-specific, for they also occur in monkeys treated with slow-release amphetamine pellets or following several months of frequent high doses of amphetamine (Ellinwood and Kilbey, 1977). The behavioral measures are suggestive of hallucinatory behaviors, since they also occur following LSD or mescaline injections. Like amphetamine psychosis, this syndrome rapidly disappears following the discontinuation of amphetamine.

3. Unique Biochemical Effects Induced by the Amphetamine Pellet

When we conducted studies of catecholamine fluorescence in animals implanted with our slow-release amphetamine pellet we found a remarkable effect (Ellison et al., 1978). In normal animals the caudate nucleus has a relatively diffuse fluorescence with extremely fine terminals and axons. In animals sacrificed from two to three days after amphetamine pellet implantation, the caudate takes on a cloudy or filmy appearance, as though much of the dopamine has diffused into extracellular space. However, in animals sacrificed five days after pellet implantation, during the period of paranoid-like social disruptions, a highly unusual type of fluorescence is observed in the caudate nucleus. The background fluorescence is considerably less than that in normal animals, and large, distinct axons with multiple and extremely large, swollen enlargements can be observed. These alterations in fluorescence microscopy persist long after the pellet is removed, at least until 17 days after pellet implantation and 10 days after pellet removal. Extremely large, stump-like objects can be observed, as can thick and long axons. These axons are larger in diameter than those observed in normal tissue in any brain area, and they sometimes have varicosities of an equally swollen nature. Similar swollen and enlarged fluorescent axons in the ventrolateral caudate have been observed following partial aspiration lesions of the caudate and are presumed to reflect an accumulation of amines in the remaining axons. These observations therefore imply that damage to dopamine terminals in the caudate can be produced by the continuous amphetamine administration achieved by pellet implantation.

This implication was validated using regional assays of amines and tyrosine hydroxylase (TH) in several brain regions at various times after amphetamine pellet implantation. TH activity and dopamine and norepinephrine levels were measured in regional brain homogenates (cortex, caudate, accumbens, hypothalamus, and brainstem). The largest changes in TH activity during pellet implantation were in the caudate nucleus, where it fell to 50% of control levels. Dopamine content of the caudate was also significantly reduced, as were norepinephrine levels during pellet implantation in most brain regions. These results are presumed to reflect the depletions produced by amphetamine-stimulated release of catecholamines. However, although catecholamine levels had recovered to near control levels in the animals sacrificed 110 days after pellet removal, caudate TH activity remained significantly less than in controls, although it was elevated in the other brain regions.

In further studies, we attempted to determine whether these persisting decreases in caudate TH activity were related to the total amount of amphetamine released by the pellet in seven days, or to the continuous nature of the amphetamine release. One group was implanted with amphetamine pellets for seven days and given a 60-day recovery period; a second group was injected with an amount of amphetamine equivalent to that released by the pellet in seven days (25 mg), but this was given in seven daily injections of (3.7 mg of amphetamine each). TH activity in the caudate 60 days after the cessation of amphetamine injections or pellet removal was again significantly reduced relative to controls in the amphetamine-pellet-implanted animals, but not in the injection groups, implying that continuous amphetamine administration has a unique capability for inducing long-lasting structural and biochemical alterations in dopamine terminals in the caudate nucleus. This general conclusion has been validated in a number of other laboratories, using different techniques, such as amphetamine delivered by minipumps or by very frequent injections (for example, Nwanze and Jonsson, 1981; Ricaute et al., 1980; Steranka and Sanders-Bush, 1980). Fuller and Hemrick-Luecke (1980) found that amphetamine administered in combination with drugs that slow its metabolism becomes neurotoxic to caudate dopamine terminals.

As the caudate becomes depleted and damaged, the nucleus accumbens becomes more active. When rats were injected with labeled 2-deoxyglucose (2DG) and sacrificed at various times after amphetamine-pellet implantation (Eison et al., 1981) we found that, whereas acute injections of amphetamine induced increased 2DG uptake in several brain regions (many of which are correlated with sensory functions), the "late-stage" pellet-implanted animal shows a distinctively different pattern of glucose uptake. Most notable in this animal is an increased radioactivity in the nucleus accumbens and related mesolimbic structures. This conclusion was validated in a related study using similar microdissection techniques (Eison et al., 1981), in which it was found that the regional distribution of ^{3}H-labeled amphetamine-derived radio-activity is maximal in the accumbens of pellet-implanted animals. These changes do not occur for serotonin and its metabolite, 5HIAA (Ellison and Ratan, 1982).

4. Continuous Amphetamines or Repeated Injections: Which Models Psychosis?

We have repeatedly found that continuous amphetamine has effects on behavior that are strikingly different from those of intermittent amphetamine. When rats were pretreated with a seven day amphetamine pellet and then the pellet was removed and a drug-free period given, the animals displayed a completely different reaction to injections of either amphetamine or apomorphine, both just after drug treatment and 30 days after drug treatment, compared to animals given daily injections of amphetamine (Nelson and Ellison, 1978). Whereas "daily-injection" animals treated for seven days (of strong injections) or 30 days (of weak injections) demonstrate heightened motor stereotypies compared to controls (showing the inverse tolerance effect), animals tested immediately after seven days of the amphetamine pellet are, conversely, less reactive to amphetamine, using sterotypies as a measure. This presumably reflects the downregulation in dopamine D2 receptors that occurs at this time (Nielsen et al., 1980b). We subsequently (Ellison and Morris, 1981) replicated these findings and studied the relationship of these changes to in vivo spiroperidol accumulation. Again, in these experiments, the seven day pellet induced changes opposite to those displayed by animals pretreated with daily injections of amphetamine. This experiment also included a 28-day pellet group, which received continuous amphetamine at a much lower daily dose, one that we have found to be nontoxic to dopamine innervations of caudate. In this study, it was also found that this pellet was extraordinarily different from the daily injection in its effects on behavior and spiroperidol accumulation. Intermittent, daily injections of amphetamines leads to inverse tolerance for motor stereotypies, with little change in receptor number, whereas continuous amphetamine initially leads to an inverse tolerance for motor stereotypies, but then to a late-stage in which there is a motor-stereotypy tolerance and down-regulation of receptors.

These findings are important because Robinson and Becker (1986) have presented a very different interpretation of the effects of chronic amphetamine administration on brain and behavior. They conclude that it is the sensitization produced by intermittent amphetamine that is the best model of amphetamine psychosis, rather than the "neurotoxicity" produced by continuous amphetamine. We do not agree with this conclusion for several reasons.

1. Many of the arguments presented by Robinson and Becker are not completely accurate. They argue the amphetamine-pellet experiments involve very high doses of amphetamine—much higher than addicts actually self-administer. The doses used by chronic human addicts can become immense, especially as binges progress. Doses of several grams of methamphetamine in a day are not uncommon. The brain levels of amphetamine induced by the pellet are not high (the equivalent of a 1.7 mg/kg ip dose). One simply cannot, as they attempt to do, compare human doses with rat doses and conclude that the dose required to produce amphetamine neurotoxicity is many times higher than that required to produced an amphetamine psychosis. The doses of continuous amphetamine found sufficient to induce a late-stage

in monkeys were quite low compared to the doses employed in rats and are close to the human range.

2. Robinson and Becker equate continuous amphetamine with neurotoxicity, yet clearly amphetamine can be administered in a continuous fashion, at lower doses, without producing neurotoxic effects. One testable question is whether what we have called a late-stage will appear with these milder continuous-amphetamine regimes. Later in this chapter, we will present data on the effects of continuous cocaine that clearly invalidate the conclusion that the dopamine neurotoxicity is a necessary prerequisite for the late-stage.
3. Robinson and Becker argue that former amphetamine addicts show an enduring hypersensitivity to amphetamine and cite our research as saying that animals given continuous amphetamine, then withdrawn from amphetamine, and later challenged with an acute injection are not supersensitive to amphetamine. In fact our published report (Ellison and Morris, 1981) shows that they do become hypersensitive to the drug, but that this supersensitivity develops over time.
4. Most important, amphetamine (and cocaine) addicts simply do not take the drug once daily. This regimen, perhaps more applicable to the way the drug is administered to hyperkinetic children, leads to a daily "crash" that actually outlasts the action of the drug. Any nonbiased reading of the literature on drug abuse will lead to this conclusion.

The most important issue is not which is the "better" model, but whether either, or both, of these two very different effects induced by amphetamine can shed light on schizophrenia and paranoia. Motor stereotypies are clearly present in schizophrenics, and actually predominate in many, and the intermittent, daily-injection regimen clearly induces this type of behavior remarkably well. An animal in a motor stereotypy shows a complete inhibition of social behaviors (as did our colony animals during the early phases of the speed run), never shows any sudden spontaneous fleeing behaviors, and actually shows a suppression of the distinctive behaviors induced by hallucinogens.

Yet some of these latter behaviors, especially aberrant social interactions, are a hallmark of paranoia.

5. Cocaine Psychosis

Fortunately, the argument that dopamine neurotoxicity is a hallmark of the "late-stage" behaviors can now be dismissed as a result of more recent developments. We have recently begun to extend our continuous-drug model to the study of chronic cocaine administration. If similar stages appear with cocaine administration, this would be a highly significant finding, for it may permit one to dissociate unique effects of these two drugs and allow one to better determine the necessary and sufficient conditions for producing the late-stage paranoid-like effects.

On the one hand, it is quite remarkable that virtually every one of the above findings with amphetamine has also been reported to be true for cocaine. This parallel is true for the production of motor stereotypies. Post (1976) found that repeated cocaine injections in monkeys induced the development of a variety of stereotypic responses and, in rats, sensitization to the motor stereotypy-inducing effects (and seizure-inducing properties of cocaine have been reported in the two-hour period following each of a series of daily injections of the same dose of cocaine [Post, 1975; Post and Rose, 1976; Stripling and Ellinwood, 1976; Kilbey and Ellinwood, 1977a]). This increased sensitivity was still seen as long as seven weeks after the end of chronic administration (Kilbey and Ellinwood, 1977b). This sensitization phenomenon has also been seen after chronic amphetamine administration (Segal and Mandell, 1974; Magos, 1969; Klawans and Margolin, 1975; Kilbey and Ellinwood, 1976). Thus, there are a number of remarkable parallels between chronic effects of cocaine and amphetamine on dopamine systems.

It is also very clear that cocaine addicts binge and become paranoid. Cocaine is noted for inducing paranoid reactions that can be quite similar to those induced by amphetamine (Lesko et al., 1982; Ellinwood, 1981). Several investigators have found that humans who abuse cocaine for prolonged periods of time often showed increased irritability, difficulty in concentration, disruption of eating and sleeping habits, perceptual disturbance (pseudohallucinations), paranoid thinking, and, occasionally, overt psychosis (Segal, 1977; Waldorf et al., 1977; Wesson and Smith, 1977). As with amphetamine, this psychosis is treated with neuroleptics (Gawin, 1986).

It is also clear that cocaine, like amphetamine, has important effects on the dopamine system. The question is whether, like amphetamine, continuous cocaine also induces a destruction of dopamine terminals, for this could lead to an important test of the continuous-amphetamine paranoid model.

Fortunately, much of this research has already been performed. A number of investigators have studied whether cocaine, when administered continuously, has the same neurotoxic properties that *d*-amphetamine has. Although Trulson et al. (1986) reported that 10 mg/kg of cocaine administered every 12 hours for ten consecutive days induced decreased TH staining axons and decreased TH activity in the caudate when the animals were studied 60 days later, this report has not been replicated by any other laboratory. Kleven et al. (1988) administered the same dose and also two extremely large doses (12.5 mg/kg eight times daily for 10 days or 100 mg/kg/day for 21 days) and reported no long lasting changes in dopamine (DA), 5-HT, or their major metabolites in striatum, hippocampus, hypothalamus, or cortex. Ryan et al. (1988), administered cocaine continuously at doses of 50–450 mg/kg/day and failed to detect axonal degeneration in neostriatum using silver stain or TH immunolabeling. In contrast, the amphetamine-induced degeneration was readily apparent. Indeed, Hanson et al. (1987) reported that cocaine coadministration could *block* the neurotoxic actions of multiple methamphetamine injections. This is presumably because, as has been well established (Scheel-Kruger, 1971; Ho et al., 1977; Whitby et al., 1960), one of the primary effects of cocaine is to block presynaptic reuptake of DA. The vast majority of studies, then, indicate not only that cocaine does not have the same neurotoxic effect on caudate dopamine terminals that *d*-amphetamine and methamphetamine have, but also that it even protects against these neurotoxic effects.

5.1. Cocaine Psychosis in Rats: The Late Stage

In collaboration with Jack Lipton and Steven Zeigler in my laboratory, we have recently developed a slow-release cocaine pellet for rats. Similar in design to the amphetamine pellet, it holds 125 mg of cocaine base, over 50% of which is released during the first five days after implantation. This is an extremely simple pellet to construct. A short length of silicone tubing is first capped on one end with silastic elastomer, the cocaine base is inserted, the pellet is filled with PEG, and the other end of the tubing is capped. This pellet is implanted subcutaneously like our previous pellets.

Rats implanted with these pellets are also initially hyperexploratory and then enter motor stereotypies, but the stereotypies are much less intense than with amphetamine, involving less biting, and more sniffing and head-bobbing. However, as shown in Fig. 3, the stereotypies induced by the two drugs equally dominate the animals' behavior during the first few days. Still, as the animals enter the "late-stage" after several days, the two drugs become even more distinct. The cocaine-implanted animals are much more vividly "hallucinatory" than amphetamine-implanted animals (Fig 4). This figure demonstrates both the problem of observing these "hallucinatory behaviors" in amphetamine animals (their very low frequency when the animals are simply observed for ten minutes daily), and the extraordinary level to which these behaviors appear in the cocaine-implanted animals. Although these late-stage cocaine-implanted animals eventually develop an extremely strong syndrome of limbflicks and wet-dog shakes, other behaviors they exhibit are also particularly interesting. "Panic attacks" are seen, in which the animals suddenly, explosively, bounce off the sides of the cage, as though fleeing an invisible enemy. These sudden, spontaneous startle attacks, which can be elicited by a slight noise, but which also occur in the absolute absence of any obvious eliciting stimulus, are reminiscent of the explosive fleeing episodes observed in monkeys given continuous amphetamines and the spontaneous startle responses shown by the amphetamine rats

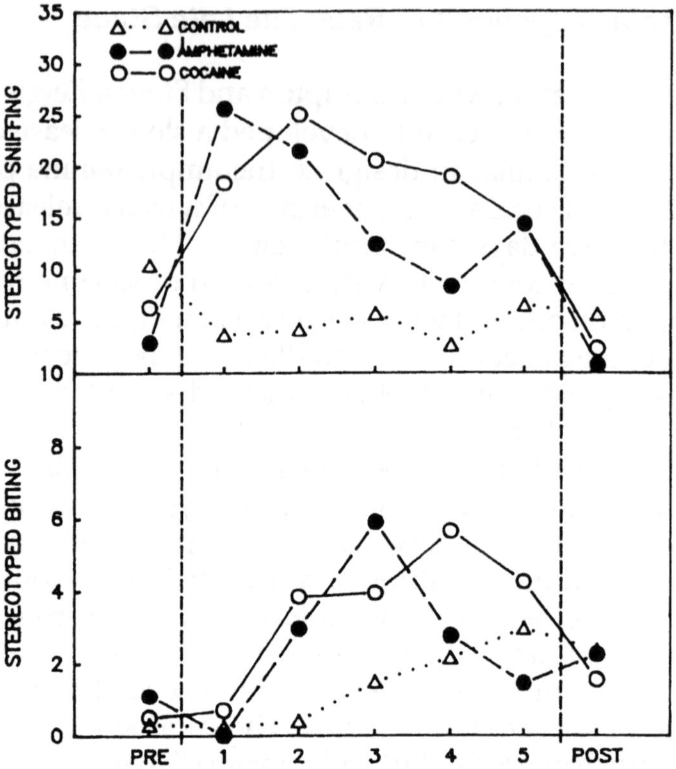

Fig. 3. During the early stage of continuous stimulants, both the amphetamine and the cocaine animals show a high level of stereotypies.

in the colony and in activity cages. The cocaine rats have also been observed showing a variety of other behaviors very suggestive of hallucinatory behaviors. Some are observed reaching out repeatedly with the paws to grasp invisible objects in space; others also show much more parasitotic grooming than controls. It is clear that this cocaine pellet is an extraordinary tool for eliciting late-stage behaviors.

Until very recently we thought that these findings showed that our previous hypothesis linking dopamine neurotoxicity to the "late-stage" of hallucinatory behaviors was incorrect. Yet while the cocaine pellet clearly does not produce the dopamine neurotoxicity which the amphetamine pellet does, it turns out that it does induce another, and possibly related neurotoxicity. This was discovered when various groups of rats were implanted

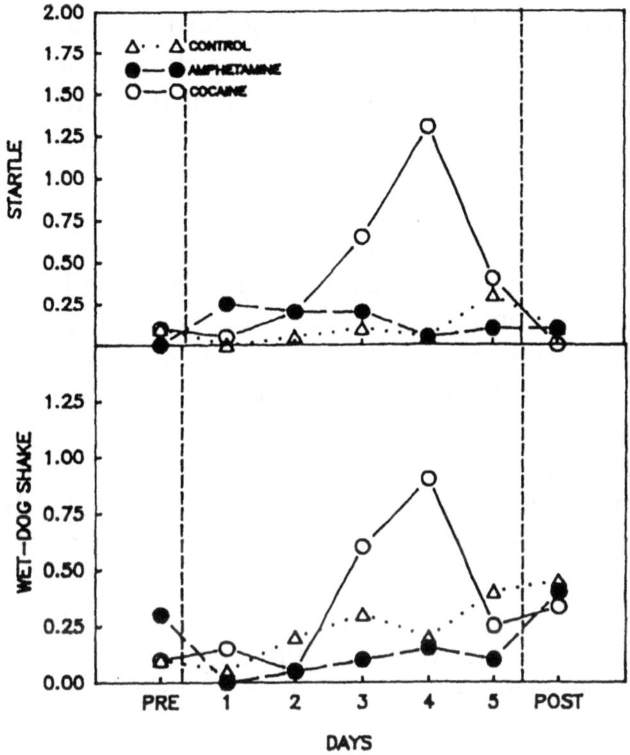

Fig. 4. After several days of continuous cocaine, substantial amounts of spontaneous "startle" responses and "biting" grooming of the skin occur. This latter behavior may reflect parasitotic grooming, in that it also occurs in rats infected with parasites such as mites.

with amphetamine pellets, or cocaine pellets, or given intermittent injections of cocaine for 5 days, or given control injections (Ziegler et al., 1991). All animals were then given a prolonged recovery period and tested in open field. A clear dissociation between the groups was obtained. In open field, the "cocaine injection" animals were hyperactive, showing the well-known inverse tolerance effect, whereas the cocaine pellet animals behaved very differently, acting as though they were extremely frightened.

In vitro autoradiography was then utilized to examine whether there were persisting changes in receptor binding for D2 ([^{3}H]spiperone), D1 ([^{3}H]SCH23390), benzodiazepine

($[^3H]$flunitrazepam), 5HT-1 ($[^3H]$5HT), 5HT-2 ($[^3H]$ketanserin), and muscarinic ACh recptors ($[^3H]$QNB). The amphetamine pellet animals showed large, persisting changes in D1 and D2 receptors, replicating the dopamine neurotoxicity effect, and these were not present with chronic cocaine, either when administered intermittently or continuously. This was as expected. But the continuous cocaine (pellet) animals showed a completely different pattern of brain neurotoxicity, with appreciable increases in $[^3H]$flunitrazepam binding in dapamine-rich areas, cortex, and amygdala but decreased $[^3H]$QNB binding in dopamine-rich areas, hippocampus, and amygdala. This means that continuous cocaine has enduring effects on a completely different neurochemical systems from continuous amphetamine: continuous cocaine damages acetylcholine and GABA systems in dopamine-rich areas. It seems that the uptake blockade of dopamine produced by the cocaine protects dopamine terminals from neurotoxicity, but leads to toxic effects on those cells receiving dopamine input. This might indicate that a functionally related kind of neurotoxicity occurs with both continuous amphetamine and continuous cocaine.

6. Continuous Stimulants, Dopamine Neurotoxicity, and Psychosis: A Reevaluation

From these observations, we conclude that it is some unique property of several days of continuous stimulants, not just the damage to dopamine terminals, that eventually leads from the motor-stereotypy phase to the distinctive "late-stage" behaviors. Continuous-cocaine animals dissociate the two phenomena: They are quite spectacular in their late-stage behaviors yet they apparently have minimal, if any, neurotoxic effects on dopamine terminals. This necessitates a reevaluation of what these continuous-stimulant models of paranoid psychosis have to tell us.

The answer to this question clearly does not lie in a simplified "intermittent injection, inverse tolerance, and motor stereotypy" vs "continuous administration, neurotoxic effects, and distinctive late-stage behaviors" controversy. Some facts just don't fit this simple dichotomy. For example, continuous stimulants, in their initial phases, are quite potent in inducing the inverse tolerance effect for motor stereotypies. If rats are given continuous amphetamine for 24 hours (and, judging from other evidence, much of the damage to dopamine terminals has already occurred), the *d*-amphetamine is discontinued for 12 hours, and a probe, low-dose injection of *d*-amp is given (Ellison and Morris, 1981), then these animals will show a very strong "inverse tolerance" potentiation of motor stereotypies to the *d*-amp. Clearly, both the early stages of continuous amphetamines and the intermittent-injection paradigm induce a state that models, or reproduces in some way, the motor-stereotypy aspect of chronic schizophrenia. In fact, the mechanisms underlying this effect probably are part of the reason why the "dopamine theory of schizophrenia" became so predominant several years ago. This effect presumably provides information on the processes that, in the early stages of a psychotic break, lead to a compulsive focusing on certain issues or behaviors and that, later in the disorder, can persist and even proliferate, resulting in the motor stereotypies of chronic schizophrenics that have been extraordinarily well documented.

Yet this view of psychosis is only a partial glimpse of the disorder, and especially of the later stages of the psychotic break. An animal in a motor stereotypy, or a human in a motor stereotypy, is completely asocial and is certainly a very different creature from the animal or human in the late-stage of paranoid schizophrenia or of the paranoid amphetamine or cocaine psychosis, when spontaneous bizarre, even violent, social behaviors occur. Following several days of continuous stimulants, an entirely new phenomenon develops, one that in some ways is the exact opposite of the motor-stereotypy phase insofar as dopaminergic mechanisms are concerned. While pure motor stereotypies diminish and caudate dopaminergic control wanes, a new class of behaviors emerges, accompanied by increased dopamine activity and glucose metabolism in the nucleus accumbens.

Yet stereotypies, albeit less motoric, persist in this late-stage, for the fight or flight behaviors become focused on one individual enemy, and this is clearly a form of social stereotypy. One common aspect of both amphetamine and cocaine binges is the continuous sympathomimetic arousal to which the brain is subjected, and an abundant literature relates this to dopaminergic hyperstimulation.* What are the neurochemical underpinnings of this state? Clearly there is an apparently highly important, and perhaps decisive, aspect in the incessant hyperstimulation of dopamine systems, which eventually leads to a destruction of dopamine circuitry (in the case of amphetamine) or of related circuitry (in the case of cocaine), and, perhaps more important, a depletion and downregulation of dopamine receptors in both amphetamine and cocaine psychosis. An animal showing "late-stage" behaviors can always be immediately brought out of these by one of two opposite manipulations: a higher dose of a dopamine agonist (which will immediately induce motor stereotypies and a cessation of all social and hallucinatory behaviors), or discontinuation of either the stimulant or administration of a dopamine blocker (which will eliminate all behaviors). This leads to the hypothesis that the "paranoid stimulant late-stage model" is a unique reflection of this cumulative stage of intermediate dopaminergic neurotransmission—when dopaminergic systems are becoming depleted and receptor number is falling, but their stimulation continues.

However, in the broader conception of this disorder, it seems quite inappropriate to focus on one neurotransmitter as being especially important. Stimulants, and especially incessant stimulants, induce a paranoid-like state because they alter an immense number of neurochemical systems. There is a fascinating possibility that has never been tested: Could hallucinatory metabolites of amphetamine or cocaine, or even endogeneous hallucinogens, perhaps derivatives of dopamine, be produced during this late-stage? There is a crucial test of this interesting possibility that has never been conducted. As is well known,

*It may be fruitful to consider when, in an evolutionary scheme, such events might occur. In nature, such a prolonged state of catecholamine stimulation would result only during persisting life-threatenting situations, such as when an animal is on the run, with no place to hide and highly vulnerable. The paranoid searching for enemies would be a highly adaptive reaction to this, and the cumulative sleep deprivation would contribute to the syndrome. It may be fruitful to think of the paranoid individual as being in such a state, for this might simplify the theoretical issues.

repeated administration of the hallucinogens LSD and mescaline induce a profound tolerance for these drugs. It is possible that animals in the late-stage of continuous stimulants might be almost completely tolerant to injections of these hallucinogens, a cross-tolerance that should occur only when animals enter the late-stage.

Whatever neurochemical alterations ultimately prove to be responsible, it seems clear that the late-stage of continuous stimulants, such as amphetamine or cocaine, represents a highly distinctive state of great relevance to paranoid schizophrenia, and a novel model of hallucinatory behaviors.

References

American Psychiatric Association. DSM-III: Diagnostic and Statistical Manual of Mental Disorders. 3rd Ed. Washington, DC: The Association, 1980.

Angrist M. and Gershon S. (1970) The phenomenology of experimentally induced amphetamine psychosis—preliminary observations. *Biol. Psychiat.* **2,** 95–107.

Angrist B., Lee H. K., and Gershon S. (1974a) The Antagonism of amphetamine-induced symptomatology by a neuroleptic. *Am. J. Psychiat.* **131,** 817–819.

Angrist B., Sathananthan G., Wilk S., and Gershon S. (1974b) Amphetamine psychosis: Behavioral and biochemical aspects. *J. Psychiatr. Res.* **11,** 13–23.

Bell D. S. (1965) Comparison of amphetamine psychosis and schizophrenia. *Am. Psychiat.* **111,** 701–707.

Bell D. S. (1973) The experimental reproduction of amphetamine psychosis. *Arch. Gen. Psychiat.* **29,** 35–40.

Bowers M. (1968) Pathogenesis of acute schizophrenic psychosis. *Arch. Gen. Pychiatr.* **19,** 348–355.

Connell P. (1958) *Amphetamine Psychosis,* Maudsley Monographs No. 5. Oxford University Press, London.

Eison M. S., Ellison G., and Eison A. S. (1981) The regional distribution of amphetamine in rat brain is altered by dosage and prior exposue to the drug. *J. Pharmacol. Exp. Ther.* **218,** 237–241.

Ellinwood E. H. (1981) Assault and homicide associated with amphetamine abuse. *Am. J. Psychiat.* **127,** 1170–1175.

Ellinwood E. H. and Kilbey M. (1977) Chronic stimulant intoxication models of psychosis, in *Animal Models in Psychiatry and Neurology* (Hanin I. and Udin E., eds.), Pergamon Press, New York, NY, pp. 61–74.

Ellinwood E. H. and Sludilovsky A. (1973) Chronic amphetamine intoxication: Behavioral model of psychoses, in *Psychopathology and Psychopharmacology* (Cole J. O., Freedman A. M., and Friedhoff A. J., eds.), Johns Hopkins University Press, Baltimore, MD, p. 51–71.

Ellinwood E. H. Jr. (1967) Amphetamine psychosis: I. Description of the individuals and the process. *J. Nerv. Ment. Dis* **144,** 273–283.

Ellison G. and Morris W. (1981) Opposed stages of continuous amphetamine administration: Parallel alterations in motor sterotypies and in vivo spiroperidol accumulation. *Eur J. Pharmacol.* **74,** 207–214.

Ellison G. and Ratan R. (1982) The late-stage following continuous amphetamine administration to rats is correlated with altered dopamine but not serotonin metabolism. *Life Sci.* **31,** 771–777.

Ellison G., Eison M., Huberman H., and Daniel F. (1978) Structural and biochemical alterations in dopaminergic innervation of the caudate nucleus following continuous amphetamine administration. *Science* **201,** 276–278.

Ellison G., See R., Levin E., and Kinney J. (1987) Tremorous mouth movements in rats administered chronic neuroleptics. *Psychopharmacol.* **92,** 122–126.

Ellison G., Staugaitis S., and Crane P. A. (1981b) A silicone delivery system for producing binge and continuous intoxication in rats. *Pharmacol. Biochem. Behav.* **14,** 207–211.

Ellison G. D., Nielsen E. B., and Lyon M. (1981) Animal models of psychosis: Hallucinatory behaviors in monkeys during the late-stage of continuous amphetamine intoxication. *J. Psychiatr. Res.* **16,** 13–22.

Ellison G. D. and Eison M. S. (1983) Continuous amphetamine intoxication: An animal model of the acute psychotic episode. *Psychol. Med.* **13,** 751–761.

Erickson C. K., Stavchanaky S. A., Koch K. I., and McGinity J. W. (1982) A new subcutaneously-implantable reservoir for sustained release of nicotine in the rat. *Pharmacol. Biochem. Behav.* **17,** 183–185.

Fuller R. and Hemrick-Luecke S. (1980) Long-lasting depletion of striatal dopamine by a single injection of amphetamine in iprindole-treated rats. *Science* **209,** 305,306.

Gately P. F., Segal D. S., and Geyer M. A. (1987) Sequential changes in behavior induced by continuous infusions of amphetamine in rats. *Psychopharmacology* **91,** 217–220.

Gawin F. H. (1986) Neuroleptic reduction of cocaine-induced paranoia but not euphoria? *Psychopharmacology* **90,** 142,143.

Gold M. S. and Bowers M. Jr. (1978) Neurobiological vulnerability to low-dose amphetamine psychosis. *Am. J. Psychiat.* **35,** 1546–1548.

Griffith J., Cavanaugh J., Held N., and Oates J. (1972) *d*-Amphetamine: Evaluation of psychotomimetic properties in man. *Arch. Gen. Psychiat.* **26,** 97–100.

Hanson G. R., Matsuda L., and Gibb J. W. (1987) Effects of cocaine on methamphetamine-induced neurochemical changes: Characterization of cocaine as a monoamine uptake blocker. *J. Pharmacol. Exp. Ther.* **242,** 507–513.

Ho B. T., Taylor D. T., Estevez V. S., Englert L. F., and McKenna M. L. (1977) Behavioral effects of cocaine: Metabolic and neurochemical approach, in *Cocaine and Other Stimulants* (Ellinwood E. H. and Kilbey M. M., eds.), Plenum, New York, pp. 229–240.

Hotchkiss A. and Gibb J. (1980) Long-term effects of multiple doses of methamphetamine on tryptophan hydroxylase and tyrosine hydroxylase activity in rat brain. *J. Pharmacol. Exp. Ther.* **214,** 257–262.

Huberman H., Eison M., Byran K., and Ellison G. (1977) A slow-release silicone pellet for chronic amphetamine administration. *Eur. J. Pharmacol.* **45**, 237–242.

Jacobs B. L. and Trulson M. E. (1979) Long-term amphetamine treatment decreases brain serotonin metabolism: Implications for theories of schizophrenia. *Science* **205**,1295–1297.

Jacobs B. L., Trulson M. E., and Stern W. C. (1976) An animal model for studying the actions of LSD and related hallucinogens. *Science* **194**, 741–743.

Kilbey M. M. and Ellinwood E. H. (1977a) Reverse tolerance to stimulant-induced abnormal behavior. *Life Sci.* **20**, 1063–1076.

Kilbey M. M. and Ellinwood E. H. (1977b) Chronic administration of stimulant drugs: Response modification, in *Cocaine and Other Stimulants* (Ellinwood E. H. and Kilbey M. M., eds.), Plenum, New York, pp. 409–430.

Klawans H. L. and Margolin D. I. (1975) Amphetamine-induced dopaminergic hypersensitivity in guinea pigs. *Arch. Gen. Psychiat.* **32**, 725–732.

Kleven M. S., Woolverton W., and Seiden L. (1988) Lack of long-term monoamine depletions folowing repeated or continuous exposure to cocaine. *Brain Res. Bull.* **21**, 233–237

Kramer J. C. (1969) Introduction to amphetamine abuse. *J. Psychedelic Drugs* **2**, 8–13.

Kramer J. C., Gischman V., and Littlefield D. (1967) Amphetamine abuse: Pattern and effects of high doses taken intravenously. *J. Am. Med. Assoc.* **201**, 89–93.

Lesko L. M., Fischman M., Javaid J., and Davis J. (1982) Iatrogenous cocaine psychosis. *N. Engl. J. Med.* **307**, 1153.

Lipton J., Zeigler S., Wildins J., and Ellison G. (1991) Silicone pellet for continuous cocaine administration: Heightened late-stage behaviors compared to continuous amphetamine. *Pharmacol. Biochem. Behav.*, inpress.

McGuire M. T., Raleigh R. J., and Brammer G. L. (1982) Sociopharmacology. *Ann. Rev. Pharmacol. Toxicol.* **22**, 643–661.

Magos L. (1969) Persistence of the effect of amphetamine on stereotyped activity in rats. *Eur. J. Pharmacol.* **6**, 200,201.

Nelson L. and Ellison G. (1978) Enhanced sterotypies after repeated injections but not continuous amphetamines. *Neuropharmacology* **17**, 1081–1084.

Nielsen E. and Ellison G. D. (1980) A silicone pellet for long-term continuous administration of amphetamine. *Commun. Psychopharmacol.* **4**, 17–20.

Nielsen E., Lee T., and Ellison G. D. (1980a) Following several days of continuous administration *d*-amphetamine acquires hallucination-like properties. *Psychopharmacology* **68**, 197–200.

Nielsen E. B., Neilsen M., Ellison G., and Braestrup E. (1980) Decreased spiroperidol and LSD binding in rat brain after continuous amphetamine. *Eur J. Pharmacol.* **66**, 149–154.

Nwanze E. and Jonsson G. (1981) Amphetamine neurotoxicity on dopamine nerve terminals in the caudate nucleus of mice. *Neurosci. Lett.* **26**, 163–168.

Post R. M. (1975) Cocaine psychoses: A continuum model. *Am. J. Psychiat.* **132,** 225–231.
Post R. M. (1976) Clinical aspects of cocaine: Assessment of acute and chronic effects in animals and man, in *Cocaine: Chemical, Biological, Clinical, Social and Treatment Aspects* (Mule S. J., ed.), CRC, Cleveland, pp. 203–215.
Post R. M. and Rose H. (1976) Increasing effects with repetitive cocaine administration in the rats. *Nature* **260,** 731, 732.
Potthoff A. D., Ellison G., and Nelson L. (1983) Ethanol intake increases awareness during continuous administration of amphetamine and nicotine, but not several other drugs. *Pharmacol. Biochem. Behav.* **18,** 489–493.
Ricaurte G. A., Schuster C. R., and Seiden L. S. (1980) Long-term effects of repeated methylamphetamine administration on dopamine and serotonin neurons in the rat brain: A regional study. *Brain Res.* **193,** 153–163.
Robinson T. and Becker J. (1986) Enduring changes in brain and behavior produced by chronic amphetamine administration: A review and evaluation of animal models of amphetamine psychosis. *Brain Res. Rev.* **11,** 157–198.
Ryan L. J., Martone M., Linder J., and Groves P. M. (1988) Cocaine, in contrast to *d*-amphetamine, does not cause axonal terminal degeneration in neostriatum and agranular frontal cortex of long-evans rats. *Life Sci.* **43,** 1403–1409.
Scheel-Kruger J. (1971) Behavioral and biochemical comparison of amphetamine derivatives, cocaine, benztropine and tricyclic antidepressant drugs. *Eur. J. Pharmacol.* **18,** 63–73.
Segal D. S. and Mandell A. J. (1974) Long-term administration of *d*-amphetamine: Progressive augmentation of motor activity and stereotypy. *Pharmacol. Biochem. Behav.* **2,** 249.
Segal R. K. (1977) Cocaine: Recreational use and intoxication. In *Cocaine 1977* (Petersen R. C. and Stillman R. C., eds.), US Government Printing Office, Washington, DC.
Slater E. (1959) Amphetamine psychosis. *Br Med. J.* ***i,*** 488.
Smith R. C. (1969) The world of the Haight-Ashbury speed freak. *J. Psychedelic Drugs* **2,** 77–83.
Snyder S. H., Bannerjee S., Yamamura H., and Greenberg D. (1974) Drugs, neurotransmitters and schizophrenia: Phenothiazines, amphetamine and enzymes synthesizing psychotomimetic drugs and schizophrenia research. *Science* **184,** 1243–1253.
Steranka L. and Sanders-Bush E. (1980) Long-term effects of continuous exposure to amphetamine on brain dopamine concentration and synaptosomal uptake in mice. *Eur. J. Pharmacol.* **65,** 439–443.
Stripling J. S. and Ellinwood E. H. (1976) Cocaine: Physiological and behavioral effects of acute and chronic administration, in *Cocaine: Chemical, Biological, Chemical, Social and Treatment Aspects* (Mule S. J., ed.), CRC, Cleveland.
Trulson M. E., Babb S., Joe J., and Raese J. (1986) Chronic cocaine administration depletes tyrosine hydroxylase immunoreactivity in the rat brain

nigral striatal system: Quantitative light microscopic studies. *Exp. Neurol.* **94,** 744–756.
Waldorf D., Murphy S., Reinarman C., and Joyce B. (1977) Doing Coke: An Ethnography of Users and Sellers (Drug Abuse Council, Washington, DC).
Wesson D. R. and Smith D. E. (1977) Cocaine: Its use for central nervous system stimulation including recreational and medical uses, in *Cocaine 1977* (Petersen R. C. and Stillman R. C., eds.), US Government Printing Office, Washington, DC.
Whitby L. G., Hertting G. G., and Axelrod J. (1960) Effect of cocaine on the disposition of noradrenaline labeled with tritium. *Nature* **187,** 604,605.
Wyatt R. (1978). Is there an endogenous amphetamine? A testable hypothesis of schizophrenia, in *The Nature of Schizophrenia* (Wynne L., Cromwell R., and Matthysse S., eds.), Wiley, New York, pp. 116–125.
Young D. and Scoville W. B. (1938) Paranoid psychosis in narcolepsy and the possible danger of Benzedrine™ treatment. *Med. Clin. North Am.* **22,** 637–646.
Zeigler S., Lipton J., Toga A., and Ellison G. (1991) Continuous cocaine administration produces persisting changes in brain neurochemistry and behavior. *Brain Res.*, in press.

Appendix 1: Silicone Pellet Technology

Many drugs have distinctively different effects when presented in a continuous manner in than when they are given intermittently, which in most laboratories means daily injections every seven days. Almost any drug can be delivered by inexpensive, highly nontoxic silicone pellets; the only requirement is that the drug be in the nonpolar, base form. The alternative—osmotic minipumps—are expensive, but there are some cases in which minipumps are unacceptable on other grounds. Minipumps have an upper limit to the amount of drug that can be administered (and very high doses of amphetamine base have a pH that is not tolerated by the pumps) and they cause skin necrosis with cocaine. These considerations rarely hold for silicone pellets, which can be made of any size and release rate.

Drugs such as amphetamine base are often purchased as the salt form (typically, amphetamine sulfate or hydrochloride). The amphetamine sulfate is dissolved in water, and 10% ammonium hydroxide is added until the mixture becomes cloudy, meaning that precipitation of the base has occurred. A volatile substance, such as chloroform, is added, the mixture is shaken, and the volatile lower portion drained off using a separation flask. This extraction is repeated several times until the pH of the water layer does not change when the extraction is performed (typically, at about pH 9.0). The water layer is discarded and a new batch extracted. Finally, the total volatile extract is placed in a flask warmed with water (40°C) and nitrogen gas is bubbled through the mixture, evaporating until only the pure free base remains. This can be recognized by the consistency and color, by the lack of further loss to nitrogen, and by the odor. The amphetamine base is then dissolved in polyethylene glycol 300 (J. T. Baker Chemical Co., Chicago, IL) to a final concentration of 231 mg/mL [free base].

Many compounds can be administered using these capsules. The size of the capsule is determined by the total amount of drug to be delivered; the release rate is determined by the amount of silicone wall given access to the drug. Nonconducting "sleeves" made from polyethylene or cut from 1-cc tuberculin syringes are placed inside the silicone tubing to cut down the release rate, leaving only small windows of silicone to transfer drugs. *D*-Amp, PCP, and mescaline have been administered using silastic slow-release pellets as described by Nielson and Ellison (1980). A 45-mm-long polyethylene cylinder, with an inside diameter of 5 mm and two 1.3-mm-diameter holes near its middle, was enclosed in 1.59-mm-thick silastic tubing, filled with 0.7 mL of drug base in vehicle, and capped at each end with silastic polymer. In the experiments of Potthoff et al. (1983), continuous slow-release *D*-amphetamine, PCP, and mescaline pellets contained 65, 71, and 165 mg of drug base, respectively, suspended in 0.7 mL polyethylene glycol (PEG) and released at an average of 1.1, 0.98, and 0.55 mg/day, respectively, over 30 days. The secobarbital and haloperidol pellets represent a modification of the prototype pellet, in which the polyethylene cylinder has been omitted. Thus, 45-mm-long, 1.59-mm-thick, 4.76-mm-id, 7.94-mm-od silastic tubing was filled and capped at each end with silastic polymer. The secobarbital pellet contained 800 mg of base suspended in 0.8 mL of PEG. The haloperidol pellet contained 100 mg in 0.8 mL of lactic acid. The secobarbitol and haloperidol pellets released 6.52 and 1.24 mg/day, respectively, over 30 days. We have more recently used haloperidol "pillows" made from very thin silicone sheeting to release even more haloperidol (Ellison et al., 1987). Even alcohol can be administered with these devices (Ellison et al., 1981b). Nicotine can be administered via implantable glass reservoirs constructed according to methods described by Erickson et al. (1982). Glass capsules, 20-mm long, 4-mm id, filled with 0.2 mL of nicotine base (Sigma), and capped with silastic polymer, were immersed in distilled water for 72 hours, and then implanted subcutaneously. These nicotine pellets release 3–4 mg/day. Caffeine has been administered via a refillable silastic pillow as described by Ellison et al. (1981b). A 64- × 76-mm pouch of 0.51-mm-thick silastic sheeting was sealed with silicone type A adhesive and connected to a 0.16-mm-id silastic tube, which served as a spout. Joints were coated with silastic polymer in order to minimize skin irritation. The pillows were implanted under sodium pentabarbital general anesthesia, filled with water, and plugged with a stainless steel bong. At the beginning of the experimental period, the plug was removed and the water withdrawn via the spout using a 12-cc syringe. The water from the pillow was replaced with caffeine solution and the plug replaced. The pillows were then emptied and refilled daily during the experimental period. The drug-treatment period was limited to 21 days because of severe infections in some animals. The caffeine pillows released 200 mg/day.

Various pellet designs can initially be screened in vitro by measuring the amount of drug released into saline during successive 24-hour incubations at 37°C. When many of pellets are to be made, junctions between the sleeves are critical, and a machined tool aids in cutting to assure a tight joint when the pieces are fitted together. The pellet is then sealed at the remaining end with catalyzed silastic polymer. This was originally available as Dow-Corning Medical Grade Silastic Elastomer, but other companies now have adequate substitutes. Because the drug will begin to diffuse into the silicone, and then into the air, upon filling, control (containing PEG only) and drug pellets should be filled and sealed within a few hours prior to use.

Drug-pellet implantation is extremely simple, taking only several minutes per animal. Rats are shaved on one side of the back (just off the midline, where they cannot reach the wound), given either local (procaine) injections or general volatile or barbiturate anaesthesia and restrained, and a 15-mm incision is made in the skin of the back. The connective tissue under the skin is cut and rapidly separated using a blunt dissector, forming a 3 × 2-cm cavity.

Appendix 2: Colony Procedures

Typically, a colony includes about 27 male hooded (Long-Evans) rats weighing 160–180 g. Albinos yield disappointing social behaviors, and mixed male-female colonies become unwieldy. This is because fighting, including killing, among the males gradually increases, and then babies, who can escape from almost any enclosure, begin to appear.

In a typical experiment, these males are allowed at least two months to grow and establish stable dominance hierarchies. They are then captured, divided into equal groups based on body weight, and a surgical or pharmacological manipulation is performed.

For example, central injections of neurotoxins can be administered providing the animals are suitably prepared. They can be anesthetized with sodium pentothal and bilateral skull holes drilled over the lateral ventricles (L 1.5, 0.2 mm posterior to bregma). The scalp is then sutured and the animals marked with a fur dye (Nyanza Inc., Lawrence, MA) by painting a stripe down the animal's front left and shoulders, belly and back, or hind legs and rump. The animals are then reintroduced into the colony enclosure for two weeks; during this time their behavior is observed and prelesioning baseline data collected. All animals are then recaptured and, on three successive days, are anesthetized with ether, placed on an inclined head holder, and a 25-gage hypodermic needle is inserted through the scalp and a previously drilled hole to a depth of 6 mm (determined by a guide on the needle shaft). The neurotoxin is injected through this needle and the animals are immediately returned to the colony enclosure and their behavior during the next 50 days is observed.

Alternatively, upon capture, the animals can be marked with the fur dye, returned to the colony enclosure for prelesioning data, and then, two weeks later, recaptured briefly anaesthetized with a volatile anaesthetic, and implanted with a slow-release pellet. In other experiments, rats are merely observed while free access to drugs, such as alcohol, is present.

The colony enclosure is a large enclosure, especially constructed for rat colonies and divided into separate areas for burrows, a free-behavior arena, and a feeding area. The arena is a 3-m × 4-m × 2-m room completely enclosed by concrete, wire mesh and screening, and glass. On the walls are multiple ledges and climbing structures, and in the center is a large multitiered structure with climbing ramps and thick ropes leading to the ceiling and wall ledges. The more space is available, the less fighting will occur, but there must be a clear behavioral-arena floor where most social interactions occur. This floor is covered with straw, which need only be changed monthly, since the animal's will avoid soiling this area. A "behavioral sink" can be established by providing two watering fountains on the foot level of this structure. On the straw-covered floor of the arena we also typically have activity wheels and several play objects. Outside this arena are completely enclosed observer's blinds. This arena is kept on a reverse day–night lighting cycle with overhead fluorescent lighting during the light phase (10:00PM–10:00AM) and dim red lighting with three 15-W white bulbs during the dark phase (10:00AM–10:00PM).

A burrows area must be provided, for the animals readily recognize the evolutionary significance of this, and treat it accordingly. On burrows we constructed at one end of the arena. Twelve holes (8 cm in diameter) are drilled in the wall at just above floor level, each separated by an average of 20 cm. These holes lead through tubes 15–45 cm long to 12 straw-lined cages (50 cm long, 23 cm in diameter) that serve as the burrows. The burrows area, which is completely enclosed, is kept in constant dim red light, and is seldom entered by humans during the experiment. Certain animals, especially the "king rat," adopt a certain burrow and defend it on all occasions; others move from burrow to burrow, living four to five to an enclosure.

Because of a peculiarity in the original room that enclosed our first colony, we built a separate feeding area. This proved to be a remarkable asset, for it provided a clear physical location where feeding behavior could be studied, and it also meant that once, in the middle of each dark cycle, all animals left the burrows and could be observed. Fighting and other social behavior in the colony just before and after the feeding hour became our most important data, as did order through the tube linking the arena to the feeding arena.

Thus, in the arena wall opposite the burrows was a hole that led through a large valve and a 1-m-long tube to a separate 2.5- × 2.5- × 2-m enclosure. In the center of this area was an elevated feeding stand with a wide ramp leading to it. Each day at 4:00PM (in the middle of the dark cycle) a tray filled with food was placed on the food stand, chimes were rung, and the valve was opened. The rats were allowed 1 hour of access to the food; they were then chased from the feeding area and the valve was closed. This is not unlike the present ecological niche of rats, who have become scavengers of the garbage of humans.

For an optimal colony, we use leftovers from the luncheon table at the UCLA faculty club. This food is a mixture of fresh table scraps (meat, vegetables, and fresh greens) to which Purina ground laboratory chow mixed with water and Purina Chow pellets have been added. An average of 3.0 kg of food was available each day, this was always more than was consumed. When studying drugs for which caloric intake must be strictly controlled (such as alcohol), we have used colonies given only Purina chow. However, observing these colonies is much less enjoyable.

During the period before the experiment begins, the animals must be habituated to the presence of observers sitting quietly outside the colony enclosure. During the experiment the animals we observed daily during four standard periods: just after the beginning of the dark period (10:00AM–12:00 noon), during the two hours just before feeding (2:00–4:00PM), during the hour of feeding (4:00–5PM), and after feeding (5:30–7:30PM). This covers the periods of maximal activity in the colony, in that few animals are out of the burrows during the light period. Where two or more observers were present during any given period, their observations were combined and averaged. During the hour of feeding, three observers are generally present (observing the arena, tube, and feeding area, respectively).

Throughout the experiment, all observers must be kept unaware of the animals' experimental treatment (i.e., which marking group had received which type of injection) and the marking code should be changed between replications. The observers score the following behaviors:

Every half hour a census is taken of the arena, noting how many animals from each group are (a) on the arena floor or at the watering place, (b) in the rafters of the ceiling or on ledges, and (c) in the burrows. The last figure is obtained by subtracting the total animals in locations a + b from the total number in the colony, *N*.

The categories of behavior we have employed are:

- "Approaching humans." Rat comes to the screening at the edge of the colony enclosure in proximity to the observer and peers at the observer.
- "Wheel running." Rat enters the activity wheel and runs for at least 15 seconds. An additional mark is recorded for each two minutes of incessant activity.
- "Mounting" and "mounted." Only complete mounts are scored, and the experimental groups of both the "mounter" and the "mounted" animal are recorded.
- "Broadsiding." In broadsiding (or "threat posture"), one rat approaches another, turning its flank sideways as it does so, and then pushes against its opponent with its side and hindquarters. The experimental group of the pusher is recorded.
- "Stand and box." An aggressive posture involving two or more rats standing erect and facing each other. This is usually accomplished by pawing movements, but sometimes the two rats stand motionless with snouts in close proximity for several seconds. The experimental groups of both animals are recorded.
- "Wrestling." Two rats roll together on the ground, clutching and striking at each other with their paws, and sometimes attempting to bite. After several seconds of wrestling, one rat usually "pins" the other and the two rats remain motionless for several seconds before disengaging. At this

time the experimental group of the rat ending up on top is recorded as "top dog," and the experimental group of the rat ending up on his back (underneath the victor) is recorded as "bottom dog." In those matches in which the two rats alternately pin each other, only the final victor and bottom rat are recorded. When no victor can be determined, both rats are recorded as "top dog."

- " Violent fighting." These we fights of extreme intensity, initially signaled by loud squeals, rustling of the straw, and by an encounter that usually involves vicious biting and is always ended by one of the rats suddenly breaking off the fight and fleeing. The experimental group of the biter, or of the nonfleeing rat, is recorded.
- "Tube behavior." is observed at the tube connecting the arena with the feeding area.
- "Initial order." About 30 minutes before the time of feeding, rats gather at the tube entrance, pushing and vying for position. The food is placed on the feeding tray, chimes rung, and the valve opened. With rare exceptions, all rats enter the feeding area within a few minutes, pushing or pulling at the animal ahead. Order into the feeding area is recorded by noting each animal's experimental group as he emerges from the tube into the feeding area.
- "Hoarding." The experimental group of each rat that returned from the feeding area in a path directed to the burrows with food in his mouth is recorded. Returns without food we not recorded.

Censuses are taken at the 10th, 20th, and 30th minute after the beginning of feeding, recording how many animals from each experimental group were on the feeding tray and how many were on the floor of the feeding area.

Finally, at the bottom of each rating sheet, the observers must be given a chance to note the occurrence of unusually meaningful or rare behaviors. Entries in this category have sometimes provided rare insights into the actual events occurring in the colony as a whole. These have included insights into paranoia, such as the first observations of the fixations and goading of "late-stage" amphetamine rats of other particular enemies and the related terrorization of the entire colony by a late-stage amphetamine "king rat." However, we have also seen some extraordinary examples of what were clearly altruistic attempts by control animals to help lesioned animals into the feeding area during the hour of feeding. The degree of homosexual pairing that can develop in all-male rat colonies is striking, as is the kidnapping of rat pups by females who have not yet become pregnant in mixed colonies. Also, once, just after Thanksgiving, one of my students brought an entire turkey carcass to the feeding area, rang the feeding bell, and opened the valve. The rats streamed in and, as a group, attempted to drag the entire carcass back through the two-inch valve to the burrows.

Animal Models for the Symptoms of Mania

Melvin Lyon

1. Introduction and Working Hypothesis

Methods for modeling a human illness must have relevance to the clinical description of the abnormal condition, and they should be capable of generating an experimental condition that responds to the treatments commonly used with that disorder. In the case of mania, there are some immediate problems with definition of the syndrome. The following discussion is based on the American Psychiatric Association's DSM-III (1980) classifications of mania and depression. However, it is suggested that this classification may be deficient in some important respects (Lyon, 1990), and any critique of neuromethods for inducing an animal state resembling mania must take these shortcomings into account.

Mania is typically held to be a separate disorder from schizophrenia, even though some symptoms overlap in practice, and it is usually classified as one pole of a manic-depressive continuum, with some patients, however, having essentially only a depressive set of symptoms. It is noticeable in DSM-III (1980) that only mania is strongly associated with hyperactivity, and also with highly repetitive, or constantly switching, response patterns. However, as noted in Chapter 2 by Lyon, on animal models of schizophrenia, these symptomatic features are probably also a major factor in producing the seemingly different syndrome of schizophrenia.

From: *Neuromethods, Vol. 18: Animal Models in Psychiatry I*
Eds: A. Boulton, G. Baker, and M. Martin-Iverson

It is probably more than coincidental that neuroleptics are used in conjunction with lithium in the early treatment of hypermanic patients. The neuroleptics are said to have a more immediate calming effect, whereas the lithium seems to exert its effect somewhat later, in damping the otherwise dangerously strong swings in mood. Since neuroleptic drugs principally act to block dopaminergic (DA) activity, this suggests that an overly active DA system is one of the typical signs of mania (though not of depression).

In agreement with this, neuroleptics usually do not have any beneficial effect on purely depressive symptoms, and may make them worse. This suggests that the initial feature that dominates in the manic phase is overstimulation of DA systems, with stimulation of the nigrostriatal DA system affecting mainly the *response choice,* which indirectly also affects the incidence of "switching" and/or repetition of responses (Lyon and Robbins, 1975). The mesolimbic DA system has a greater effect on the *rate of responding* and, directly or indirectly, on the autonomic excitation (emotional tone) accompanying the behavior (Iversen, 1977; Stevens, 1975).

The above actions on DA systems must be considered against the background of circadian rhythms and environmental stimuli affecting the organism at the time of treatment (Martin-Iverson et al., 1988; *see also* Chapter 4 by Martin-Iverson, this volume). Intermittent vs continuous treatment with DA agonists can also strongly affect the behavioral results of the treatment (Nielsen, 1981; Eison et al., 1983), and thus alter the modeling value for mania.

The activity of DA systems can also be modified by blocking the systems directly with neuroleptics, or by changing the dopamine/serotonin (DA/5-HT) balance, which is normally in favor of DA during activity and in favor of 5-HT during rest. In a similar manner, other neurotransmitters (GABA, ACh) and neuromodulators (β-endorphins) are also seen as affecting the DA/5-HT relationship. This viewpoint is used as a "working hypothesis" for the present discussion of neuromethods for mimicking the effects of mania in humans.

1.1. Clinical Manifestations of Mania and the Working Hypothesis

Mania is frequently accompanied by a spurt of creative energy, accompanied by a strong feeling of elation, and some artists and writers accomplish their best work during a manic phase (Andreasen and Powers, 1975). As the mania becomes more intense, the rapidly changing behavior either results in a superficial sampling of continuously changing ideas, or in a form of perseveration, where the same idea occurs so repeatedly that it blocks further progress. The manic individual becomes "hung up" in a narrowly focused set of behaviors, and this stage frequently is followed by irritability and depression.

Certainly, two aspects of manic hyperactivity tend to separate it from other types. One difference lies in the way the manic hyperactivity tends to evolve into a *specific and limited behavioral form,* in which the individual wholeheartedly pursues a single activity for such long periods that it evokes protest from others, and is oblivious to social pressures. For instance, a guest in a house, on discovering one dull kitchen knife, insists on sharpening every knife in the house, including the host's fine table silverware, and proceeds to do this over the host's open protest. Another example is that of the author who locks herself in her room, typing continuously and refusing food, water, social contact, or sleep until she can finish her manuscript. These examples illustrate the "focused" nature of the hyperactivity, which distinguishes it from simple hyperactive exploration, or excessive energy devoted to whatever is at hand or is socially appropriate to the situation.

The second difference in manic hyperactivity is its *excessive emotional tone,* which is out of all proportion to the actual situation. For instance, there is tremendous elation if the activity is allowed to continue without disturbance, as well as a quick change to irritation and rage if this activity is prevented.

In terms of neurotransmitters, the depressive stage most commonly resembles the behavioral state produced by a continuous overstimulation of the noradrenergic (NE) and DA systems, which reduces the supply of available dopamine. This may explain why DA stimulant drugs often produce a temporary, but noticeable, upward swing in activity and mood in depressive patients (Beckmann and Heinemann, 1976; Brown and Mueller, 1979). However, given that the nervous system in this illness continues to overstimulate these systems, the exhaustion and depression soon returns. Surely, this is too simple a hypothesis to explain all that happens in depression, but it does indicate that DA activity reduction is, at least, a concomitant feature of depression.

1.2. How Animal Models Relate to the Present Working Hypothesis

The above reasoning suggests that animal models for mania should probably include a direct or indirect stimulation, or overbalance, in favor of the dopaminergic systems, and that models of depression will tend to be found with methods producing DA and NE depletion. However, long-term stimulation of DA systems can lead to severe depletion of DA resources and behavior resembling the post-amphetamine "crash" syndrome, and the like, that displays many elements of behavioral depression. It has even been suggested that the aftereffects of large doses of amphetamine be used to model depression directly (Gerner et al., 1976). Certainly, the phenomenon of "crashing" fits well with the supposition that the early phases of intense mania are principally related to DA overstimulation. The purpose of this chapter is to examine how well this initial working hypothesis bears up under scrutiny from both pharmacological and behavioral viewpoints.

1.3. Problems of Definition for Animal Models

There is a great temptation to place too much emphasis on hyperactivity in defining the symptom characteristics for animal models for mania. Robbins and Sahakian (1980) reviewed the animal literature related to mania, and they suggested that irritability and elation should also be included among the basic attributes of such models. Lyon (1990) has gone further in attempting to define elation and irritability in terms of behavioral features recognizable in an animal model. Table 1 summarizes in an extremely brief fashion some of the essential criteria that can be seen in animal models to have close parallels with human symptomatology in mania. In proposing any methods for inducing manic-like states in animals, it may be helpful to look more closely at this table. The most important point for any animal model must be that it does not simply mimic one of these potential symptoms, but several of them, in the same way that diagnostic systems such as DSM-III require the simultaneous appearance of several indicators.

A few points deserve special mention. *Elation* is often supposed to be a uniquely human feature of emotional experience, with few measurable parallels among animals. The present chapter takes the view that an excessive valuation placed on secondary reinforcements is a proper measure of elation. Secondary reinforcements are seen as the cognitively associated reinforcing value, which is not so directly mixed with the satisfaction of basic physiological needs as are primary reinforcers, such as food and water. Hence, the value of a secondary reinforcer, like that of beauty, lies "in the eye of the beholder," and consequently, can be under- or overvalued, according to the emotional content of the moment in which it appears. An excessive emotional reaction of a positive sort is presumably what we call elation. Since the overvaluation of secondary reinforcers can be measured in animals, and, if carefully done, separated from the mere stimulation of response activity (Hill, 1970; Robbins, 1976), we have a relatively distinct measure of elation in animals.

The *rapid shifting of mood* that occasionally appears in mania, where depressive episodes intrude on moments of elation, and vice versa, is also a feature that needs to be modeled in animal studies. However, direct measurement of shifts in mood can be difficult to assess behaviorally in animals. *Irritability* is one feature of the shifting mood function, and it is relatively easy to measure in animals since they shift readily to aggressive behaviors when irritated. Very few studies, however, have taken the time to investigate the level of irritability exhibited, or to compare it properly with control levels. A related phenomenon may be the kindling, or *sensitization* by electroconvulsive shock or

Table 1
Potential Animal Correlates for Human Symptoms of Mania

Human symptoms	Measurable animal correlates
Elation	Highly increased value of secondary reinforcement
Inflated self-esteem	Assumes inappropriate dominant role responses
Irritability	Intense vocalization Increased startle response Easily provoked aggression
Rapid mood switching	Sudden and inappropriate changes in vocalizing, bodily posture, and response to stimuli Abrupt changes in secondary reinforcement value Abrupt changes in thresholds for startle and aggression
Increased intensity of social contact	Increased social vocalization More approach, contact, and grooming of others
Increased aggression	Increased threat by voice or posture Unprovoked attack and biting
Increased frequency of sexual contacts	Increased mating calls, presenting postures Increased sexual investigation, foreplay, copulation
Hyperactivity	Hyperactivity in vocalization, locomotion, and responding in general
Pressure of speech	Increased rate of vocalization or any other form of direct social communication
Flight of ideas	Increased and inappropriate switching between social, aggressive, and sexual behaviors
Decreased sleep	Decreased sleeping time
Distractibility	Behavior easily disrupted by irrelevant stimuli
Impaired judgment and impulse control	Avoidance learning loss with escape and memory intact Loss of passive avoidance in GO/NO–GO task
Perseverative and/or stereotyped responses	Constant repetitions of behavioral patterns or perseveration in a single response

repeated low-dose drug effects, which cause sudden shifts to occur in mood and activity functions when some threshold is exceeded (Post et al., 1984). What is most relevant here is that following the initial kindled episode, the threshold for further stimulation is markedly reduced, thus leading to more rapid shifting of reactions when a new stimulus is applied. *Emotional lability* is the clinical name for such abrupt shifting, and it appears to find here a close animal parallel.

Finally, much criticism of animal models of mania is directed at the supposedly patent absurdity of measuring *accelerated thought processes* in animals, as opposed to human subjects. The present view, however, is that accelerated thought processes are actually very effectively communicated by animals in the speed of their reaction on tasks requiring serial reaction that do not require exact repetition of a response sequence. For instance, operant conditioning techniques using several manipulanda that must all be operated, but in any desired order (Thompson, 1973; Vogel and Annau, 1973). Such responding tends to develop into fixed patterns for each individual, which represent a series of thought processes (Schwartz, 1986), and rate changes in the execution of these patterns can be directly related to acceleration of thought processes.

Furthermore, there is an increasing number of psychophysiological techniques that allow measurement of the speed of reactive brain processes as "thinking" proceeds (Hiramatsu et al., 1984). The speed with which evoked potentials have their effects, and the rapid shifting of mood mentioned above, are both measurable by direct physiological techniques.

1.4. Plan for the Present Chapter

The models are subdivided somewhat arbitrarily, according to their emphasis on causal factors, into the following groups: catecholaminergic, indoleaminergic, lithium-response, opioid, cholinergic, glutamic acid/GABA, brain lesions, intracranial self-stimulation, and kindling/sensitization. There is, of course, considerable overlap between many of these model types, and only certain of them have been selected here for presenting methodological detail.

2. Animal Models of Mania

2.1. Catecholaminergic Models

Amphetamine models are the oldest and most investigated of all animal models of mania, but in recent years, a number of other drugs (i.e., cocaine, methylphenidate, apomorphine) with actions affecting dopamine (DA) have been added to this group (for reviews, *see* Murphy, 1977; Robbins and Sahakian, 1980; Iversen, 1986; Lyon, 1990). Methodologically, these models can be separated into treatment categories of

1. Large acute injections;
2. Once-daily injections for several days;
3. Several treatments per day repeated over days;
4. Subcutaneous mini-pumps, or implanted capsules, with slow-release lasting for several days, or even weeks.

The single, large injection methods yield, for the most part, only evidence of hyperactivity, which is an important, but not definitive, symptom of mania. However, when combined with operant conditioning methods to test for changes in secondary reinforcement (*see* Table 1), even these treatments can produce signs of elation, defined here as a sharp increase in secondary reinforcement values. Amphetamine is not preeminent in the latter effect, with pipradrol and methylphenidate producing more obvious effects (Hill, 1970; Robbins, 1975,1976).

In such operant experiments, careful attention must be paid to separating the effects of these drugs on simple response rate from the effects owing to changes in the secondary reinforcing value proper. This problem, of separating simple rate changes from other measures of reinforcing value, is ubiquitous in experiments with DA stimulant drugs (Lyon and Robbins, 1975).

Behaviorally, studies of amphetamine, methylphenidate, pipradrol, and other catecholaminergically active agents produce many of the symptoms of mania, including hyperactivity, elation (as above), increased irritability and aggression, increased startle reactions, interference with sexual activity, and insomnia. As in humans, many of these symptoms, especially in the early stages, lead to greater social interactions, and even to improvements in productive behavior on learned tasks. However, with continuing stimulation, the behaviors invariably tend toward excessive responding, strongly repetitive actions, and loss of normal social interactions.

Pharmacologically, amphetamine and similar drugs are acting principally on the DA systems in producing these effects. There appears, however, to be a functional separation within the DA systems of the brain, such that hyperactivity changes are most closely related to the basal ganglia, whereas the emotional tone (mood) accompanying the symptoms is apparently more closely related to the mesolimbic DA system, including the nucleus accumbens (NAC) (Stevens, 1975). DA stimulation of the NAC appears to increase the rapidity with which responses are made. Behaviorally, this would appear as overly rapid reactions, quickly shifting attention, behavioral switching, and the emotional concomitants of irritability, and rapid escalation to aggression. At the same time, the DA stimulation of the basal ganglia would be presumed to be leading in the direction of repetition of responding and perseveration in behavior, with the end result being stereotyped activity (Iversen, 1977).

In the initial phases of mania, the characteristics of behavior would be increased switching of behavior, and in the later phases, a repetitive and stereotyped ("hung-up") type of response that is also characteristic of schizophrenia (*see* Lyon et al., 1986; Lyon and Gerlach, 1988). The behavioral changes may also be related to changes in neurotransmitter balance, rather than to DA alone. For instance, lithium (Li), which has an excellent antimanic effect, tends to dampen hyperactivity produced by amphetamine, yet there is little evidence that Li reduces DA *per se.* What apparently happens is that Li increases 5-HT levels, and this acts to counterbalance the overstimulation of DA systems (Gerson and Baldessarini, 1980). This fits well with the knowledge that 5-HT depletion with parachlorophenylalanine (PCPA) leads to hyperactivity, which has been compared with manic hyperactivity, and this type of depletion is aided by the effect of Li.

2.1.1. Examples of Methods Using Catecholaminergic Changes

The first method chosen to illustrate the measurement of variables relevant to mania was not originally intended as such (Hill, 1970; Robbins 1975,1976,1978; Robbins et al., 1983). However, this method provides one of the few available assessments of what is defined here as elation. If elation is indeed closely related to sudden and intense increases in secondary reinforcing strength, then this animal model could be extremely important in relation to mania.

2.1.1.1. Measuring Secondary Reinforcement as a Parallel to Elation

Robbins et al. (1983) have described an excellent method for measuring secondary reinforcement. The method does not depend on responses during extinction of an operant, but rather on the acquisition of a totally new response in which secondary reinforcement is the only source of reward.

The following is a brief description of this method.

Rats were trained in an operant conditioning chamber to push open a panel in order to reach a water feeder. Training consisted of a 0.5-s light flash above the dipper feeder, followed by a 7.5-s presentation of the dipper with water. Water was given independently of responding, every 30 s on a fixed, and later on a random time schedule of reinforcement. To emphasize discrimination of the "light + dipper noise" combinations, the house light was turned off and the reinforcement schedule stopped if the animal poked its head into the dipper feeder when no light flash had been given. This was followed by a 3-s pause to prevent spurious "dipper + no-light" associations from forming.

There were also two levers present in the box, but during training, depression of the levers had no effect beyond being recorded. Training required about six sessions, arranged on alternate days. At the end of training, all animals consumed the water offered, and ignored the levers.

During the test trials for secondary reinforcement effects, no water was delivered, although the empty dipper and the flashing light were presented when correct (to be rewarded) responses were made on one of the levers. One of the levers, counterbalanced among animals, was chosen as the correct lever, and produced flash + dipper closure on a random interval (RI) 5-s schedule, whereas responding on the other lever had no effect except being recorded. What was measured was the degree of acquisition of responding on the lever producing flash + dipper closure (secondary reinforcement). This preference was tested using, among others, the following drugs: pipradrol hydrochloride (1.5–13.5 mg/kg dissolved in a 1:2 mixture of polyethylene glycol and distilled water), *d*-amphetamine sulfate (0.25–2.9 mg/kg dissolved in 0.9% saline), cocaine hydrochloride (1.7–41.4 mg/kg dissolved in 0.9% saline), apomorphine hydrochloride (0.03–3.0 mg/kg dissolved by slow warming in 0.2% ascorbate in 0.9% saline and given subcutaneously in the flank), α-flupenthixol dihydrochloride (0.01–1.0 mg/kg dissolved in 0.9% saline), morphine hydrochloride (1.0–10.0 mg/kg dissolved in 0.9% saline).

Drugs were injected ip about 15 min prior to testing (which may not have been ideal for *d*-amphetamine's peak action), except morphine and α-flupenthixol, which were given 30 min before sessions. During the pause, the injected animal was placed in a special holding cage. Sessions with varying drug doses and control treatments were counterbalanced using Latin-Square designs.

Pipradrol and *d*-amphetamine clearly produced increased responding on the correct (conditioned reinforcement) lever, whereas cocaine had little effect. Morphine and α-flupenthixol decreased responding on both levers, whereas apomorphine increased responding on both levers, thus not indicating an "elational" increase, but simply an "activity" increase. The results indicate the differential nature of responding under the drugs, and indicate that drugs such as *d*-amphetamine and pipradrol may be more closely related to this type of secondary reinforcement than cocaine and morphine, both of which have distinctly analgesic properties. As noted later, morphine may be reinforcing in a different manner than are DA stimulants *per se*. However this may be, the above method seems to provide an excellent tool for examining differences in elational properties.

2.1.1.2. Direct Infusion into the Nucleus Accumbens

The second method chosen for analysis here is that of infusion of dopaminergic agents directly into the nucleus accumbens (NAC). This method has the special merit that it affects, in a localized manner, one of the rostral structures connected with the mesolimbic system. As mentioned above, the NAC has been implicated in some behavioral functions having possible relationships to mania. A key finding was that of Kelly et al. (1975), showing that there is an apparent separation of neural control concerning hyperactivity (rate of response initiation) in itself, vs stereotyped, or intensely repetitive, behaviors. Iversen (1977) summarized some of this work, and suggested that the NAC was more closely related to response rate functions, whereas the nigrostriatal system was more closely related to stereotyped behavior (Randrup and Munkvad, 1968; Randrup et al., 1981). Since that time, Costall et al. (1984a,b) have shown that direct infusion of DA agonists into the NAC affects some behaviors in a cyclical fashion, and may demonstrate exceedingly lasting effects (>1 yr) on DA-related behaviors.

The latest work with infusion of DA agents into the NAC goes much more into depth with the behavioral changes produced, and provides evidence of DA effects on both conditioned and unconditioned (species-specific) behaviors (Taylor and Robbins, 1984; Jones et al., 1987,1989). Thus, it may provide clues relevant to both the learned and unlearned (biologically constraining) factors that are involved in mania.

The following is a brief description of the daily injection vs minipump infusion methods used by Costall et al. (1984a,b). It is suggested that with some methodological revisions, the comparison of these two methods, particularly as it reflects on the mesolimbic DA system under different treatment conditions, may come to provide some valuable insights into manic behavior.

Adult, male rats were used as subjects, and they were selected as either high (60–80 counts/5 min) or low (10–25 counts/5 min) in photocell activity cages. It should be noted that only about one-half of the animals fitted into these two categories. Selected animals were stereotaxically implanted with guide cannulas for bilateral injection into the nucleus accumbens. Correct placement of cannulas was checked by killing at least three animals at the time of surgery and examining frozen sections of the brain. No mention is made of postmortem anatomical checks on the tested animals. The guide cannulas were of stainless steel and 0.65 mm diameter, and were held in place with perspex holders fastened to the skull. Guides ended 3.5 mm above the NAC and were kept open by means of a stylet protruding 0.5 mm below the end of the guide.

Drug treatment was given to one group by direct daily injections by microsyringe for 13 successive days. Doses of DA hydrochloride ranging from 1.56–50 μg were delivered in 1 μL of a 0.1% sodium metabisulphite solution bubbled with nitrogen. A second group was implanted with a 14-d minipump (Alzet®) attached to a short piece of polyethylene tubing fitted to a stainless steel injection unit inserted into the guide cannula and extending 3.5 mm beyond its tip. The minipumps were removed on d 13 to avoid a potential fall-off in drug infusion at the design limit of the pumps.

For subcutaneous injection, the dopamine agonist (-)*N-n*-propylnorapomorphine ((-)NPA) was dissolved in 0.1% sodium bisulphite, and for ip injections, methysergide hydrogen malenate, piperoxan hydrochloride, and (+)propanolol were prepared with water. Haloperidol and (-)sulpiride were dissolved in "minimum" quantities of lactic acid and hydrochloric acid, respectively. All peripheral injections were calculated as the base and given in a vol of 1 mL/kg body wt.

Behavior was measured by activity in the photocell cages, individually, and by Automex® activity meters in social groups of five animals. Unfortunately, no other behavioral measures are reported, although the activity measures were continued up to 98 d after DA withdrawal. This method provides two potentially useful parallels to mania.

The first parallel lies in a cyclic change in hyperactivity produced in both high and low base activity animals at a one-week interval with the minipump infusion method. The initial phase is hyperactive followed by a three-day return to baseline activity levels, and then followed by a second peak *exactly* seven days after the first one. The peak in activity comes abruptly on the third and the tenth day of treatment, and activity is significantly above baseline levels for 3–4 d. This strongly suggests (although the authors do not mention it) that the hyperactivity is correlated with the general activity in the laboratory, including the two weekends in the 13-d test period. Since control animals did not show this peak, it appears that quiet/active periods in the laboratory are associated with a significant change in the incidence of hyperactivity. Furthermore, this is only a function of the minipump infusion method, since the injection procedure seems to show only one prolonged peak of hyperactivity lasting ten days or more. The exact causal factors in this apparent sensitivity of the infusion paradigm to laboratory activity suggests an important parallel to the effects of daily activity stress on mania. The infusion method is also one of the better methods (but *see* comparison with the implanted capsule method, Eison et al., 1983) for mimicking the relatively stable DA stimulation, which must occur in the natural course of manic illness. The injection method does not offer this parallel so exactly since peak drug levels are constantly changing between injections. It is unfortunate that stereotyped behavior, and some form of conditioned behavior were not examined by Costall et al., since these might have provided still further evidence of parallels to mania.

The second potentially important parallel to mania in these studies is that Costall et al. (1984b) have shown that hypersensitivity to DA challenge is retained more than one year after intracerebral treatment with haloperidol. This suggests that some changes resulting from this injection treatment have more or less permanent effects on sensitivity to DA stimulant drugs. Such an effect resembles the apparently permanent tendency to relapse in unmedicated manic patients, even after periods of many months without the illness.

That it is the change in DA levels in the accumbens and striatum that is the relevant variable for the changes seen is strongly supported by the studies of Jones et al. (1987) and Kuczenski and Segal (1989), using in vivo dialysis of these regions to demonstrate DA activity changes. It is of great interest that the effects seen are so long-lasting, but one severe detriment to these studies is the failure to examine other behaviors than simple motor activation (activity counts). However, the method itself provides an excellent means of investigating the potentially manic effects of localized NAC injections of DA stimulants.

2.1.2. Other Catecholamine-Related Models

Most of the other potential models for manic behavior in this section are based on such relative modulation of neurotransmitters, but it is contended that the basic feature of change is always most closely related to dopamine. For instance, it will be seen that lithium, which acts more on serotonin than on dopamine, has little effect directly on amphetamine-induced behavior (Fessler et al., 1982), but indirectly, lithium's effect on serotonin may allow dopamine to achieve an increased dominance in certain types of behavioral initiation.

2.1.2.1. Alpha-methyl-para-tyrosine (AMPT)

Agents that interfere with dopamine formation, such as AMPT, which interferes with the process of tyrosine conversion to L-dopa, may reduce the hyperactivity induced by amphetamine (Davies et al., 1974). Some antimanic properties have been reported for AMPT in humans (Brodie et al., 1971). In addition to those drugs directly affecting dopamine, some agents primarily affect norepinephrine, acetylcholine, and serotonin levels, with corresponding charges in the relative effect of dopamine stimulation.

2.1.2.2. Amphetamine/Chlordiazepoxide Model

U'Prichard and Steinberg (1972) suggested a potential model for prolonged hyperactivity induced by amphetamine injections combined with the benzodiazepine chlordiazepoxide. The behavioral methods applied (*see also* Davies et al., 1974) included measures of locomotor activity, and studies of hole-board activity (Poitou et al., 1975). Although it was initially suggested that amphetamine stereotypy (repetitive sniffing, licking, and head movements) was dampened by the chlordiazepoxide pretreatment, later studies did not support this distinction (Poitou et al., 1975). The latter authors suggested that the combination of chlordiazepoxide and amphetamine caused an imbalance in the normal relative levels of norepinephrine and serotonin, and they cited the partial correction of this behavioral effect by lithium to support their view.

2.1.2.3. Catecholamine Super- or Subsensitivity Models for Mania

2.1.2.3.1. Supersensitivity. There has also been speculation that super- or subsensitivity induced in DA receptors might be the main feature of mania. Robbins and Sahakian (1980), after reviewing the evidence for DA receptor supersensitivity, as induced by localized NAC lesions with 6-OHDA, suggested that this model should be further investigated. There are several points arguing against this direct hypothesis of induced supersensitivity. Chronic pretreatment with amphetamine does not produce supersensitivity as measured by constant hyperactivity or stereotypy, over the entire period of treatment (Huberman et al., 1977; Nielsen, 1981; and Jackson et al., 1981, all in rats; Mehrabian, 1986, in humans). Instead, there appears to be an initial phase of increased activity and perseverative behavior, which is accompanied by lack of sleep and increased irritability. This phase is followed by one in which the activity and stereotypy is reduced, while feeding and resting behaviors gradually increase, though not to normal levels in most cases.

2.1.2.3.2. Subsensitivity. Antelman and Chiodo (1981) have suggested that the principal feature in the induction of amphetamine psychosis, which has many symptomatic features resembling mania, is dopamine autoreceptor subsensitivity, rather than postsynaptic receptor supersensitivity. According to this view, the autoreceptors, which normally function as "shutoff valves" for further DA release into the synapse, are subsensitive to the released DA, and therefore do not prevent further release. The result would be increased activity, and at least, a temporary respite from depressive symptoms. Since tricyclic antidepressants and electroconvulsive shock both appear to induce subsensitivity of DA autoreceptors, this seems to support this possibility (Chiodo and Antelman, 1980).

The truly chronic treatment methods, especially those with slow-release capsules (Huberman et al., 1977), mini-pumps

(Nielsen, 1981), and chronic infusion (Fischman and Schuster, 1974), show that continuous treatment with central stimulants produces a cessation or weakening of stereotypic motor acts, and less overall bodily activity. However, this may be caused simply by temporary exhaustion and the effects of almost continuous physiological stress induced by DA / NE overstimulation. In short, the outward behavior may have changed, but the sensitivity of the DA system may largely be intact.

2.1.2.4. Monoaminergic/Cholinergic Sensitivity Balance Model

Dilsaver and Greden (1984) have suggested that there may be an important interaction between cholinergic and monoaminergic systems in the production of mania and the subsequent depression that so often accompanies it. The idea is still at the theoretical stage, but may be worth considering in developing methods to mimic the switching between mania and depression that is so characteristic of these disorders.

Basically, the concept of Dilsaver and Greden is that when the brain neurotransmitter balances are disturbed, the opposing systems will attempt to restore balance by changing the sensitivity of pre- and postsynaptic receptors. For instance, cholinergic overstimulation may lead to an increased synthesis of tyrosine hydroxylase in noradrenergic and dopaminergic neurons, with resulting overstimulation of these systems, which, indeed, seems to be the case in mania. However, overestimation of the DA and NE systems would then gradually lead to counteracting changes in the cholinergic system, with subsensitive autoreceptors on presynaptic endings, and supersensitive postsynaptic receptors. This would lead to another cyclic switch from mania to depression. The concept seems relevant, but methodology for separating these effects is not yet available.

2.1.3. Summary of the Catecholaminergic Models

The catecholaminergic models provide a very important set of parallels with manic behavior. The combined stimulation of the DA and NE systems does not seem to be necessary, as most of the symptoms can be produced by models that are essentially dependent on only the DA systems.

The principal features modeled here are hyperactivity, elation, and switching of mood tone. Secondary reinforcement strength is used as a prime measure of elation, and direct stimulation of the nucleus accumbens of the mesolimbic system, and the corpus striatum of the nigrostriatal system have been suggested as models for the induction of manic symptoms.

2.2. Indoleamine Related Models

Serotoninergic systems have often been implicated in models of mania, but careful attention must be paid to methodology, since many of these models include pretreatment with monoamine oxidase inhibitors (MAOIs), which automatically increases the level of available monoamines as well. This criticism applies particularly to models purporting to demonstrate manic behaviors produced by increasing the amount of normally available serotonin (Grahame-Smith, 1971; Green and Grahame-Smith, 1974; Jacobs, 1976; *see also,* Summary in Iversen, 1986). Further doubt is cast on the likelihood that excessive serotonin is the prime factor in mania by the finding that ICSS rates are reduced, rather than increased, by treatment of rats with the serotonin precursor L-tryptophan, together with the MAOI pargyline (Herberg and Franklin, 1976). This suggests that elation effects are reduced rather than increased by this treatment, although the caveat regarding interpretation of ICSS rate decreases as purely motivational in causation must be recalled.

However, it should be noted that with high doses of amphetamine, 5-HT levels are increased in the corpus striatum of rats (Kuczenski and Segal, 1989). This is a dose-dependent effect, and several other researchers (*see review* by Gerson and Baldessarini, 1980) have concluded that the increased 5-HT levels are associated mainly with the stereotyped behavior phase of amphetamine stimulation. In any case, serotoninergic actions must be seen against the background of the DA stimulation, which they appear to modulate, rather than control fully.

Similar criticisms may be applied to some lysergic acid diethlamide (LSD) models of mania in cats, such as that reported by Jacobs et al. (1976), which is said to depend on the antagonism of 5-HT receptors by this drug. The symptoms described following the LSD treatment include features known to be typical of dopaminergic overstimulation, such as fractionation of behaviors (Schiorring, 1971; Sudilovsky, 1975; Ellinwood and Kilbey, 1977) and continuous switching between a limited number of forms of activity (Evenden and Robbins, 1983). Since Kelly and Iversen (1975) have shown that LSD may also work directly as a DA agonist on mesolimbic structures, it seems doubtful to attribute these effects to serotoninergic systems alone.

In contrast to the questionable nature of serotonin (5-HT) increases as causative factors in mania, there appears to be good evidence that *depletion* of 5-HT by para-chlorophenylalanine (PCPA) is followed by hyperactivity, increased frequency of social and aggressive acts, and increased irritability to novel stimuli (Fibiger and Campbell, 1971; Ellison and Bresler, 1974; Marsden and Curzon, 1976). Whereas PCPA does have effects on other neurotransmitter systems (Miller et al., 1970), it has been shown that the behavioral effects mentioned here can be reversed by treatment with tryptophan or 5-HT itself. Furthermore, the PCPA-induced hyperactivity is not readily reduced by the specific DA receptor blocker pimozide, whereas the increased aggression can be blocked by lithium. These facts tend to increase the likelihood that specific 5-HT effects are at work in addition to any DA overbalance that might occur as a result of the depleted serotonin following PCPA treatment. Because this model includes changes in social behavior, including aggression and irritability, as well as the more commonly reported hyperactivity, it is a model deserving further research.

2.2.1. Summary of Indoleamine Related Models

These models suffer from the fact that reductions in 5-HT are unfailingly followed by a relative overbalance of the opposing monoaminergic system (DA + NE). Whereas some features of behavior are clearly influenced by the 5-HT levels alone, it is virtually impossible to separate the indoleaminergic from the monoaminergic effects. However, as several studies have suggested, the DA/5-HT balance may be a critical factor in mania, and this may be the principal cause of the manic-like symptoms produced by 5-HT depletion models.

2.3. Opioid Related Models

2.3.1. Morphine Models

The opioid related models of mania have a special interest in comparison with the DA related models discussed above. The reasons are that behavioral changes produced by opioid stimulation include, but are not limited by, increases in bodily activity, and also, that opioid effects appear to be more specifically related to elational and behaviorally reinforcing actions (Barchas et al., 1985).

Morphine was long thought to produce diverse behavioral effects in different species, with cats being prone to "feline mania" under its influence, although rats were frequently sedated. Recent work has shown that most of this diversity was simply the result of dose response differences along a steeply climbing curve of dose effect (Ayhan and Randrup, 1973; Villablanca et al., 1984). However, a second difference exists. Morphine tends to stimulate the frequency of abrupt changes in the types of activity being engaged in, with sudden shifts in the morphine-treated rat between such diverse activities as locomotion, social interaction, eating, and grooming (Ayhan and Randrup, 1973; Schiørring and Hecht, unpublished). The relatively rapid shifting between these more complicated behaviors appears somewhat different from the increased shifting between simpler motor acts, which is so characteristic of the DA overstimulation models (Lyon and Robbins, 1975). The result is that the DA models yield manic-like behaviors that are more limited in scope and become more rapidly stereotypical and repetitive, whereas the opioid models disrupt behavior by producing disorder in the normal sequence of events. The common feature of both DA and opioid models is that behavior becomes fractionated, and eventually inappropriate, as the shifting of behavioral patterns increases. Behavior under opioids may also become stereotypical and repetitive, but it seems that this end result is reached more slowly.

Besides the above differences between opioid and DA models, there are also differences in the strength of elational and reinforcing effects, which have traditionally been supposed to be extremely high with opiates. However, it is also known that the DA systems are closely related to reinforcing effects, with the mesolimbic DA system being most important in this context. ICSS in the hypothalamic regions containing many mesolimbic fiber projections can be used as a reinforcing effect in its own right, although DA cells may be only indirectly involved (Gallistel, 1986).

The relationship of these reinforcement measures to those involving morphine euphoria is also difficult to assess, since lithium itself has a slightly dysphoric effect in humans, and does not counteract morphine-induced euphoria (Jasinski et al., 1977). This finding also militates against the acceptance of morphine stimulation as a model for the euphoric/elational aspects of mania, which apparently can be counteracted by Li.

If morphine is able to induce a state similar to mania by its effect on opioid receptors sites in the brain, then morphine receptor antagonists, such as naloxone and naltrexone, should be capable of producing at least a temporary reduction of manic symptoms. However, the effects of naloxone on mania are equivocal, with some positive response, but more frequently, no improvement (Berger and Barchas, 1982).

2.3.2. Endogenous Opioid Models

Morphine may not be the best of the opioid substances for modeling the effects of mania. Katz (1982) has suggested that endogenous opioids may be the actual source of manic disorder. His experiments with mice tend to bear this out, with *d*-Ala2-substituted amides of Leu and Met enkephalins producing stereotyped running very similar to that seen following morphine treatment. Kyotorphin, which inhibits the catabolic enzyme enkephalinase, given intraventricularly in <30 s, tends to potentiate enkephalin activity and also produces increases in rearing on the hind legs at lower doses (ca. 75 μg in 5 μL), and stereotyped intensive running at high doses (ca. 150 μg in 5 μL). At doses higher than 150 μg, there is a period of intense stereotypical responding lasting no longer than 30 min and followed by convulsions and death.

Probably the most interesting of the models employing endogenous opioids is that of Schwartz et al. (1982), who pretreated rats with morphine or lactose via implanted subcutaneous Silastic® capsules, and then tested them during the hypersensitivity induced following removal of the morphine capsules by injecting beta-endorphins into the ventral tegmental area. During the supersensitive postmorphine period, morphine, but not amphetamine, challenge was followed by greatly increased activity. Furthermore, gamma-endorphins themselves can apparently induce a haloperidol-like supersensitivity, yet amphetamine and apomorphine challenge treatments indicate that these changes are independent of changes in striatal DA function. The following is a brief resume of the methods used by Schwartz et al. (1982).

Adult male Wistar rats were anesthetized with Chloropent® ip, and implanted with stainless steel guide cannulas bilaterally aimed at the ventral tegmental area (VTA). The cannulas were cemented to the skull and blocked with a stylet during surgical recovery of 1 wk. All animals were then tested with a single 0.5 μg infusion of β-endorphin (0.3 nmol total dose) and locomotor activity was recorded in a photocell box for 2 h. Rats were then paired with respect to similar locomotor activity scores and one member of each pair assigned to receive morphine or placebo pellet implants.

Silastic® pellets were prepared (McGinity and Mehta, 1978) that contained either 100 mg morphine sulfate or lactose, and left implanted sc for 3 d. Two additional pellets were then implanted for 3 more d. At the end of this procedure, all pellets were removed and animals received 0.4 mg/kg of the opioid antagonist naloxone hydrochloride, sc. The presence of ptosis, wet-dog shakes, and diarrhea were confirmed during the next 90 min in the morphine group, but not in the placebo group.

Microinfusion through the cannulas into the VTA of β-endorphin was performed using 1 μL of solution infused over a period of 105 s by a syringe pump. The injection cannulas were left in place for 1 min after infusion, then stylets were placed in the cannulas and activity counts were taken over a 2-h period.

One day after pellet removal and naloxone treatment (as above), animals were subjected to either 0.5 or 1.0 μg/μL β-endorphin, and on d 4 post-pellet, they received the opposite dose.

On d 6, animals were given saline infusions, on d 7 *d*-amphetamine was injected sc, and on d 8, they received 5.0 mg/kg morphine sc.

The results showed that pretreatment with morphine resulted in a change from hypo- to hyperresponsiveness over the next 3 d, whereas lactose had no effect. Schwartz et al. (1982) suggest that this change models the switch from hypoactivity to mania, and that the morphine pretreatment followed by β-endorphin infusion into the mesolimbic pathway at the VTA provides an improved model for the induction of mania. It should be noted that this model implicates the mesolimbic DA system in an opioid-controlled behavior. Exactly how this is accomplished is not yet clear. Joyce and Iversen (1979) demonstrated, however, that injections of morphine into the substantia nigra pars compacta did not cause the hyperactivity that resulted when the VTA was treated. The further use of this model should provide answers to the question of DA/opioid interactions in mania.

2.3.3. Summary of Opioid Related Models

It is possible that in evaluating opioid models of mania, that the mesolimbic DA system from ventral tegmental area to nucleus accumbens should be considered separately from the nigrostriatal DA system that is responsible for amphetamine stimulation of motor activity (Schwartz et al., 1982; Van Ree et al., 1982). It is also possible, as Stevens (1975,1979) has often suggested, that GABA stimulation of limbic structures is more important than DA stimulation of the striatum in producing the typical symptoms of mania.

The most problematic feature of the opioid related models, so far, is that the evidence is conflicting on whether naloxone, which blocks opiate receptor activity, can produce any improvement in mania (Berger and Barchas, 1982). This suggests that the effects of morphine, and possibly of the endogenous opiates as well (de Wied, 1979), are mainly owing to effects produced in other neurotransmitter systems than those directly controlled by the opiate receptors. The problem of defining which opioid receptor types might be involved in mania remains to be solved.

2.4. Acetylcholinergic Factors in Mania

Janowsky and J. Davis and their colleagues (Janowsky et al., 1972a,b) were among the first to advance a NE/ACh hypothesis for mania and depression. According to the original form of this hypothesis, high NE and low ACh would be associated with mania, whereas low NE and high ACh would accompany depression. Whereas present views give more emphasis to DA as well as NE in this relationship, there is no doubt that ACh, at the very least, can act as a modulator for the monoaminergic effects. Some of the earliest studies with manics treated with the cholinesterase inhibitor physostigmine reported a complete reversal of mania, even resulting in depression in 8/8 patients tested (Janowsky et al., 1973). However, later studies of ACh and mania have shown that physostigmine acts principally on the rate increases in thought production and bodily activity induced by mania, and it had much less effect, if any, on the thought *content* or emotional *tone* (J. Davis et al., 1978). Physostigmine tended to make euphoric manic patients feel tired, listless, and ill or nauseated, with clearly depressive overtones, yet it failed to alter the state of irritable manic patients, even resulting in increased irritability on cessation of the physostigmine infusion (K. Davis et al., 1978).

Animal studies with physostigmine seem to support the above description, with an initial depressant effect on behavior, followed by excessive overactivity. On the other hand, the ACh antagonist, atropine, tends to produce a more perseverative form of behavior ("behavioral trapping"), with an apparent lack of habituation that seems to mimic partially the effects of DA agonists (De Vietti et al., 1985).

2.4.1. The Effects of Lithium (Li) on Cholinergic Functions

Lithium was originally reported to have little significant effect on cholinergic systems (Boissier, 1975), but more recent work indicates that it may be important in allowing the brain to retain more choline, or to increase the concentration of brain ACh. Such an effect would explain the effectiveness of Li in preventing the return of the massive mood swings in bipolar affective illness, not by a direct effect on the DA/NE systems (for which there is less clear evidence anyway), but by insuring the constant presence of adequate amounts of ACh when this neurotransmitter must act to balance the presumed changes in the DA/NE systems.

Putting this and other evidence together, Dilsaver and Greden (1984) have proposed an adrenergic/cholinergic model similar to that originally proposed by Janowsky and J. Davis *(see above)*, but in this case, attempting to explain how the bipolar mood changes are initiated. Adrenergic excess leads to mania, whereas cholinergic excess favors behavioral inhibition, and possibly leads to depression. Therefore, Dilsaver and Greden suggest that since anticholinesterases and cholinomimetic drugs both work to increase available ACh, these agents will, in turn, antagonize the monoaminergic systems. The particular importance of the Dilsaver and Greden theory, however, lies in the attempted explanation of the mania/depression switching, which they hypothesize is caused either by autoregulatory receptor reactions at pre- and postsynaptic monoaminergic sites, which would be accompanied by an increase in cholinergic receptor sensitivity (leading to the mania/depression switch), or by increased production of tyrosine hydroxylase by cholinergic hyperfunction (leading to the depression/mania switch).

2.4.2. Summary of ACh Related Models of Mania

There is little doubt that ACh in combination with the catecholamines provides part of the balance that is disturbed in mania. However, it is a telling point in comparing this model to human mania, that the emotional component of behavior is not easily altered by large changes in ACh. In short, there seems to be a separation of the motor activation from the elational aspects of mania. Nevertheless, as a comparison model with those involving combined changes in these variables, the manipulation of ACh levels may be quite useful. What is needed for progress in this area is a viable model of the Dilsaver/Greden hypothesis, using agents with both adrenergic and cholinergic influences in combination.

2.5. Glutamic Acid (Glu) and Gamma-Amino-Butyric Acid (GABA) Factors in Mania

Models involving the adrenergic/cholinergic interactions may be useful for studying the switching effect from mania to depression, and they may also have direct relevance to the next grouping of models that implicate the excitatory effects of glutamic acid and the inhibitory action of GABAergic systems (Fonnum, 1984; Scheel-Kruger, 1986) as they are reflected in behavior.

The potential and known interactions of the brain's most widely distributed excitatory and inhibitory neurotransmitters (Glu and GABA, respectively) are too complex to be dealt with here. Only a few relevant findings will be considered in relation to mania.

GABA is formed from glutamate by the removal of a carboxyl group. It is perhaps the most ubiquitous of all neurotransmitters, with over one-third of all nerve terminals in the CNS staining for its presence (Fonnum, 1984). GABA is an inhibitory neurotransmitter at the postsynaptic junction, but the end result can be excitatory if it inhibits a normally inhibitory pathway.

2.5.1. Excitatory Amino Acids and the Hippocampus

Several interesting relationships have been suggested between the effects of excitatory neurotransmitters, such as glutamic acid (Glu) and aspartate (Asp), and the neural activity of the hippocampus (Cooper et al., 1986). The NMDA receptors in this limbic region seem to be of particular importance, since they have relationships to long-term potentiation, which may be important in learning and short-term retention, yet these same receptors are vulnerable to a neurotoxic effect if overstimulated by glutamate in the absence of calcium (Ca^{2+}) channel blocking by either a blocking agent (such as MK-801), or magnesium (Mg^{2+}) within a specific temperature range. If the calcium channel is not blocked sufficiently often, a process begins that may end with the death of the cell.

These facts may be relevant to mania for a number of reasons. First, there is a large amount of independent evidence linking cellular damage in these same regions of the hippocampus with the major psychoses (Bogerts et al., 1985; Kovelman and Scheibel, 1984). Second, lithium-induced increases in Mg^{2+} tend to be correlated with that drug's effectiveness in reducing mania (Pavlinac et al., 1979), and a close relationship between Mg^{2+} deficiency and catecholamine functions and aggressivity has been documented by Kantak (1988). These changes could be related to the NMDA receptors, but further investigation is needed to establish the possible roles of not only Mg^{2+}, but also of Ca^{2+}, zinc, and glycine at the NMDA site.

Cocaine also magnifies kindling effects produced by electrical stimulation of the amygdala and hippocampus, which have been suggested to be dependent on excitatory amino acid (Glu) functions. If intensified by repeated stimulations over time, these local effects can lead to the kindling of seizures within the limbic system (Post et al., 1984). The kindling itself leads to a state in which the initial neurochemical stimulation is no longer necessary to invoke the disturbance of hippocampal electrical activity (*see also* Kindling and Sensitization Models).

To prevent the development of these seizures, Post tried various antiepileptogenic compounds, and discovered that carbamazepine blocked both cocaine and lidocaine-induced hippocampal seizure activity. On this basis, Post suggested that carbamazepine should also act as an antimanic medication. This has proved to be correct, at least for a subgroup of manic patients (Rubin and Zorumski, 1985).

2.5.2. The Role of GABA

Originally, it was thought that the relationship of GABA to the DA systems was mainly as a direct inhibitory effect of striatonigral GABA neurons inhibiting the action of the nigrostriatal fibers arising in the pars compacta of the substantia nigra. However, this is not what happens at normal levels of GABA activation. Instead, the striatonigral GABA neurons act principally on the cells of the pars reticulata of the substantia nigra, and indirectly affect thalamocortical relationships (Scheel-Kruger et al., 1981; Scheel-Kruger, 1986; Carlsson, 1988), which then, in turn, affect the corpus striatum once more. Furthermore, stimulation with the GABA agonist muscimol of the VTA results in different reactions from the rostral and caudal ends of that structure (Arnt and Scheel-Kruger, 1979). GABA stimulation of the caudal VTA results in:

> ...a dose-dependent increase in non-explorative locomotion, a strong aggression shown as immediate attack of everything presented and violent fighting after placing two rats in a single cage...increased food intake in satiated rats...no stereotypies (sniffing, head movements, licking or gnawing) after bilateral injection or turning after unilateral injections...(Scheel-Kruger et al., 1980, p. 263).

On the other hand, muscimol injected into the rostral VTA causes a decrease in spontaneous activity that eventually becomes sedative, with a hunch-back posture and some rigidity of the limbs. There is no aggressive behavior or induced eating.

These behavioral descriptions fit reasonably well with the symptoms of mania, including the fact that limited behavioral stereotypies are not the most prominent symptom as they tend to be in schizophrenia (Bleuler, 1950). Stevens (1975) reported extreme emotional responses of fear, hypervigilance, and apparent "hallucinations" following GABA blockade of the VTA with the GABA antagonist bicuculline in the cat. Scheel-Kruger et al. (1980) found that bicuculline and picrotoxin in the rostral VTA of the rat induced strong hypermotility and vertical activity (rearing on the hind legs, an investigatory movement in uncertain situations). From this, one might guess that Stevens' injections mainly reached the rostral VTA. Stevens' conclusion was that disturbances of the mesolimbic system had much to do with schizophrenia, but it may be that lesser doses would produce the equivalent of manic behavior.

It is also extremely interesting that the only two *long-axon* pathways involving GABA neurons are the striatonigral path already mentioned, and the output axons of Purkinje cells in the cerebellum, which terminate in the basal cerebellar and vestibular nuclei (Fonnum, 1984). These pathways are of interest because the first of these pathways is quite possibly involved in mania, as mentioned above, and the cerebellar path has been suggested to have a potential role in the abnormal eye movements seen in schizophrenia (Karson et al., 1990).

2.6. Brain Lesion Models

Robbins and Sahakian (1980) made an attempt to review thoroughly the various ways in which hyperactivity following brain lesions might be seen as a model for mania. They concluded appropriately that hyperactivity alone may not be a sufficient index of manic-like behavior in animals. This is because such a wide variety of agents affecting the brain can cause excessive activity. However, in the following, the various lesion models have been grouped roughly according to the structures that they affect, rather than according to the degree of hyperactivity they produce.

Before assessing the effectiveness of various lesioning procedures in modeling mania, rather than for producing simple behavioral activation, it would be wise to recall the two major aspects of manic hyperactivity as discussed in the introduction. Manic hyperactivity (1) *tends to evolve into a specific and limited behavioral form* after a period of intensely active switching between such forms; and (2) *exhibits an excessive emotional tone*, whether of elation, irritability, or aggression.

Using the above differences in manic hyperactivity as criteria, many of the supposed parallels to manic hyperactivity in animals are not convincing. Heavy metal poisoning often involves a strong deposition of the metal in the hippocampus, a structure frequently suggested to be implicated in psychosis. However, the hyperactivity produced by poisoning with lead, cadmium, or rubidium (Silbergeld and Goldberg, 1974; Rastogi et al., 1977; Meltzer et al., 1969) is "paradoxically" reversible by treatment with amphetamine or methylphenidate, which is not the typical case in mania, although occasional decreases in manic symptomatology are reported in the literature following central stimulant drug treatment. In any case, the present lack of knowledge about the fundamental effects and localization of heavy metal toxicity makes it difficult to use these treatments as a source for modeling.

The same criticisms apply to methods used by Norton et al. (1976). They compared the behavioral effects of X-irradiation at gestational d 14–15, carbon monoxide exposure on postnatal d 5, and direct electrolytic lesions of the globus pallidus. However, we do not yet know enough about the effects of agents limited to d 14–15 of gestation in the rat, or to the exact effects of postnatal d-5 treatment. Destruction of the globus pallidus, which contains the major descending output of the corpus striatum, would be expected to remove a large part of the inhibitory influence on motor reactions, and thus, to effectively raise the level of muscular activity, and hasten the rate of change between various activities by allowing a more rapid initiation of new responses. This is exactly what Norton et al. (1976) found, using these methods. Explorational behavior, measured by videotape analyses of behavioral acts at 1-s intervals, occurred more frequently, but with briefer durations, and there appeared to be a more random, and more frequent, switching between behaviors. It should be noted that at the 14th gestational day, the cells in the cortex of the parahippocampal gyrus had not fully migrated to their cortical layering, and disturbances at this anatomical position have been found in the brains of schizophrenic patients (Jakob and Beckmann, 1986). Postnatal hypoxia induced by the carbon monoxide treatment on d 5 may also affect the hippocampus, as well as the neocortex, thalamus, and sensory pathways of the midbrain (inferior colliculus, and so on), since these regions are still under steady development at this time. Finally, the cerebellum still contains immature and only partially migrated cells destined for the lateral cerebellar cortex, including their input to the basal cerebellar nuclei (Gilles et al., 1983; Lyon and Barr, in press). Although the behavioral effects of lesions at these developmental stages are similar to mania, we are once again without precise localization. Hence, the methods used by Norton et al. (1976) are not well designed for basic modeling of mania, although their methods for measuring behavior are excellent for this task.

2.6.1. Lesion Methods Related to Serotoninergic Systems

Several methods have been applied to the destruction of 5-HT systems, and two of these will be mentioned here: selective destruction of 5-HT cells by neurotoxins, and electrolytic lesions of the raphe nuclei in the brain stem (Lorens et al., 1976).

The selective neurotoxins 5,6- and 5,7-dihydroxytryptamine (DHT) can be used to destroy 5-HT cells in the CNS. This can be done either by microinjections into selected areas, or by periventricular injections, which affect a much broader periventricular area (Diaz et al., 1974; Green and Grahame-Smith, 1974). This method should be compared with PCPA injections, which do not destroy the 5-HT cells, but do lower the amount of 5-HT available. The lesion method with 5,6-DHT has a dose-related effect. A lower dose, injected intraventricularly in rats, produces an initial increase in whole body locomotion and a reduction in rearing, whereas a higher dose reduces whole-body locomotion but increases rearing on the hind legs.

This effect is reminiscent of the effects reported with low and high doses of amphetamine in rats responding on a shock avoidance task, in which such increased locomotion and increased rearing disturb, respectively, passive and active avoidance (Lyon and Randrup, 1972). As mentioned previously, such parallels suggest that perhaps changes in DA balance, rather than 5-HT activity *per se,* is the relevant variable in producing these changes. However, the activity changes may not be as long-lasting in the case of the DHT lesions. Mailman et al. (1981) reported that 5,7-DHT lesions given 2-d postnatally, resulted in hyperactivity 14 d later, but only hypoactivity 28 d posttreatment. Once again, this is consistent with the suggestion that the ac-tivity changes can be compensated for by DA systems over time.

Lesions of the raphe nuclei have been studied with both electrolytic and neurotoxic lesions (Lorens et al., 1976). As with many other systems, the effects of these two lesion methods do not produce the same behavioral effects. Electrolytic lesions were followed by increases in whole body locomotion, but not by increases in rearing or sniffing at the floor of the cage. In addition, running wheel, and open-field, activity also were increased. These effects are similar to those seen with low doses of *d*-amphetamine, and once again, suggest an indirect relationship with DA system overbalance produced by decreasing the normal 5-HT levels of activity (Gerson and Baldessarini, 1980; Balsara et al., 1979). However, 5-DHT lesions of the median raphe did not produce the same symptoms, which suggests that part of the reason for the increased locomotion may be the result of fibers coursing through the region of the median raphe, but without direct connection with 5-HT systems.

Furthermore, it appears that the dorsal nucleus of the raphe is more closely connected to the caudate-putamen, whereas the median raphe nucleus has more extensive connections with the hippocampus, medial preoptic area, suprachiasmatic nucleus, and anterior hypothalamus (Van de Kar and Lorens, 1979). These connections suggest that the dorsal nucleus may have more to do with the rapid repetition of responding seen in mania, whereas the median nucleus is more closely related to the potential kindling effects in the hippocampus (*see below* under Kindling and Sensitization Models).

2.6.2. Lesion Methods Related to DA Systems

Selective lesioning of specific neurotransmitter systems began with the use of 6-hydroxydopamine (6-OHDA), a neurotoxin that appears to be relatively specific to DA systems. It is perhaps one of the most widely used of all neurotoxins, and several of its uses result in behavioral symptoms with an apparent relation to mania. In the following, a few of these experiments have been selected in order to show some of the internal neural arrangements that may be dysfunctional in mania. Once again, the two basic DA systems that are considered for such a role are the nigrostriatal, and the mesolimbic systems.

Lesions of the nigrostriatal system involving DA cells lead to a loss of motor activity, principally through the lack of initiation of movement by the corpus striatum and associated structures. Lesions caused by disease result in Parkinsonism, and more recently, the abused drug 1-methyl-4-phenyl-1,2,3,6-tetrahydropyridine (MPTP) has been shown to produce the same losses of DA cells in the substantia nigra, and the same symptoms as Parkinsonism (Heikkila et al., 1984). The only parallels with mania lie in the supersensitivity to DA of the remaining nigral cells, which may temporarily result in overly repetitive actions, either as facial or limb tics, uncontrollable tremors of the extremities, or simply inordinately perseverative repetitions of some otherwise normal behavioral sequences. However, since the basic course of the disease leads to less and less active responding, any further parallels with mania are difficult to find. Furthermore, it is particularly symptomatic of Parkinsonism that facial expressions and normal social interactions are diminished or lost. In short, there is reduced reaction to environmental stimuli, and increasing impairment in the initiation of acts.

On the other hand, lesions of the mesolimbic system, dependent on where in this broader system they occur, tend to produce other effects than do the nigrostriatal lesions. Selective destruction of DA neurons in the nucleus accumbens of the rat is followed by a decrease in amphetamine-induced hyperactivity, but to increases in locomotor reactions produced by apomorphine, a direct DA receptor agonist. However, the timing of the behavioral test is very important in this context. For a few days after the 6-OHDA treatment of the nucleus accumbens, there may be a decrease in activity initiated by stress or amphetamine (Evetts et al., 1970), but several weeks later, such stimulation may increase activity again.

There is also a possibly confounding factor in activity analyses following accumbens lesions. Lesions involving damage to the lateral ventricles and septal region, as most NAC lesions are bound to do, sometimes result in a period of hyperesthesia, which was originally described as "septal hyperemotionality." However, the effect has nothing to do with specific septal nuclei (Harrison and Lyon, 1957), and may depend on a transient denervation supersensitivity. This might explain, in some cases, why apomorphine *increases* activity in 6-OHDA accumbens lesioned rats.

This suggestion fits well with the observation that mania does not necessarily lead to excessive *motor* activity (although it may do so). What is more characteristic of mania is the increased irritability, hypersensitivity to external stimuli, increased aggressive responses, and increased vocalization. These are all important symptoms of manic behavior, and the question now is, how are they related to periventricular brain structures?

Petty and Sherman (1981) introduced the intraventricular injection of 6-OHDA as a possible animal model for mania. They pointed out that 6-OHDA injected into the ventricles was known to deplete NE sharply, without severe effects on 5-HT and GABA.

Furthermore, animals treated in this manner showed "...strikingly increased irritability, aggression, hyperemotionality, hyperreactivity, and vocalization" (Nakamura and Thoenen, 1972; Coscina et al., 1973). These behavioral changes became worse if the animals received intraventricular NE, yet the behaviors were reduced by neuroleptics and by intense handling (Coscina et al., 1975). Petty and Sherman also recognized the importance of testing lithium, tricyclic antidepressants, neuroleptic drugs, and electroconvulsive shock treatments with their model. From the present point of view, we would not expect tricyclic antidepressants or shock treatment to be very effective in treating mania, since these treatments are not as commonly used, or found particularly effective, for the manic end of the bipolar mood disorder continuum. However, lithium and neuroleptic drugs are both supposed to be beneficial in mania, with lithium restoring DA/5-HT balance gradually, whereas neuroleptics act more immediately to block the hyperdopaminergia found in the early stages of mania.

The major problem with the Petty and Sherman study was the lack of differentiating behavioral tests. Only hyperreactivity, especially following an intense electric shock, was tested, using a Coulbourn modular test cage that was mounted on a BRS/LVE swinging pendulum activity meter.

On each test day, the initial stage was a 2-min familiarization period in the activity cage, followed by measurement of activity until the animal reached a criterion of 5 s with no measurable activity. After this, ten test trials were given, in which each test trial began with a 1 mA foot-shock for 1 s, followed by activity measurement until there was again a 5 s period of no activity. A 20-s intertrial interval separated the test trials. Responses of >40 counts on the activity meter were scored as hyperreactivity.

Separate groups were tested with

1. No drug or other treatment; reactivity test;
2. 6-OHDA by intraventricular injection; groups of animals tested on days 0, 2,4,6 after 6-OHDA; saline injection only; each group tested only once;

3. 6-OHDA as above; groups tested on days 0, 2, 4, 6 after 6-OHDA; lithium lactate injections (0.75 mmol/kg) twice daily after the 6-OHDA treatment; each group tested only once;
4. 6-OHDA as above; one group tested only on d 4 after 6-OHDA; chlorpromazine (5 mg/kg, ip) given once on d 4 with testing occurring only 1 h later;
5. 6-OHDA as above; one group tested only on d 2 after 6-OHDA; imipramine (3 mg/kg, ip) given once on d 1 and 2.
6. 6-OHDA as above; one group tested only on d 6 after 6-OHDA; electroconvulsive shock (110 mA, 0.3 s) given through earclip electrodes twice daily on d 2–5. Each animal had a tonic-clonic seizure of at least 20 s duration at each treatment session. Testing occurred at least 18 h after the last convulsive shock.
7. 6-OHDA as above; one group tested only on d 6 after 6-OHDA; subconvulsive shock (30 mA, 0.3 s) given on d 2–5; no animals showed seizures at any time. Animals were tested at least 18 h after their last shock.

Petty and Sherman found no changes in general activity between treatments before testing began with electric foot-shock. Animals receiving only 6-OHDA treatments were progressively hyperreactive to the shock over days, with a 67% increase on d 6. Lithium treatment prevented this progressive increase in reactivity, and animals were essentially at the normal level by d 6. Chlorpromazine also significantly reduced the hyperreactivity to shock, whereas imipramine, as predicted above, caused a greater reactivity than 6-OHDA treatment alone on d 4. Chronic electroconvulsive shock, but not the subconvulsive shock, had some effect on the hyperreactivity to foot-shock, but did not bring about a complete return to normal reactivity. These results fit in very favorably with the suggestion that this method provides a good model for mania. What is missing, in particular, is a good measure of elation. It is suggested that such a measure be added to future studies with this model.

There are also a number of interesting anatomical points concerning this model. It should be noted that among the structures separated from CSF only by the ependymal lining of the ventricles (in the rat at least) are the nucleus accumbens, corpus striatum, hypothalamus, hippocampus, and periaqueductal gray matter. Theoretically, agents in the ventricular fluid that are able to pass the CSF/brain barrier will easily affect all of these structures. Thus, injection of 6-OHDA into the ventricles has the potential to affect directly DA transmission in both the mesolimbic and the nigrostriatal systems. *In both cases, the rostral part of the systems is most clearly affected.* The substantia nigra and the ventral tegmental area, from which the neurons arise that will most affect DA functions rostrally, are spared from the direct effect of intraventricular injection. From a methodological point of view, it is interesting to note that the nucleus accumbens and the hypothalamus, of the mesolimbic system, lie in the most ventral regions of the ventricles, so that fluids with a heavier mol wt might tend to gravitate to these regions following injection into the dorsal part of the ventricular system.

Perhaps of special significance for the Petty and Sherman model is the fact that the 6-OHDA treatment resulted in a lasting hypersensitivity to electric shock, and that this hypersensitivity *increased* in strength of a period of several days. Such an effect shows important parallels with the kindling model suggested by Post et al. (1984). Furthermore, this hypersensitivity could be reduced by lithium and, to some extent, by chlorpromazine and electroconvulsive shock, and was increased by the tricyclic antidepressant imipramine. These features make this model one of the most attractive of all the models for mania yet introduced.

2.7. Intracranial Self-Stimulation (ICSS) Models

ICSS has often been suggested as a good model for the elational effects of mania, on the assumption that the systems yielding a positive increase in responding to ICSS are also those most intimately connected with reinforcement. However, this assumption does not always bear up under close scrutiny. Electrical stimulation of the brain leads to an increasing motor activation, more frequent repetitions of the response preceding stimulation, and so on, but this does not totally negate the motor stimulatory effects of the brain stimulation, which also exist (Phillips et al., 1976). The problem then becomes one of judging whether the rate increases typically measured in ICSS paradigms can be directly compared with elational effects. Certainly, ICSS thresholds provide a more secure measure of reinforcing strength, than do simple rate measures (Robbins and Sahakian, 1980). The definition of elation in terms of animal models probably requires a secondary reinforcement paradigm, in which the experimenter has control over the conditioning strength of the positive (elation-provoking?) alternatives (Lyon, 1990).

Leith and Barrett (1980) suggested that ICSS in combination with prolonged amphetamine treatment could be used to model the changes from mania to depression. They were able to show that as the amphetamine treatment continued, the ICSS rate was initially stimulated and then later depressed. This, they suggested, was owing to the initial DA stimulating effect of amphetamine, which is eventually followed by a depression in rate as the monoamine neurotransmitter resources are used up by the continual DA stimulation (*see also* Gallistel, 1986). The logic of this model is reasonable to a certain extent, and the fact that it addresses the problem of bipolar illness is unusual, but it is unclear from this how the full cycle, including the depression/mania change, would be achieved. This, added to the uncertainties surrounding the ICSS rate measure as a yardstick for elation, leaves the model with a suggestive, but not convincing, parallel to mania.

Perhaps ICSS could be used in combination with a model, such as that suggested by Dilsaver and Greden (1984) discussed above, which attempts to account for both directions of switching between the manic and depressive states. At this time, the ICSS model appears only to relate to single aspects of the manic syndrome, and even these are not entirely clear.

2.8. Kindling and Sensitization Models

Kindling refers to the gradual growth of sensitivity to periodic subconvulsive electrical stimulation of brain structures, until finally a convulsive episode occurs to the previously ineffective stimulus. Furthermore, if convulsions are elicited a number of times by this method, an independent convulsive cycle may be established, such that convulsions may be elicited without the priming effect of the subconvulsive stimulus (Goddard et al., 1969).

There is excellent evidence that the hippocampus, amygdala, and the nearby transitional cortex are very susceptible to the induction of long-term potentiation and kindling effects (Post et al., 1984). Increasing dose levels, as well as frequent repetition of small doses, leads to cocaine sensitization, and eventually to seizures and death (Post et al., 1988). The sensitization is partially context-dependent, which agrees with the context dependent behavior frequently found with *d*-amphetamine and other DA agonists (Fischman and Schuster, 1974).

Recognition of this phenomenon has led to a reevaluation of the causes for convulsive episodes, and also to further study of the periodicity in these events. Post et al. (1981) showed that periodic effects of kindling could be produced by electrical stimulation of the amygdala in the rat, and Wake and Wada (1975) demonstrated essentially the same phenomenon in kindling of the frontal cortex in the cat. Perhaps even more important in the context of mania and depression, was the fact that these researchers were able to show a cyclic variation in sensitivity to the electrical stimulation, with a periodicity of several days. Thus, this method seemed to reproduce the basic cyclic nature of the bipolar mood disorders. However, the cycling time is short, compared with the human manic/depressive cycle, and the behaviors produced during the sensitization to amygdaloid stimulation are not exactly like those that best characterize mania.

In a second paradigm, Post (1977) showed that repeated low-dose treatment with cocaine in monkeys resulted in a phase of behavioral activation followed by an "inhibitory syndrome" consisting of reduced motor activity, catalepsy, and abnormally strong staring and visual preoccupation with minute details in the environment. This resembles the results obtained by Lyon and Nielsen (1979), and by Nielsen et al. (1983) on the effects of continuous amphetamine treatment of vervet monkeys with implanted amphetamine capsules (*see* Chapter by Lyon on Models of Schizophrenia in this book for details of method). It is not surprising perhaps that cocaine and amphetamine treatments can provide similar results, but the fact that a very low-dose treatment can kindle, over time, a sort of high-dose effect is intriguing. It implies that within the amygdala, frontal cortex, and hippocampus, and perhaps elsewhere, there is some form of long-term effect that is perhaps characterized by the hippocampal effect known as long-term potentiation (LTP) (Brown et al., 1988).

As with LTP, the kindling effect fades over time if it is not elicited, and in this way, is dissimilar to long-term *memory*, which remains intact over such periods. However, it is reasonable to suppose that lengthy periods of overstimulation of Glu-modulated systems, such as those involving the NMDA receptors in the hippocampus and elsewhere, can result in the dysfunction and eventual destruction of the cells, with a relatively permanent hypersensitivity as a result.

3. Summary

The present review of models for mania suggests that very important behavioral effects resembling mania can be found after manipulation of DA and Glu related systems. It appears that the DA systems obtain an overbalance as mania increases, and that the resultant effects can be modulated by other neurotransmitter systems, such as those related to ACh, 5-HT, and some neuropeptides. The origin of the DA overbalance may lie in the LTP effects of excessive stimulation of the NMDA receptors in the hippocampus and/or amygdala, with an end result of relatively permanent hypersensitivity (kindling effect). The cycling of mania and depression appears, from this view, to be resulting from the natural corrective tendencies of the brain in response to an overbalance of a particular neurotransmitter influence. However, this mechanism can have only a temporary effect, and on repeated insult, severe and long-lasting depression becomes increasingly probable.

The models for mania that seem most promising at this time are therefore those related to DA overstimulation of mesolimbic structures, as demonstrated by the dialysis method of Kuczenski and Segal, and by the intraventricular infusion method of Petty and Sherman. For studying the modulating effects of 5-HT on this basic DA model, PCPA treatment provides an excellent model. The role of opioid receptors and the relationship of endogenous opioids to mania can best be studied by using morphine stimulation compared with specific beta-endorphins.

Lithium antagonism of an agent's effect should not be taken as the ultimate measure of a particular treatment's relationship to mania. On the other hand, both lithium and neuroleptic antagonism of model effects should be considered with regard to NMDA and DA receptor mechanisms.

Acknowledgments

This work was prepared while the author was on leave of absence from Copenhagen University, Denmark, and held an appointment as Visiting Professor and Research Associate at the Social Science Research Institute, University of Southern California. Thanks are owing to SSRI director Ward Edwards and all his colleagues and staff for providing a stimulating milieu in which to work. Special thanks to William O. McClure for both inspiration and valuable discussions.

References

American Psychiatric Association (1980) *DSM-III: Diagnostic and Statistical Manual of Mental Disorders,* 3rd Ed. American Psychiatric Association, Washington, DC.

Andreasen N. C. and Powers P. S. (1975) Creativity and psychosis. An examination of conceptual style. *Arch. Gen. Psychiat.* **32,** 70–73.

Antelman S. M. and Chiodo L. A. (1981) Dopamine autoreceptor subsensitivity: A mechanism common to the treatment of depression and the induction of amphetamine psychosis? *Biol. Psychiat.* **16,** 717–727.

Arnt J. and Scheel-Kruger J. (1979) GABA in the ventral tegmental area: Differential regional effects on locomotion, aggression and food intake after microinjection of GABA agonists and antagonists. *Life Sci.* **25,** 1351–1360.

Ayhan I. and Randrup A. (1973) Behavioural and pharmacological studies on morphine-induced excitation of rats. Possible relation to brain catecholamines. *Psychopharmacol. (Berl.)* **29,** 317–328.

Balsara J. J., Jadhav J. H., Muley M. P., and Chandorkar A. G. (1979) Effect of drugs influencing central serotonergic mechanisms on methamphetamine-induced stereotyped behavior in the rat. *Psychopharmacol.* **64,** 303–307.

Barchas J. D., Evans C., Elliott G. R., and Berger P. A. (1985) Peptide neuroregulators: The opioid system as a model. *Yale J. Biol. Med.* **58,** 579–596.

Beckmann H. and Heinemann H. (1976) *d*-Amphetamin beim manischen Syndrom. *Arzneim-Forschungen (Drug Res.)* **26(6),** 1185,1186.

Berger P. A. and Barchas J. D. (1982) Studies of beta-endorphin in psychiatric patients, in *Opioids in Mental Illness: Theories, Clinical Observations, and Treatment Possibilities,* Vol. 398 (Verebey K, ed.) *Ann. NY Acad. Sci.,* NY, pp. 448–459.

Bleuler E. (1950) *Dementia Praecox.* International University Press, NY.

Bogerts B., Meertz E., and Schonfeldt-Bausch R. (1985) Basal ganglia and limbic system pathology in schizophrenia. *Arch. Gen. Psychiat.* **42,** 784–791.

Boissier J. -R. (1975) L'utilisation des sels de lithium en therapeutique. *Annales pharmaceutiques francaises* **33(8–9),** 447–458.

Brodie H. K. H., Murphy D. L, Goodwin F. K., and Bunney W. E., Jr. (1971) Catecholamines and mania: The effect of alpha-methyl-para-tyrosine on manic behavior and catecholamine metabolism. *Clin. Pharmacol. Therapy* **12,** 218.

Brown T. H., Chapman P. F., Kairiss E. W., and Keenan C. L. (1988) Long-term synaptic potentiation. *Science* **242,** 724–728.

Brown W. A. and Mueller B. (1979) Alleviation of manic symptoms with catecholamine agonists. *Am. J. Psychiat.* **136(2),** 230, 231.

Carlsson A. (1988) The current status of the dopamine hypothesis of schizophrenia. *Neuropsychopharmacol.* **1(3),** 179–186.

Chiodo L. A. and Antelman S. M. (1980) Electroconvulsive shock: Progressive dopamine autoreceptor subsensitivity independent of repeated treatment. *Science* **210,** 799–801.

Cooper J. R., Bloom F. E., and Roth R. H. (1986) *The Biochemical Basis of Neuropharmacology,* 5th Ed., Oxford University Press, NY, p. 400.

Coscina D. V., Seggie J., Godse D. D., and Stancer H. C. (1973) Induction of rage in rats by central injection of 6-hydroxydopamine. *Pharmacol. Biochem. Behav.* **1,** 1–6.

Coscina D. V., Goodman J., Godse D. D., and Stancer H. C. (1975) Taming effects of handling on 6-hydroxydopamine induced rage. *Pharmacol. Biochem. Behav.* **3,** 525–528.

Costall B., Domeney A. M., and Naylor R. J. (1984a) Locomotor hyperactivity caused by dopamine infusion into the nucleus accumbens of rat brain: specificity of action. *Psychopharmacol.* **82,** 174–180.

Costall B., Domeney A. M., and Naylor R. J. (1984b) Long-term consequences of antagonism by neuroleptics of behavioural events occurring during mesolimbic dopamine infusion. *Neuropharmacol.* **23,** 287–294.

Davies C., Sanger D. J., Steinberg H., Tomkiewicz M., and U'Prichard D. C. (1974) Lithium and alpha-methyl-*p*-tyrosine prevent 'manic' activity in rodents. *Psychopharmacol. (Berl.)* **36,** 263–274.

Davis J. M., Janowsky D., Tamminga C., and Smith R. C. (1978) Cholinergic mechanisms in schizophrenia, mania and depression, in *Cholinergic Mechanisms and Psychopharmacology* (Jenden D. J., ed.), Plenum, NY, pp. 805–815.

Davis K. L., Berger P. A., Hollister L. E., Domaral J. R., and Barchas J. D. (1978) Cholinergic dysfunction in mania and movement disorders, in *Cholinergic Mechanisms and Psychopharmacology* (Jenden D. J., ed.) Plenum, NY, pp. 755–779.

DeVietti T. L., Pellis S. M., Pellis V. C., and Teitelbaum P. (1985) Previous experience disrupts atropine-induced stereotyped "trapping" in rats. *Behav. Neurosci.* **99,** 1128–1141.

de Wied D. (1979) Schizophrenia as an inborn error in the degradation of beta-endorphin—a hypothesis. *Trends in Neurosci.* **2,** 79–82.

Diaz J., Ellison G., and Matsuoka D. (1974) Opposed behavioral syndromes in rats with partial and more complete central serotonergic lesions made with 5,6-dihydroxytryptamine. *Psychopharmacol. (Berl.)* **37,** 67–79.

Dilsaver S. C. and Greden J. F. (1984) Antidepressant withdrawal-induced activation (hypomania and mania): Mechanism and theoretical significance. *Brain Res. Rev.* **7,** 29–48.

Eison M. S., Eison A. S., and Iversen S. D. (1983) Two routes of continuous amphetamine administration induce different behavioral and neurochemical effects in the rat. *Neurosci. Lett.* **39,** 313–319.

Ellinwood E. H., Jr. and Kilbey M. M. (1977) Chronic stimulant intoxication models of psychosis, in *Animal Models in Psychiatry and Neurology,* (Hanin I., and Usdin E., eds.) Pergamon, Oxford, UK, pp. 61–74.

Ellison G. and Bresler O. (1974) Tests of emotional behavior in rats following depletion of norepinephrine, of serotonin, or both. *Psychopharmacol. (Berl.)* **34,** 275–288.

Evenden J. L. and Robbins T. W. (1983) Increased response switching, perseveration and perseveration switching following *d*-amphetamine in the rat. *Psychopharmacol.* **80,** 67–73.

Evetts K. D., Uretsky N. J., Iversen L. L, and Iversen S. D. (1970) Effects of 6-hydroxydopamine on CNS catecholamines, spontaneous motor activity and amphetamine-induced hyperactivity in rats. *Nature* **225,** 961,962.

Fessler R. G., Sturgeon R. D., London S. F., and Meltzer H. Y. (1982) Effects of lithium on behaviour induced by phencyclidine and amphetamine in rats. *Psychopharmacol.* **78,** 373–376.

Fibiger H. C. and Campbell B. A. (1971) The effect of parachlorophenylalanine on spontaneous locomotor activity in the rat. *Neuropharmacol.* **10,** 25–32.

Fischman M. W. and Schuster C. R. (1974) Tolerance development to chronic methamphetamine intoxication in the rhesus monkey. *Pharmacol. Biochem. Behav.* **2,** 503–508.

Fonnum F. (1984) Glutamate: A neurotransmitter in mammalian brain. *J. Neurochem.* **42,** 1–11.

Gallistel C. R. (1986) The role of the dopaminergic projections in MFB self-stimulation. *Behav. Brain Res.* **20,** 313–321.

Gerner R. H., Post R. M., and Bunney W. E., Jr. (1976) A dopaminergic mechanism of mania. *Am. J. Psychiat.* **133,** 1177–1179.

Gerson SC, and Baldessarini R. J. (1980) Minireview: Motor effects of serotonin in the central nervous system. *Life Sci.* **27,** 1435–1451.

Gilles F. H., Leviton A., and Dooling E. C. (1983) *The Developing Human Brain, Growth and Epidemiological Neuropathology.* Wright PSG, Boston, MA.

Goddard G. V., McIntyre D. C., and Leech C. K (1969) A permanent change in brain function resulting from daily electrical stimulation. *Exptl. Neurol.* **25,** 295–301.

Grahame-Smith D. G. (1971) Studies in vivo on the relationships between brain tryptophan, brain 5-HT synthesis, and hyperactivity in rats treated with a MAO inhibitor and *l*-tryptophan. *J. Neurochem.* **18,** 1053–1066.

Green A. R. and Grahame-Smith D. G. (1974) The role of brain dopamine in the hyperactivity syndrome produced by increased 5-hydroxytryptamine synthesis in rats. *Neuropharmacol.* **13,** 949–959.

Harrison J. M. and Lyon M. (1957) The role of the septal nuclei and the components of the fornix in the behavior of the rat. *J. Comp. Neurol.* **108,** 121–137.

Heikkila R. E., Hess A., and Duvoisin R. C. (1984) Dopaminergic neurotoxicity of 1-methyl-4-phenyl-1,2,5,6-tetrahydropyridine in mice. *Science* **224,** 1451–1453.

Herberg L. J. and Franklin K. B. J. (1976) The 'stimulant' action of tryptophan-monoamine oxidase inhibitor combinations: Suppression of self-stimulation. *Neuropharmacol.* **15,** 349–351.

Hill R. T. (1970) Facilitation of conditioned reinforcement as a mechanism for psychomotor stimulation, in *Amphetamines and Related Compounds,* (Costa E. and Garattini S., eds.) Raven, NY, pp. 791–795.

Hiramatsu K., Kameyama T., Saitoh O., Niwa S., Rymar K., and Itoh K. (1984) Correlations of event-related potentials with schizophrenic deficits in information processing and hemispheric dysfunction. *Biol. Psychol.* **19,** 281–294.

Huberman H. S., Eison M. S., Bryan K., and Ellison G. (1977) A slow-release pellet for chronic amphetamine administration. *Eur. J. Pharmacol.* **45,** 237–240.

Iversen S. D. (1977) Striatal function and stereotyped behaviour, in *The Psychobiology of the Striatum* (Cools A. R., Lohman A. H. M., and van der Bercken J. H. L., eds.), North-Holland, Amsterdam, pp. 99–118.

Iversen S. D. (1986) Animal models of schizophrenia, in *The Psychopharmacology and Treatment of Schizophrenia* (Bradley P. B. and Hirsch S. R., eds.), Oxford University Press, Oxford, UK.

Jackson D. M., Bailey R. C., Christie M. J., Crisp E. A., and Skerrit J. H. (1981) Long-term *d*-amphetamine in rats: Lack of change in postsynaptic dopamine receptor sensitivity. *Psychopharmacol.* **73,** 276–280.

Jacobs B. L. (1976) An animal behaviour model for studying central serotonergic synapses. *Life Sci.* **19,** 777–786.

Jacobs B. L., Trulson M. E., and Stern W. C. (1976) An animal behavior model for studying the action of LSD and related hallucinogens. *Science* **194,** 741–743.

Jakob H. and Beckmann H. (1986) Prenatal developmental disturbances in the limbic allocortex in schizophrenia. *J. Neurotrans.* **65,** 303–326.

Janowsky D. S., El-Yousef M. K., Davis J. M., Hubbard B., and Sekerke H. J. (1972a) A cholinergic adrenergic hypothesis of mania and depression. *Lancet* **2,** 632–635.

Janowsky D. S., El-Yousef M. K., Davis J. M., and Sekerke H. J. (1972) Cholinergic antagonism of methylphenidate-induced stereotyped behavior. *Psychopharmacologia (Berl.)* **27,** 295–303.

Janowsky D. S., El-Yousef M. K., Davis J. M., and Sekerke H. J. (1973) Parasympathetic suppression of manic symptoms by physostigmine. *Arch. Gen. Psychiat.* **28,** 542–547.

Jones G. H., Hernandez T. D., and Robbins T. W. (1987) Isolation-rearing impairs the acquisition of schedule-induced polydipsia. *Abst. Soc. Neurosci.* **13,** 405.

Jones G. H., Marsden C. A., and Robbins T. W. (1989) Hypersensitivity to reward-related stimuli and to intra-accumbens *d*-amphetamine following social deprivation in rats. *Soc. Neurosci. Abst.* Vol. **15 (1),** p. 412.

Joyce E. M. and Iversen S. D. (1979) The effect of morphine applied locally to mesencephalic dopamine cell bodies on spontaneous motor activity in the rat. *Neurosci. Lett.* **14,** 207–212.

Kantak K. M. (1988) Magnesium deficiency alters aggressive behavior and catecholamine function. *Behav. Neurosci.* **102,** 303–311.

Karson C. N., Dykman R. A., and Paige S. R. (1990) Blink rates in schizophrenia. *Schiz. Bull.* **16(2),** 345–354.

Katz R. J. (1982) Morphine- and endorphin-induced behavioral activation in the mouse: Implications for mania and some recent pharmacogenetic studies, in *Opioids in Mental Illness,* vol. 398 (Verebey K, ed.), *Ann. NY Acad. Sci.,* NY, pp. 291–300.

Kelly P. H. and Iversen L. L. (1975) LSD as an agonist at mesolimbic dopamine receptors. *Psychopharmacologia (Berl.)* **45,** 221–224.

Kelly P. H., Seviour P. W., and Iversen S. D. (1975) Amphetamine and apomorphine response in the rat following 6-OHDA lesions of the nucleus accumbens septi and corpus striatum. *Brain Res.* **94,** 507–522.

Kovelman J. A. and Scheibel A. B. (1984) A neurohistological correlate of schizophrenia. *Biol. Psychiat.* **19,** 1601–1621.

Kuczenski R. and Segal D. (1989) Concomitant characterization of behavioral and striatal neurotransmitter response to amphetamine using *in vivo* microdialysis. *J. Neurosci.* **9(6),** 2051–2065.

Leith N. J. and Barrett R. J. (1980) Effects of chronic amphetamine or reserpine on self-stimulation responding: Animal model of depression? *Psychopharmacol.* **72,** 9–15.

Lorens S. A., Guldberg H. C., Hole K., Kohler C., and Srebro B. (1976) Activity, avoidance learning and regional 5-hydroxytryptamine following intra-brain stem 5,7-dihydroxytryptamine and electrolytic midbrain raphe lesions in the rat. *Brain Res.* **108,** 97–113.

Lyon M. (1990) Animal models of mania and schizophrenia, in *Behavioural Models in Psychopharmacology* (Willner P., ed.), Cambridge University Press, Cambridge, UK.

Lyon M. and Barr C. E. (in press) Possible interactions of obstetrical complications and abnormal fetal brain development in schizophrenia, in *Fetal Neural Development and Adult Schizophrenia* (Mednick S. A., Cannon T. D., Barr C. E., and Lyon M., eds.), Cambridge University Press, NY.

Lyon M. and Nielsen E. B. (1979) Psychosis and drug induced stereotypies, in *Psychopathology in Animals: Research and Clinical Implications* (Keehn J. D., ed.), Academic, NY, pp. 103–142.

Lyon M. and Randrup A. (1972) The dose-response effect of amphetamine upon avoidance behavior in the rat seen as a function of increasing stereotypy. *Psvchopharmacologia (Berl.)* **23,** 324–347.

Lyon M. and Robbins T. W. (1975) The action of central nervous system stimulant drugs: A general theory concerning amphetamine effects, in *Current Developments in Psychopharmacology,* Vol 2. (Essman W. and Valzelli L., eds.), Spectrum, NY, pp. 79–163.

Lyon N. and Gerlach J. (1988) Perseverative structuring of responses by schizophrenic and affective disorder patients. *J. Psychiat. Res.* **23,** 261–277.

Lyon N., Mejsholm B., and Lyon M. (1986) Stereotyped responding in schizophrenic outpatients: Cross-cultural confirmation of perseverative switching on a two-choice task. *J. Psychiat. Res.* **20,** 137–150.

Mailman R. B., Lewis M. H., and Kilts C. D. (1981) Animal models related to developmental disorders: Theoretical and pharmacological analyses. *Appl. Res. Ment. Retard.* **2,** 1–12.

Marsden C. A. and Curzon G. (1976) Studies on the behavioural effects of tryptophan and *p*-chlorophenylalanine. *Neuropharmacology* **15,** 165–171.

Martin-Iverson M. T., Stahl S. M., and Iversen S. D. (1988) Chronic administration of a selective dopamine D-2 agonist: factors determining behavioral tolerance and sensitization. *Psychopharmacol.* **95,** 534–539.

McGinity J. W. and Mehta C. S. (1978) Preparation and evaluation of a sustained morphine delivery system in rats. *Pharmacol. Biochem. Behav.* **9,** 705.

Mehrabian A. (1986) Arousal-reducing effects of chronic stimulant use. *Motivation and Emotion* **10(1),** 1–10.

Meltzer H. L., Taylor R. M., Platman S. R., and Fieve R. R. (1969) Rubidium: A potential modifier of affect and behaviour. *Nature* **223,** 321–322.

Miller F. P., Cox R. H., Snodgrass W. R., and Maikel R. P. (1970) Comparative effects of *p*-chlorophenylalanine, *p*-chloroamphetamine, and *p*-chloro-*N* methylamphetamine on rat brain norepinephrine, serotonin, and 5-hydroxyindole-3-acetic acid. *Biochem. Pharmacol.* **19,** 435–442.

Murphy D. L. (1977) Animal models for mania, in *Animal Models in Psychiatry and Neurology* (Hanin I. and Usdin E., eds.), Pergamon, NY, pp. 211–222.

Nakamura K and Thoenen H. (1972) Increased iritability—A permanent behavior change induced in the rat by intraventricular administration of 6-hydroxydopamine. *Psychopharmacologia (Berl.)* **24,** 359–372.

Nielsen E. B. (1981) Rapid decline of stereotyped behavior in rats during constant one week administration of amphetamine via implanted ALZET® osmotic minipumps. *Pharmacol. Biochem. Behav.* **15,** 161–165.

Nielsen E. B., Lyon M., and Ellison G. (1983) Apparent hallucinations in monkeys during around-the-clock amphetamine for seven to fourteen days. Possible relevance to amphetamine psychosis. *J. Nerv. Ment. Dis.* **171,** 222–233.

Norton S., Mullenix P., and Culver B. (1976) Comparison of the structure of hyperactive behavior in rats after brain damage from X-irradiation, carbon monoxide and pallidal lesions. *Brain Res.* **116,** 49–67.

Pavlinac D., Langer R., Lenhard L., and Deftos L. (1979) Magnesium in affective disorders. *Biol. Psychiat.* **14(4),** 657–661.

Petty F. and Sherman A. D. (1981) A pharmacologically pertinent animal model of mania. *J. Affect. Dis.* **3,** 381–387.

Phillips A. G., Carter D. A., and Fibiger H. C. (1976) Dopaminergic substrates of intracranial self-stimulation in the caudate-putamen. *Brain Res.* **104,** 221–232.

Poitou P., Boulu R., and Bohuon C. (1975) Effect of lithium and other drugs on the amphetamine chlordiazepoxide hyperactivity in mice. *Experientia* **15(1),** 99–101.

Post R. M. (1977) Approaches to rapidly cycling manic-depressive illness, in *Animal Models in Psychiatry and Neurology* (Hanin I. and Usdin E., eds.), Pergamon, NY, pp. 201–210.

Post R. M., Squillace K. M., Sass W., and Pert A. (1981) The effect of amygdaloid kindling on spontaneous and cocaine-induced motor activity and lidocaine seizures. *Psychopharmacology* **72,** 189–196.

Post R. M., Rubinow D., and Ballenger J. (1984) Conditioning, sensitization, and kindling: Implications for the course of affective illness, in *Neurobiology of Mood Disorders* (Post R. M. and Ballenger J., eds.), Williams and Wilkins, Baltimore, pp. 432–466.

Post R. M., Weiss S. R. B., and Pert A. (1988) Cocaine-induced behavioral sensitization and kindling: Implications for the emergence of psychopathology and seizures, in *The Mesocorticolimbic Dopamine System* (Kalivas P. W. and Nemeroff C. B., eds.), *Ann. NY Acad. Sci.*, NY, pp. 292–308.

Randrup A. and Munkvad I. (1968): Behavioural stereotypies induced by pharmacological agents. *Pharmakopsychiatri und Neuropsychopharmakologi* **1,** 18–26.

Randrup A., Munkvad I., and Fog R. (1981) Mental and behavioural stereotypies elicited by stimulant drugs. Relation to the dopamine hypothesis of schizophrenia, mania, and depression, in *Recent Advances in Neuropsychopharmacology* (Angrist B., Burrows G. D., Lader M., Lingjaerde O., Sedvall G., and Wheatley D., eds.), Pergamon, Oxford, UK, pp. 63–74.

Rastogi R. B., Merali Z., and Singhal R. L. (1977) Cadmium alters behaviour and biosynthetic capacity for catecholamine and serotonin in neonatal rat brain. *J. Neurochem.* **28,** 789–794.

Robbins T. W. (1975) The potentiation of conditioned reinforcement by psychomotor stimulant drugs: A test of Hill's hypothesis. *Psychopharmacologia (Berl.)* **45,** 103–112.

Robbins T. W. (1976) Relationship between reward-enhancing and stereotypical effects of psychomotor stimulant drugs. *Nature (Lond.)* **254,** 57–59.

Robbins T. W. (1978) The acquisition of responding with conditioned reinforcement: Effects of pipradrol, methylphenidate, *d*-amphetamine, and nomifensine. *Psychopharmacol.* **58,** 79–87.

Robbins T. W. and Sahakian B. J. (1980) Animal models of mania, in *Mania: An Evolving Concept* (Belmaker R. and van Praag H., eds.), Spectrum, NY, pp. 143–216.

Robbins T. W., Watson B. A., Gaskin M., and Ennis C. (1983) Contrasting interactions of pipradrol, *d*-amphetamine, cocaine, cocaine analogues, apomorphine and other drugs with conditioned reinforcement. *Psychopharmacol.* **80,** 113–119.

Rubin E. H. and Zorumski C. F. (1985) Limbic seizures, kindling and psychosis: A link between neurobiology and clinical psychiatry. *Comp. Ther.* **1,** 54–58.

Scheel-Kruger J. (1986) Dopamine-GABA interactions: Evidence that GABA transmits, modulates and mediates dopaminergic functions in the basal ganglia and the limbic system. *Acta Neurol. Scand. (Suppl.)* **107,** 1–54.

Scheel-Kruger J., Arnt J., Magelund G., Olianas M., Przewlocka B., and Christensen A. V. (1980) Behavioural functions of GABA in basal ganglia and limbic system. *Brain Res. Bull.* **5(2),** 261–267.

Scheel-Kruger J., Magelund G., and Olianas M. C. (1981) Role of GABA in the striatal output system: Globus pallidus, nucleus entopeduncularis, substantia nigra and nucleus subthalamicus, in *GABA and the Basal Ganglia, Advances in Biochemistry and Psychopharmacology,* Vol. 30 (Di Chiara G. and Gessa G. L., eds.), pp. 165–186.

Schiørring E. (1971) Amphetamine induced selective stimulation of certain behaviour items with concurrent inhibition of others in an open-field test with rats. *Behaviour* **39,** 1–17.

Schwartz B. (1986) Allocation of complex sequential operants on multiple and concurrent schedules of reinforcement. *J. Exp. Anal. Behav.* **45,** 283–295.

Schwartz J. M., Ksir C., Koob G. F., and Bloom F. E. (1982) Changes in locomotor response to beta-endorphin microinfusion during and after opiate abstinence syndrome—A proposal for a model of the onset of mania. *Psychiat. Res.* **7,** 153–161.

Silbergeld E. K. and Goldberg A. M. (1974) Lead-induced behavioral dysfunction: An animal model of hyperactivity. *Exptl. Neurol.* **42,** 146–157.

Stevens J. R. (1975) GABA blockade, dopamine and schizophrenia: Experimental activation of the mesolimbic system. *Int. J. Neurol.* **10,** 115–127.

Stevens J. R. (1979) Schizophrenia and dopamine regulation in the mesolimbic system. *Trends Neurosci.* **2,** 102–105.

Sudilovsky A. (1975) Effects of disulfiram on the amphetamine-induced behavioral syndrome in the cat as a model of psychosis. *National Institute on Drug Abuse Research, Monograph Series 3,* pp. 109–135.

Taylor J. R. and Robbins T. W. (1984) Enhanced behavioral control by conditioned reinforcers following microinjections of *d*-amphetamine into the nucleus accumbens. *Psychopharmacology* **84,** 405–412.

Thompson D. M. (1973) Repeated acquisition of response sequences: Effects of *d*-amphetamine and chlorpromazine. *Pharmacol. Biochem. Behav.* **2,** 741–746.

U'Prichard D. C. and Steinberg H. (1972) Selective effects of lithium on two forms of spontaneous activity. *Br. J. Pharmacol.* **44,** 349,350.

Van de Kar L. D. and Lorens S. A. (1979) Differential serotonergic innervation of individual hypothalamic nuclei and other forebrain regions by the dorsal and medial midbrain raphe nuclei. *Brain Res.* **162,** 45–54.

Van Ree J. M., Verhoeven W. M. A., de Wied D., and van Praag H. M. (1982) The use of synthetic peptides y-type endorphins in mentally ill patients, in Opioids in *Mental Illness: Theories, Clinical Observations, and Treatment Possibilities,* Vol. 398 (Verebey K., ed.), *Ann. NY Acad. Sci.,* NY, pp. 487–495.

Villablanca J. R., Harris C. M., Burgess J. W., and de Andres I. (1984) Reassessing morphine effects in cats: I. Specific behavioral responses in intact and unilaterally brain-lesioned arimals. *Pharmacol. Biochem. Behav.* **21,** 913–921.

Vogel R. and Annau Z. (1973) An operant discrimination task allowing variability of reinforced response patterning. *J. Exp. Anal. Behav.* **20,** 1–6.

Wake A. and Wada J. (1975) Frontal cortical kindling in cats. *Can. J. Neurol. Sci.* **2,** 493–496.

Animal Models in Tardive Dyskinesia

Helen Rosengarten, Jack W. Schweitzer, and Arnold J. Friedhoff

1. History and Symptomatology of Tardive Dyskinesia

1.1. Introduction

The introduction of neuroleptics in 1952 to clinical practice was followed by numerous reports in the literature describing a syndrome of abnormal involuntary movements that appeared with increasing frequency during treatment or during withdrawal from neuroleptics. This syndrome was first described in the late 1950s (Hall et al., 1956; Kullenkampf and Tarnow, 1956; Schoenecker, 1957; Sigwald et al., 1959) and these reversible and irreversible drug-related dyskinesias were subsequently named tardive dyskinesia (Uhrband and Faurbye, 1960). The role of neuroleptics as the cause of tardive dyskinesia remained controversial for many years; subsequently, a number of investigators concluded that chronic neuroleptic treatment may precipitate TD only in those patients who are already predisposed to the development of this disorder (Smith and Baldessarini, 1980; Itil et al., 1981; Kane and Smith, 1982; Marsden, 1985; Waddington et al., 1988; Toenniessen et al., 1985; Hansen et al., 1988).

By definition, tardive dyskinesia (TD) is a neuroleptic-induced condition and its prevalence has been increasing during the past 25 years. Although the pathophysiology of this syn-

From: *Neuromethods, Vol. 18: Animal Models in Psychiatry I*
Eds: A. Boulton, G. Baker, and M. Martin-Iverson ©1991 The Humana Press Inc.

drome is still unclear, the traditional hypothesis is that increased dopaminergic activity in the basal ganglia is the cause of TD (Baldessarini and Tarsy, 1980). The variability of forms of dyskinetic movements is well known and the syndrome of abnormal involuntary movement disorder most frequently involves facial, buccal, lingual, and masticatory muscles, and often extends to the upper and lower extremities, fingers, toes, trunk, and neck (Gardos and Cole, 1980). Movements of the mouth, tongue, and face include chewing, licking, sucking, tongue protrusion, and side-to-side tongue movements. These movements can be rhythmic or irregular, repetitive, choreiform, myoclonic, athetotic, dystonic, rolling, or tic-like and they can be suppressed by voluntary activity or activated by stress. The most common form, the bucco-linguo-masticatory triad is seen predominantly among patients receiving chronic neuroleptic treatment and often appears for the first time after cessation of neuroleptic therapy (Uhrband and Faurbye, 1960; Klawans, 1973; Marsden et al., 1973; Jus et al.,1979; Baldessarini and Tarsy, 1980; Carpenter et al., 1980; Gardos and Cole, 1980; Smith and Baldessarini, 1980; Seeman, 1981; Wegner and Kane, 1982). The type of neuroleptic used, presence or absence of Parkinsonian side effects, and presence of organic brain changes have been considered as possible contributory factors in the development of this disorder (Hansen, 1988).

1.2. Prevalent Views on the Pathogenesis of Tardive Dyskinesia

The role of neuroleptics as the cause of tardive dyskinesia remained controversial for many years and several investigators concluded that chronic neuroleptic treatment may precipitate TD only in those patients who are already predisposed to the development of this disorder (Marsden, 1985; Waddington et al., 1988).

Neuroleptics, potent antagonists of the dopamine D2 receptors, produce supersensitivity in rats within a relatively short time after initiation of treatment, and there have been numerous attempts to associate alterations of the D2 receptor number with this syndrome (Gianutsos et al., 1974; Baldessarini and Tarsy,

1976; Burt et al., 1977; Friedhoff et al., 1977; Muller and Seeman, 1978; Tarsy and Baldessarini, 1977) and, as well, other mechanisms that may underlie the pathology of this iatrogenic disorder (Casey and Denney, 1977; Waddington et al., 1983; Rosengarten et al., 1983,1986,1989; Gunne et al., 1980,1983,1984; Scheel-Kruger, 1985).

This controversial hypothesis involving the D2 receptor blockade and the resulting D2 receptor biochemical and behavioral supersensitivity presents a number of problems, the chief one being that in rats the supersensitivity of D2 receptors occurs in one degree or another in all rats treated with neuroleptics and appears after a relatively short period of treatment. Conversely, TD occurs only in some neuroleptic users and generally after long periods of treatment. The majority of neuroleptic drugs used in the treatment of schizophrenia are potent antagonists at D2 receptor sites (Creese et al., 1976), but less so at D1 receptor sites (Hyttel et al., 1978). As a result of this blockade, neuroleptics produce supersensitivity of the D2 receptors within a short period following initiation of treatment and the antipsychotic actions appear to be mediated by their selective action on D2 dopamine receptors (Seeman et al., 1980; Seeman, 1987; Farde et al., 1987; Sedvall et al., 1988). In this regard, it is noteworthy that functional D1 receptor supersensitivity occurs in rats after prolonged treatment that corresponds better, temporally, to the relationship between initiation of treatment and onset of TD (Rosengarten et al., 1986).

The introduction of D1 and D2 selective agonists (Tsuruta, 1981; Sibley et al., 1982; Kaiser, 1983) and antagonists (Hall et al., 1985; Hyttel, 1983; Iorio et al., 1983; Itoh et al., 1984) permitted the determination of behavioral interactions between these two dopamine receptor systems, a development that may be relevant to the mechanism of TD. It has also been demonstrated that D1 and D2 dopamine receptors in the brain interact in a complex way to elicit behavioral functions (Arnt and Scheel-Kruger, 1981; Christensen, 1980; Stoof and Kebabian, 1984; Gershanik et al., 1983; Rosengarten et al., 1983,1986,1989; Saller and Salama 1985; Salama and Saller, 1986; Braun and Chase, 1986; Fage and Scatton, 1986; Mashurano and Waddington, 1986; White and Wang, 1986;

Arnt et al., 1987; White, 1987; Meller et al., 1988; Murray and Waddington, 1989).

In view of these complexities, there is a pressing need to develop animal models of this syndrome in order to be able to understand mechanisms underlying TD pathology and to find effective treatments. Animal models of persisting tardive dyskinesia have been developed in rats and several species of monkeys.

2. Animal Models of Tardive Dyskinesia

2.1. Monkey Model of Tardive Dyskinesia

Several attempts to produce a primate model of TD have been successful. Long-term treatment of *Cebus apella* monkeys with haloperidol or fluphenazine induced symptoms of TD that persisted for a long period of time after withdrawal (Gunne and Barany, 1976; Bedard et al., 1972; Barany et al., 1979; Kovacic and Domino, 1982; Domino, 1985).

Tardive dyskinesia in monkeys represents a valid model which in many ways can serve as a replication of this syndrome in humans (Casey, 1984). Although there have been no entirely satisfactory models of abnormal involuntary dyskinetic movements of orofacial area following neuroleptic treatment, the monkey model appears to be closely related to the clinical symptomatology in humans. The use of animal models often provides an insight into human pathology and into drug-induced side effects that cannot be acquired by other means and are of particular value in the search for effective treatment. In early reports on prolonged neuroleptic treatment, it was demonstrated that dyskinetic movements of the orofacial area occurred following chlorpromazine or haloperidol administration (Bedard et al., 1972; Gunne and Barany, 1976). In 1976 Gunne and Barany reported that prolonged administration of haloperidol to *Cebus apella* and *Macaca fascicularis* monkeys at the dose of 0.5 mg/kg/d and 0.5–8 mg/kg, respectively, resulted in abnormal oral behavior. After three and six months of treatment, two of these monkeys developed bucco-lingual movements, grimacing, and tongue protrusion that were pronounced before each dose of neuroleptic and were reduced in a dose-dependent manner after each daily dose of haloperidol. Behavior similar to human TD was recorded on a TD rating scale. Biperiden, an anticholinergic drug, administered either alone or together with the regular daily dose of haloperidol reduced symptoms of acute dystonia but reinstated signs of TD (Gunne, 1976). Messiha et al. (1980) reported that chronic administration of chlorpromazine in a gradual dose increase from 100–200 mg/d over a four months period or 10–180 mg/d over a one years period resulted in the appearance of dyskinetic movements in the bucco-lingual area. Following chronic treatment monkeys demonstrated abnormal mouth and lip movements and frequent tongue protrusion, whereas short-term administration of chlorpromazine did not induce dyskinetic movements (Bedard et al., 1972; Messiha, 1980).

In a later study (Gunne, 1984), haloperidol decanoate 1–10 mg/kg or fluphenazine decanoate 4–10 mg/kg was administered to 19 *Cebus apella* monkeys at three week intervals. After varying periods of time, six of these monkeys (three treated with haloperidol and three treated with fluphenazine decanoate) developed persistent choreic movements of the trunk, neck, and extremities and one of the monkeys also developed perioral movements. Motor abnormalities persisted for a period of 1–6 years after cessation of neuroleptic treatment. Symptoms of TD manifested in *Cebus apella* monkeys are closely related to the human syndrome showing similar symptomatology, individual vulnerability, similar response to neuroleptic treatment and reversible or irreversible character of dyskinetic movements (Gunne et al., 1979; Casey, 1985).

More recently it was demonstrated in the *Cebus apella* monkey that long-term treatment with neuroleptics resulted not only in dyskinetic movements of the oro-bucco-facial area but also was accompanied by a significant decrease in the activity of nigral glutamic acid decarboxylase (GAD) activity as compared to haloperidol-treated monkeys who did not develop dyskinetic movements (Gunne et al., 1984). The emergence of spontaneous dyskinetic behavior such as vacuous chewing movements that resulted from a three to six year-long period of neuroleptic treatment, correlated well with significant depletion of GAD activity and GABA levels in substantia nigra, medial globus pallidus, and subthalamic nucleus relative to control monkeys that had been treated with neuroleptics for the same length of time but who did not demonstrate symptoms of TD (Gunne and Haggstrom, 1983; Gunne et al., 1984). This deficiency was still present when measured even two months following the discontinuation of neuroleptic treatment and was accompanied by reduced dopamine turnover in striatum and increased dopamine and low HVA levels. These findings are consistent with earlier reports that chronic neuroleptic treatment reduces GABA turnover in the substantia nigra (Mao et al., 1977) while it increases the GABA receptor density in the same area (Gale, 1980). It was demonstrated that dyskinetic monkeys exhibited an increased sensitivity to apomorphine and the indirect dopamine agonist amphetamine during and after prolonged neuroleptic treatment (Casey, 1985).

The "priming" effect of neuroleptics and the ability to reestablish dyskinetic movements in monkeys who were treated in the past with neuroleptics were reported by several investigators. Dyskinetic movements were reliably induced by a single injection of neuroleptics after a prolonged period of drug-free regimen (Weiss et al., 1977; Liebman and Neale, 1980; Porsolt and Jalfre, 1981). This "priming" effect was demonstrated in the *Cebus apella* monkey in which dyskinetic movements were fading upon initial withdrawal from neuroleptics but the symptoms could be easily reestablished upon initiation of neuroleptic therapy.

The *Macaca speciosa* monkey and the *Cebus apella* monkey manifested the closest symptomatology to human TD (Domino, 1985). In *Cebus apella* monkeys, fluphenazine enanthate was administered every other week for a period of one year in a gradually increasing dose 0.1–3.2 mg/kg/d and an attempt was made to add haloperidol 0.5 mg/kg/d. Every three weeks the animals were washed from neuroleptics. With each course of fluphenazine treatment, TD symptoms were more severe and intense following withdrawal, suggesting that the reversible form of TD may turn into irreversible TD. At the early stages of neuroleptic treatment, monkeys exhibited symptoms similar to acute dyskinetic or dystonic reaction, and these are the symptoms that can also be observed in humans following initial neuroleptic treatment. Not all of the monkeys develop these dystonic or dyskinetic symptoms. Similarly to humans, a certain percentage of subjects treated with neuroleptics for a prolonged period of time remained asymptomatic.

In a recent study, an interesting model of TD was demonstrated by Lubin and Gerlach (1988). It has been shown that D1 and D2 agonists have a direct effect on oral behavior in monkeys subjected previously to two years of haloperidol treatment. Three of the five *Cebus apella* monkeys developed oral TD manifested by tongue protrusion and/or chewing behavior. In dyskinetic monkeys administration of SKF 38393 induced or aggravated oral dyskinesia. SKF 38393-induced oral movements were inhibited by the D2 agonist LY 171555. It was also suggested that the oral dyskinesia that develops in monkeys following prolonged neuroleptic treatment is more related to D1 receptor stimulation than to D2 dopamine receptor supersensitivity (Lubin and Gerlach, 1988). This is also consistent with our hypothesis suggesting that repetitive jaw movements (RJM) in rats can be induced by D1 stimulation, enhanced by concomitant D2 receptor blockade and prolonged depot neuroleptic treatment (Rosengarten et al., 1983,1986,1987).

Serotonin–dopamine interactions were also considered as a possible factor in the TD monkey model and studies of these interactions have demonstrated that in the monkey species *Cercopithecus aethiops*, 5HT uptake inhibitors produced oral hyperkinesia that was enhanced by amphetamine and blocked by haloperidol (Kosgaard et al., 1985).

2.2. Rat Models of Tardive Dyskinesia

The first report of neuroleptic-induced perioral movements in rats (Clow et al., 1979) described enhancement of oral behavior resulting from chronic administration of chlorpromazine, trifluoperazine, or thioridazine. Rats treated with trifluoroperazine 2.5–3.5 mg/kg/d for 12 months in the drinking water exhibited increased incidence of spontaneous mouth movements while stereotypy response to apomorphine was inhibited. The authors suggested that the observed behavior has both high and low intensity components. The high intensity component is blocked by neuroleptics whereas the low intensity component, which is not blocked, is responsible for the enhanced chewing behavior.

Frontal cortex ablation in the rat has resulted in the development of prolonged chewing movements (Glassman and Glassman, 1980), which can be enhanced by chronic administration of neuroleptics (Gunne et al., 1982). Rats with frontal cortex ablations developed spontaneous jaw movements 22 weeks after the lesion and haloperidol administration further increased the rate of vacuous chewing movements (VCM). This observation prompted speculations that humans who develop oral behavior following haloperidol administration may have unsuspected frontal lesions. The authors also suggested that these movements observed in rats may have common features with human TD (Gunne et al., 1982).

In subsequent studies (Gunne et al., 1983,1984,1986), it was demonstrated that, as in monkeys, rats treated with haloperidol who manifested vacuous chewing behavior exhibited reduced GAD activity and GABA levels in substantia nigra, globus pallidus, and hypothalamic nucleus when compared to rats tested similarly who did not manifest an increase in frequency of chewing.

In another study haloperidol decanoate was administered to rats at a dose of 28.5 mg/kg every three weeks for a period of 15 weeks. After 8–10 weeks this treatment resulted in an increase of the rate of VCM, which remained elevated for a period of seven weeks after termination of drug administration (Clow et al., 1979).

Treatment of rats with fluphenazine decanoate for a six or nine-month period resulted in the manifestation of vacuous chewing behavior on apomorphine challenge whereas other dopamine mediated behaviors, including stereotypy response to apomorphine, were antagonized (Waddington and Gamble, 1980; Waddington et al., 1983). The observed chewing movements were purposeless and distinct from apomorphine-induced gnawing. Vacuous chewing behavior was evident despite con-

Table 1
Effect of Chronic Eight Month Fluphenazine Decanoate Treatment on the Rate of Spontaneous RJM and Stereotypy Response in Male Sprague Dawley Rats During Washout from Neuroleptic

Treatment	N	Stereotypy % control	p	RJM % control	p
Fluphenazine decanoate	8	38 ± 5.9	<0.001	304 ± 55	0.005
Vehicle	8	100 ± 12.6		100 ± 23	

Enhanced spontaneous RJM behavior was observed when the D2 receptor was blocked while the D1 receptor was stimulated by endogenous dopamine. Stereotypy was measured in response to sc administration of 0.5 mg of apomorphine. Student's *t*-test was used for statistical analysis of comparisons to vehicle-treated rats.

tinuous antagonism of apomorphine-induced stereotypy, and it was concluded that whereas some dopamine-mediated behaviors may be antagonized during prolonged neuroleptic treatment, others, such as perioral responses, can be enhanced.

We observed a significant increase in RJM movements in rats treated chronically with fluphenazine decanoate for an eight month period after eight weeks washout. In that study we were also able to demonstrate that the potent blockade of the D2 dopamine receptor by fluphenazine decanoate, as evidenced by a decreased stereotypy response, persisted longer than blockade of the D1 dopamine receptor. Significant RJM did not emerge earlier than after eight weeks of washout, possibly because of partial blockade of the D1 receptor by the neuroleptic (Table 1), (Rosengarten et al., 1986). The weaker affinity of fluphenazine at the D1 receptor than at the D2 receptor (Hyttel, 1983) would permit its earlier washout from the D1 receptor. If withdrawal of neuroleptics in humans leads to its earlier washout at the D1 than at the D2 receptor, this phenomenon may be responsible for withdrawal dyskinesia.

In 1987 Ellison et al., presented a rat model of tremulous mouth movements following neuroleptics (Ellison et al., 1987). In a subsequent study, Ellison and See (1989) presented a model of rat TD in which animals were treated with 21 mg of fluphenazine decanoate or haloperidol decanoate once every three weeks for two months and followed by the same dose of nondecanoate haloperidol for another two months. On withdrawal the rats manifested jaw movements. It was further determined that the behavior had a distinct frequency with a peak at 1–2 Hz. A similar frequency has been observed for extremities movements in human TD (Alpert et al., 1976; Fann et al., 1977).

2.3. Repetitive Jaw Movement (RJM) Model

The pharmacological basis for our treatment paradigm (described below) stemmed from the in vitro study by Stoof and Kebabian (1981) in which it was demonstrated that dopamine-inducible cAMP efflux from striatal slices could be augmented by simultaneous blockade of D2 receptors with the specific D2 antagonist sulpiride or diminished in the presence of the specific D2 receptor agonist, LY 141865. Thus, a D1 mediated action was capable of being modulated (reversed) by stimulation of the D2 receptor.

We, therefore, carried out a corresponding study in the intact rat. We found that RJM in rats induced with SKF 38393, a selective D1 agonist, were further enhanced by acute treatment with a D2 antagonist and inhibited by the selective D1 receptor antagonist, SCH 23390 in a dose-dependent manner (Rosengarten et al., 1983,1986,1987). We have, therefore, proposed that the D1 system may be importantly involved in the etiology of oral behavior in TD (Rosengarten et al., 1983,1986,1987,1989). The selective D1 agonist SKF 38393 induced episodes of RJM in a dose-dependent manner and the ED 50 was determined to be 33 mg/kg. Classical stereotypy (biting and licking) was never observed in the dose range from 2–60 mg/kg.

RJM and tongue protrusion constitute the two most characteristic features of this behavior. RJM occurs at a very low rate in an intact animal and increases in frequency with rat age. RJM appears in 15–20 min after subcutaneous administration of SKF 38393 and persists for ±60 min, reaching the peak frequency between 40–50 min after SKF 38393 administration.

Purposeless chewing not directed onto a distinct object and tongue protrusion is accompanied by yawning, audible tooth grinding, prominent grating, side-to-side chewing movements, and bursts of opening and closing of the rat jaws. This syndrome parallels several features characteristic of TD. Symptoms of dystonia manifested by intermittent or sustained muscle spasms and abnormal postures affecting cranial, neck, and trunk musculature are not present.

Animals: Male Sprague Dawley rats weighing 250–300 g were used for all studies. Rats were housed four to a case in an animal facility at 21 ± 1°C with a relative humidity of 55 ± 5% under a 12 h dark–light cycle and with free access to commercial food pellets and tap water.

Behavioral studies: For acute studies, drugs and dose ranges were: SKF 38393, 2–60 mg/kg; LY 141865 or LY 171555, 0.125–1.0 mg/kg. Blockade of the D2 receptor was induced by subcutaneous administration of spiroperidol in the dose of 40 kg to block the D2 receptors. Rats were placed in separate wire mesh cages and were allowed to accommodate to the behavioral testing laboratory for 30–60 min prior to drug treatment. Drugs were administered subcutaneously at the neck. Behavioral assessments were begun 10 min after drug administration and were scored by two independent observers unaware of the treatments given to specific rats. RJM scores were obtained by summing discrete bursts (2–7 second duration) of purposeless mouth movements over five one-min periods with each period separated by a 10-min interval or over a 10-min testing period 40–50 min following SKF 38393 administration. Stereotypic behavior (licking and biting) were assessed by a method modified from that reported by Costall and Naylor (1974). Sniffing was omitted from the scoring of RJM since this behavior has been determined to be inducible both by D1 and D2 agonists (Molloy and Waddington, 1984). In some studies, irreversible inactivation fo D1 or D2 receptors was carried out by administering EEDQ (6 mg/kg; ip) to rats given either spiroperidol (0.08 mg/kg) or SCH 23390 (0.4 mg/kg), respectively (Meller et al., 1985). The neuroleptics were used to protect the D1 or D2 receptors, respectively, from EEDQ and behaviors were tested 17–24 h later following administration of SKF 38393 (20 mg/kg) 15 min before. Chronic neuroleptic treatment: Rats were administered fluphenazine decanoate 25 mg/kg/im every three weeks for a total period of eight months.

Table 2
The Effect of SKF 38393 (20 mg/kg) on the Rate of RJM in Rats

	Episodes of RJM ± SEM	P
Treatment		
Vehicle	3.9 ± 0.8	
Vehicle + spiroperidol 40 μg/kg	18.4 ± 3.0	<0.005
SKF 38393	34.9 ± 3.6	<0.005
SKF 38393 + SCH 23390	1.9 ± 0.7	
SKF 38393 + LY 171555	2.1 ± 1.0	

Spiroperidol, a specific D2 receptor antagonist administered at a dose of 40 μg/kg increased the rate of spontaneous RJM in rats. SKF 38393 further increased the rate of RJM. RJM can be blocked by the specific D1 receptor antagonist SCH 23390 at the dose of 30 μg/kg. Stimulation of the D2 receptor with the specific D2 receptor agonist LY 171555 at a dose of 0.5 mg/kg significantly decreased the rate of RJM. Each set of experiments was carried out in the same testing session. Student's *t*-test was used for statistical analysis of comparisons to vehicle-treated rats.

At various times after washout, rats were tested for apomorphine inducible stereotypy and spontaneous and SKF 38393 inducible RJM.

Administration of SKF 38393 at the dose of 20 mg/kg in intact animals increased the basal RJM frequency of RJM four to five fold. Simultaneous administration of D2 blocker enhanced the RJM frequency. Administration of SCH 23390, a selective D1 antagonist prior to SKF 38393 administration was able to block the SKF 38393 inducible RJM supporting the hypothesis that RJM is mediated via the D1 receptor. LY 141865 or LY 171555 was able to decrease the SKF 38393 inducible RJM in a dose-dependent manner, supporting the view of an inhibitory role of the D2 dopamine receptor on D1 mediated RJM (Table 2). Depletion of the D1 receptor to 25–30% of their density by EEDQ administration did not change the rate of SKF 38393 inducible RJM, suggesting the presence of a functional reserve of D1 receptors for RJM behavior (Table 3).

Table 3
The Effect of SKF 38393 on RJM in Rats with a Decreased Density of D1 Receptors

Treatment	B_{max} D1 pmol/gm ± SEM	p	B_{max} D2 episodes ± SEM	p	RJM pmol/gm ± SEM
Vehicle	116.5 ± 5.5		38.9 ± 1.2		35.0 ± 3.6
EEDQ 6mg/kg	28.4 ± 2.3*	<0.005	34.1 ± 0.8		38.0 ± 2.5

Rats were pretreated with eticlopride 400 μg/kg to protect the D2 receptors prior to EEDQ administration at a dose of 6 mg/kg. In the unprotected group 74.6% of the D1 receptors were lost. SKF 38393 stimulation at a dose of 20 mg/kg produced the same degree of RJM as it did in rats with intact D1 receptors. Student's *t*-test was used for statistical analysis of comparisons to vehicle-treated rats.

Chronic eight months fluphenazine decanoate treatment resulted in a potent blockade of the D2 receptors. Spontaneous D1 mediated behavior was evident to some degree during neuroleptic treatment, whereas apomorphine induced stereotypy was completely abolished, suggesting that D2 dopamine receptors are efficiently blocked during treatment. Significant RJM emerged, however, only after eight weeks of washout from eight months fluphenazine decanoate treatment and rats showed also increased behavioral supersensitivity to SKF 38393 with no significant change in D1 receptor Bmax or Kd. D2 receptors were blocked during initial washout and remained blocked for a longer period of time than the D1 dopamine receptors. This is consistent with the fact that fluphenazine decanoate has a weaker affinity for D1 than for D2 receptors and permits their earlier washout, thus facilitating the appearance of RJM. Therefore, it appears plausible that the D1 mediated behavior may be facilitated during prolonged neuroleptic treatment by persistent D2 receptor blockade and earlier washout of neuroleptics from D1 receptors. The possibility of functional reserve of D1 receptors for RJM may further increase the frequency of this behavior even during ongoing neuroleptic administration (Table 1). Thus, enhanced oral behavior in our animal model may be expressed when the D2 receptor is blocked and 25–30% of dopamine D1 receptor is available for agonist stimulation. It is known that in humans during neuroleptic administration (Farde et al., 1987), the D1 receptor blockade is much weaker than the D2 receptor blockade and if the loss of neuroleptics from the D1 occurs earlier than from the D2 receptors, this mechanism may also be responsible for withdrawal induced dyskinesia in humans.

3. Discussion

Neuroleptic-induced involuntary oral behavior in primates and rats have similarities in their features, etiological factors, and probably a similar mechanism to human TD.

The syndrome of rapid jaw movements (RJM) was observed in rats during the acute and selective blockade of the D2 receptor with sulpiride or spiroperidol and stimulation of the D1 dopamine receptor with the selective D1 agonist, SKF 38393 (Table 2). This syndrome was also observed during washout from prolonged (eight months) fluphenazine decanoate treatment. The frequency of spontaneous repetitive jaw movements was increased, as was the response to SKF 38393 in inducing RJM (Table 1). Stereotypy response to the apomorphine challenge in these rats was abolished, suggesting potent blockade of the D2 receptors. The results of our studies indicate that blockade of D2 receptors in fluphenazine decanoate treated rats persisted much longer than the blockade of the D1 receptors. Thus, during washout the D1 receptor can be stimulated by endogenous dopamine or SKF 38393, whereas the D2 receptor is still blocked.

Under conventional neuroleptic treatment, the affinity of neuroleptic for the D2 receptor is always greater than for D1 receptors (Farde et al., 1987). Thus, different neuroleptics, depending on their differences in affinity toward these receptors (Hyttel, 1983), are capable of inducing RJM to a greater or lesser extent during washout.

We have been able to demonstrate that the status of the D2 receptors is important for the D1 receptor mediation of the RJM behavior (Rosengarten et al., 1983,1986,1987) and that the mechanism that involves opposing roles of these two receptor systems is an important factor. Although the analogy between the symptomatology of TD in humans and rat RJM should not be based on seeming similarity in appearance between oral dyskinesia and RJM, the relationship of these two syndromes to varied drug paradigms and the spontaneous occurrence of RJM in both neuroleptic treatment and in aging, is of great interest.

Despite the fact that the response of the D1 receptors to SKF 38393 challenge in rats during washout from prolonged neuroleptic treatment was increased, we were unable to find differences in the D1 receptor Bmax or Kd. Our animal model appears to be a logical extension of Stoof and Kebabian in vitro study (1981) on the opposing roles of the D1/D2 receptor systems. Our model, however, does not eliminate the possibility that other systems may also be involved and play in concert with the D1 system in the manifestation of RJM. Symptomatological similarities between oral behavior manifested during acute administration of selective D1 agonist and D2 antagonists to oral behavior manifested in rats during washout from prolonged fluphenazine decanoate treatment suggest to us similarities in the mechanism underlying both conditions. During neuroleptic treatment the degree of the D1 and D2 dopamine receptor blockade may differ, depending on the affinity of a particular neuroleptic to the D1 and the D2 receptors.

Of particular interest is the fact that 70–80% depletion of striatal D1 receptors in rats by EEDQ inactivation (Rosengarten et al., 1989) will still make the RJM behavior possible when stimulated with SKF 38393. This finding may aid in the explanation of the fact that even during potent blockade of both D2 and D1 receptors, rats manifested full RJM response to SKF 38393 challenge. From these studies, it appears that there is a functional D1 dopamine receptor reserve for RJM behavior. During potent blockade of the D2 receptors, the braking effect of the D2 system on the D1 system is eliminated, and this may also augment the D1 response. An analogy probably exists between experimental conditions in rats and humans with TD symptoms. Furthermore, it appears from PET studies (Farde et al., 1987) that in humans D2 receptor blockade is much more pronounced than that of the D1 receptors. Thus, the interaction between the D1 and D2 systems and possible functional reserve for the D1 receptor in mediating RJM may explain difficulties encountered during neuroleptic treatment designed to control RJM. Furthermore, the existence of complex interactions in CNS between dopaminergic systems and other systems in the brain may suggest that these interactions should be considered in the manifestation of this behavior and its treatment.

This article represents a description of existing animal models of TD and offers a historical perspective on the evolution of these models and their relevance to neuroleptic-induced abnormal oral behavior. An attempt was also made to stress the profound impact of the discovery of selective D1 agonists and antagonists particularly in relevance to animal models of the TD, which led to our hypothesis relating the D1/D2 receptors' involvement in animal models of TD.

Acknowledgments

Thanks to Roger J. Schoenman for helpful assistance in preparation of this manuscript. This work was supported by Grant nos. MH 35976 and MH 08618 from the National Institute of Mental Health.

References

Alpert M., Diamond F., and Friedhoff A. J. (1976) Tremographic studies in tardive dyskinesia. *Psychopharmacol. Bull.* **12,** 5–7.

Arnt J. and Scheel-Kruger J. (1980) Intranigral GABA antagonists produce dopamine independent biting in rats. *Eur. J. Pharmacol.* **62,** 51–61.

Arnt J. and Perregaard, J. (1985) Differential involvement of dopamine D1 and D2 receptors in circling behavior induced by apomorphine, SKF 38393, pergolide and LY 171555 in 6-hydroxylesioned rats. *Psychopharmacology* **85,** 346–350.

Arnt J. (1987) Behavioral studies of dopamine receptors. Evidence for regional selectivity and receptor multiplicity, in *Dopamine Receptors* (Creese I. and Fraser C. A., eds.), Liss, New York, NY, pp. 199–207.

Arnt J., Hyttel J., and Perregaard J. (1987) Dopamine D1 receptor agonist combined with the selective D2 agonist quinpirole facilitates the expression of oral stereotyped behavior in rats. *Eur. J. Pharmacol.* **133,** 137–145.

Baldessarini R. J. and Tarsy D. (1976) Mechanism underlying tardive dyskinesia. In *Basal Ganglia* (Yahr M., ed.), Raven, NY, pp. 25–36.

Baldessarini R. J. and Tarsy D. (1980) Dopamine and the pathophysiology of dyskinesia induced by antipsychotic drugs. *Ann. Rev. Neurosci.* **3,** 23–41.

Baldessarini R. J., Coke J. O., and Davis J. M. (1980) Report on the American Psychiatric Association Task Force on Late neurological effects of antipsychotic drugs, Washington, DC, American Psychiatric Association.

Barany S., Ingvast A., and Gunne L. M. (1979) Development of acute dystonia and tardive dyskinesia in cebus monkeys. *Res. Comm. Chem. Path. Pharmacol.* **25**, 269–279.

Bedard P., Larochelle L., Delean I., and Lafleur I. (1972) Dyskinesias induced by long-term administration of haloperidol in the monkey. *Physiologist* **15**, 83.

Braun A. R. and Chase T. N. (1986) Obligatory D1/D2 receptor interaction in the generation of dopamine agonist-induced behaviors. *Eur. J. Pharmacol.* **131**, 301–306.

Burt D. R., Creese I., and Snyder S. H. (1977) Antischizophrenic drugs: Chronic treatment elevates dopamine receptor binding in brain. *Science* **196**, 326–328.

Carpenter W. T., Rey A. C., and Stephens I. H. (1980) Covert dyskinesia in ambulatory Schizophrenia. *Lancet* **2**, 212–213.

Casey D. E. and Denney D. (1977) Pharmacological characterization of tardive dyskinesia. *Psychopharmacology* **54**, 1–8.

Casey D. E. (1984) Tardive Dyskinesia—animal models. *Psychopharmacol. Bull.* **20**, 376–379.

Casey D. E. (1985) Behavioral effects of long-term neuroleptic treatment in cebus monkeys, in *Dyskinesia: Research and Treatment* (Casey D. E., Chase T. N., Christensen A. V., and Gerlack J., eds.), Springer Verlag, Berlin, Heidelberg, pp. 211–216.

Clow A., Jenner P., and Marsden C. D. (1979) Changes in dopamine mediated behavior during one year's neuroleptic administration. *Eur. J. Pharmacol.* **57**, 365–375.

Costall B. and Naylor R. B. (1974) The involvement of dopaminergic systems with the stereotyped behavior patterns induced by methylphenidate. *J. Pharm. Pharmacol.* **26**, 30–33.

Creese I., Burt D. R., and Snyder S. H. (1976) Dopamine receptor binding predicts clinical and pharmacological potencies of antischizophrenic drugs. *Science* **192**, 481–483.

Ellison G., See R., Levin E., and Kinney J. (1987) Tremulous mouth movements in rats administered chronic neuroleptics. *Psychopharmacology* **92**, 122–126.

Ellison G. and See R. E. (1989) Rats administered chronic neuroleptics develop oral movements which are similar in form to those in humans with tardive dyskinesia. *Psychopharmacology* **98**, 564–566.

Fage E. and Scatton B. (1986) Effects of D1 and D2 receptor antagonists on acetylcholine levels in the rat striatum. *Eur. J. Pharmacol.* **129**, 359–362.

Fann E., Stafford J., Malone R., Frost J., and Richman B. (1977) Clinical research techniques in tardive dyskinesia. *Am. J. Psychiatr.* **134**, 759–762.

Farde L., Halldin C., Stone-Elander S., and Sedvall G. (1987) PET analysis of human dopamine receptors subtypes using 11-C-SCH 23390 and 11-C-raclopride. *Psychopharmacology* **92**, 278–284.

Friedhoff A. J., Bonnet K., and Rosengarten H. (1977) Reversal of two manifestations of dopamine receptor supersensitivity by administration of L-Dopa. *Res. Comm. Chem. Path. Pharmacol.* **116,** 411–423.

Gale K. (1980) Chronic blockade of dopamine receptors by antischizophrenic drugs enhances GABA binding in substantia nigra. *Nature* **283,** 569–570.

Gardos G., Cole J. O., and Tarsy D. (1978) Withdrawl syndromes associated with antipsychotic drugs. *Am. J. Psychiatr.* **135,** 1321–1324.

Gardos G. and Cole J. O. (1980) Problems in assessment of tardive dyskinesia, in *Tardive Dyskinesia Research and Treatment* (Fann W. E., Smith R. C., Davis J. M., and Domino E. F., eds.), Spectrum, Jamaica, NY, pp. 201–214.

Gershanik O., Heikkila R. E., and Duvoisin R. C. (1983) Behavioral correlation of dopamine receptor activation. *Neurol.* **33,** 1489–1492.

Gianutsos G., Drawbaugh R. B., Hynes M. D., and Dew Lal H. (1974) Behavioral evidence for dopaminergic supersensitivity after chronic haloperidol. *Life Sci.* **141,** 887–898.

Glassman R. B. and Glassman N. H. (1980) Oral dyskinesia in brain-damaged rats following brain lesions and neuroleptic drug administration. *Psychopharmacology* **77,** 134–139.

Gunne L. M. and Barany S. (1976) Haloperidol-induced tardive dyskinesia in monkeys. *Psychopharmacology* **50,** 237–240.

Gunne L. M., Bachus S. E., and Gale K. (1980) Oral movements induced by interference with nigral GABA neurotransmission: Relationship to tardive dyskinesias. *Exp. Neurol.* **100,** 459–469.

Gunne L. M., Growdon J., and Glaeser B. (1982) Oral dyskinesia following rat brain lesions and neuroleptic drug administration. *Psychopharmacology* **77,** 134–139.

Gunne L. M. and Haggstrom J. E. (1983) Reduction of nigral glutamic acid decarboxylase in rats with neuroleptic-induced oral dyskinesia. *Psychopharmacology* **81,** 191–194.

Gunne L. M., Haggstrom J. E., and Sjoquist B. (1984) Association with persistent neuroleptic-induced dyskinesia of regional changes in brain GABA synthesis. *Nature* **309,** 347–349.

Gunne L. M., Andersson U., Bondesson U., and Johansson P. (1986) Spontaneous chewing movements in rats during acute and chronic antipsychotic drug administration. *Pharmacol. Biochem.* **25,** 897–901.

Hall R. A., Jackson R. B., and Swain J. (1956) Neurotoxic reactions resulting from chlorpromazine administration. *J. Am. Med. Assoc.* 214–218.

Hall K., Kohler C., and Gariell L. (1985) Some in vitro receptor binding properties of ^{3}H eticlopride, a novel substituted benzamide selective for dopamine D2 receptors in the rat's brain. *Eur. J. Pharmacol.* **111,** 191–199.

Hamblin M. W. and Creese I. (1983) Behavioral and radioligand binding evidence from irreversible dopamine receptor blockade by *N*-ethoxycarbonyl-2-ethoxy-1, 2-dihydroquinoline. *Life Sci.* **32,** 2247–2255.

Hanssen T. E., Brown W. L., Wiegel R. M., and Casey D. E. (1988) Risk factors for drug-induced Parkinsonism in tardive dyskinesia patients. *J. Clin. Psychiatr.* **49,** 139–141.

Hess E. J., Battaglia G., Norman A. B., and Creese I. (1987) Differential modification of striatal dopamine receptors and effector moieties by *N*-ethoxycarbonyl 1-2-ethoxy-1, 2-dihydroquinoline in vivo and in vitro. *Mol. Pharmacol.* **31,** 50–57.

Hyttel J. (1978) Effects of neuroleptics on ^{3}H-haloperidol and ^{3}H-cis-flupenthixol binding and on adenylate cyclase activity in vitro. *Life Sci.* **23,** 551–556.

Hyttel J. (1983) SCH 23390—the first selective dopamine D1 antagonist. *Eur. J. Pharmacol.* **91,** 153–154.

Iorio L. C., Barnett A., Leitz F. H., Houser V. P., and Korduba C. A. (1983) SCH 23390, a potential benzazepine antipsychotic with unique interactions on dopaminergic systems. *J. Pharmacol. Exp. Ther.* **226,** 462–468.

Itil T. M., Reisberg B., Hugure M., and Mehta D. (1981) Clinical profiles of tardive dyskinesia. *Compr. Psychiatr.* **22,** 282–290.

Itoh Y., Beaulieu M., and Kebabian J. W. (1984) The chemical basis for the blockade of the D1 receptor by SCH 23390. *Eur. J. Pharmacol.* **100,** 119–122.

Jus A., Jus K., and Fontaine P. (1979) Long-term treatment of tardive dyskinesia. *J. Clin. Psychiatr.* **40,** 72–77.

Kaiser C. (1983) In *Stereoisomeric Probes for the Dopamine Receptors.* (Kaiser C., and Kebabian J. W., eds.), American Chemical Society, Washington, DC, pp. 223–250.

Kane J. M. and Smith J. M. (1982) Tardive dyskinesia: prevalence and risk factors 1959–1979. *Arch. Gen. Psychiatr.* **39,** 473–481.

Klawans H. L. (1973) The pharmacology of tardive dyskinesia. *Am. J. Psychiatr.* **130,** 82–86.

Kosgaard S., Gerlach J., and Christensson E. (1985) Behavioral aspects of serotonin dopamine interaction in the monkey. *Eur. J. Pharmacol.* **118,** 245–252.

Kovacic B. and Domino E. F. (1982) A monkey model of tardive dyskinesia (TD), evidence that reversible TD may turn into irreversible TD. *J. Clin. Psychopharmacol.* **2,** 305–306.

Kullenkampf C. and Tarnow G. (1956) Ein Eigentumliches Syndrom im oralen Bereich bei Megaphenaaplikation. *Nervenarzt* **27,** 178–180.

Liebman J. and Neale R. (1980) Neuroleptic-induced acute dyskinesias in squirrel monkeys. Correlation with propensity to cause extrapyramidal side effects. *Psychopharmacology* **68,** 25–29.

Lubin H. and Gerlach J. (1988) Behavioral effects of dopamine D1 and D2 receptor agonists in monkeys previously treated with haloperidol. *Eur. J. Pharmacol.* **153,** 239–245.

Mao C. C., Chenney L. M., Marco E., Revuetta A., and Costa E. (1977) Turnover times of gamma-aminobutyric acid and acetylcholine in nucleus caudatus, nucleus accumbens, globus pallidus and substantia nigra effects of repeated administration of haloperidol. *Brain Res.* **132,** 375–379.

Marsden C. D., Tarsy D., and Baldessarini R. J. (1973) Spontaneous and drug-induced movement disorders in psychotic patients, in *Psychiatric Aspects of Neurologic Disease* (Benson D. F. and Blummer D., eds.), Grunne and Stratton, NY, pp. 219–266.

Mashurano M. and Waddington J. L. (1986) Stereotyped behavior in response to the selective D2 dopamine receptor agonist RU 24213 is enhanced by pretreatment with selective D1 agonists SKF 38393. *Neuropharmacology* **25,** 947–950.

Meller E., Bohmaker K., Goldstein M., and Friedhoff A. J. (1985) Inactivation of D1 and D2 dopamine receptors by *N*-ethoxycarbonyl-2-ethoxy-1, 2-dihydroquinoline in vivo: Selective protection by neuroleptics. *J. Pharmacol. Exp. Ther.* **233,** 656–662.

Messiha F. (1980) Biochemical studies after chronic administration of neuroleptics to monkeys, in *Tardive Dyskinesia: Research and Treatment* (Fann W. E., Smith R. C., Davis J. M., and Domino E. F., eds.), Spectrum, NY, pp. 13–25.

Muller P. and Seeman P. (1977) Brain neurotransmitter receptor after long-term haloperidol. *Life Sci.* **21,** 1751–1758.

Murray A. M. and Waddington J. L. (1989) Induction of grooming and vacuous chewing by a series of selective D1 dopamine receptor agonists: Two directions of D1: D2 interaction. *Eur. J. Pharmacol.* **160,** 377–384.

Onali P., Olianas M. C., and Gessa G. L. (1984) Selective blockade of dopamine D1 receptors by SCH 23390 discloses striatal dopamine D2 receptors mediating the inhibition of adenylate cyclase in rats. *Eur. J. Pharmacol.* **99,** 127–128.

Porsolt R. D. and Jalfre M. (1981) Neuroleptic-induced acute dyskinesias in rhesus monkeys. *Psychopharmacology* **75,** 16–21.

Rosengarten H., Schweitzer J. W., and Friedhoff A. J. (1983) Induction of oral dyskinesia in naive rats by D1 stimulation. *Life Sci.* **33,** 2479–2482.

Rosengarten H., Schweitzer J. W., and Friedhoff A. J. (1986) Selective dopamine D2 receptor reduction enhances a D1 mediated oral dyskinesia in rats. *Life Sci.* **39,** 29–35.

Rosengarten H., Schweitzer J. W., and Friedhoff A. J. (1987) Prolonged fluphenazine decanoate treatment augments oral movements in rats: Mediation by D1 receptor. *Sixth Int. Catecholamine Symposium Abstracts,* p. 14.

Rosengarten H., Schweitzer J. W., and Friedhoff A. J. (1989) A full repetitive jaw movement response after 70% depletion of caudate D1 receptors *Pharmacol. Biochem. Behav.* **34,** 895–897.

Salama A. I. and Saller C. F. (1986) Functional interactive effects of D1 and D2 dopamine receptor blockade. In *Neurobiology of Central D1 Dopamine Receptors. Adv. Exp. Med. Biol.* **204,** 137–140.

Saller C. F. and Salama A. I. (1985) Dopamine receptor subtypes: in vivo biochemical evidence for functional interaction. *Eur. J. Pharmacol.* **109,** 297–300.

Seeman P. (1980) Brain dopamine receptors. *Pharmacol. Rev.* **32,** 229–313.

Seeman M. V. (1981) Tardive dyskinesia: Two-year recovery. *Compr. Psychiatr.* **22,** 189–192.

Seeman P. and Grigoriadis D. (1987) Dopamine receptors in brain and periphery. *Neurochem. Int.* **10,** 1–25.

Sedvall G., Farde L., and Hall H. (1988) D1 and D2 dopamine receptors in the human brain as studied by in vitro analysis and PET, in *Central and Peripheral Dopamine Receptors. Biochemistry and Pharmacology* (Spano P. F., Biggio G., Toffano G., and Gessa G. L., eds.), Liviana, pp. 81–86.

Scheel-Kruger J. and Arnt J. (1985) New aspects on the role of dopamine, acetylcholine and GABA in the development of tardive dyskinesia, in *Dyskinesia Research and Treatment. Psychopharmacology* 2 (Suppl.), Springer Verlag, Berlin, Heidelberg, pp. 45–56.

Schoeneker M. (1957) Ein Eigentumlichen syndrome in oralen Bereich bei Megaphen Applikation. *Nervenarzt* **28,** 35–37.

Sibley D. R., Leff S. E., and Creese I. (1982) Interaction of novel dopaminergic ligands with D1 and D2 dopamine receptors. *Life Sci.* **31,** 637–645.

Sigwald J., Bouttier D., and Raymondeaud C. (1959) Quatre cas de dyskinesie faciobucco-linguo-masticatrice a l'evolution prolongee secondaire a un traitemant par les neuroleptiques. *Rev. Neurol.* **100,** 751–755.

Smith J. M. and Baldessarini R. J. (1980) Change in prevalence severity and recovery of tardive dyskinesia with age. *Arch. Gen. Psychiatr.* **37,** 1368–1373.

Stoof J. C. and Kebabian J. W. (1981) Opposing roles for the D1 and D2 dopamine receptors in efflux of cyclic AMP from rat striatum. *Nature* **294,** 366–368.

Tarsy D. and Baldessarini R. J. (1977) The pathophysiologic basis of tardive dyskinesia. *Biol. Psychiatr.* **12,** 431–449.

Toenniesen L. M., Casey D. E., and McFarland, B. H. (1985) Tardive dyskinesia in the aged: duration and treatment relationship. *Arch. Gen. Psychiatr.* **42,** 278–284.

Tsuruta K., Frey E. A., Grewe C. W., Cote T. E., Eskay R. L., and Kebabian J. W. (1981) Evidence that LY141865 specifically stimulates the D2 dopamine receptor. *Nature* **292,** 463–465.

Uhrband L. and Faurbye A. (1960) Reversible and irreversible dyskinesia after treatment with perphanazine, chlorpromazine, reserpine and electroconvulsive therapy. *Psychopharmacology* **1,** 408–418.

Waddington J. L., Molloy A. G., Boyle K. M., and Youssef H. A. (1986) Spontaneous and drug-induced dyskinesias in rodents in relation to aging and long-term neuroleptic treatment: Relationship to tardive dyskinesia, in *Biol. Psych.,* (Shagass C., et al., eds.) Elsevier, NY, pp. 1151–1153.

Waddington J. L. and Gamble S. L. (1980) Neuroleptic treatment for a substantial proportion of adult life: Behavioral sequelae of 9 months haloperidol administration. *Eur. J. Pharmacol.* **67,** 363–364.

Waddington J. L., Cross A. J., Gamble S. J., and Bourne R. C. (1983) Spontaneous orofacial dyskinesia and dopaminergic function in rats after 6 months of neuroleptic treatment. *Science* **220,** 530–532.

Waddington J. L. and Gamble S. J. (1980) Emergency of vacuous chewing during 6 months continuous treatment with fluphenazine decanotate. *Eur. J. Pharmacol.* **68,** 387–388.

Wegner J. T. and Kane J. M. (1982) Follow-up study on the reversibility of tardive dyskinesia. *Amer. J. Psych.* **139,** 368–369.

Weiss B., Santelli S., and Lusink G. (1977) Movement disorders induced in monkeys by chronic haloperidol treatment. *Psychopharmacology* **53,** 289–293.

White F. J. (1987) D1 receptor stimulation enables the inhibition of nucleus accumbens neurones by D2 receptor agonist. *Eur. J. Pharmacol.* **135,** 101–105.

White F. Y. and Wang R. Y. (1986) Electrophysiological evidence for the existence of both D1 and D2 dopamine receptors in rat nucleus accumbens. *J. Neurosci.* **6,** 274–280.

Activity Anorexia

An Animal Model and Theory of Human Self-Starvation

W. David Pierce and W. Frank Epling

1. Introduction

Activity anorexia is a biobehavioral process that occurs in animals. Laboratory rats are fed once per day, they lose weight, but quickly adjust to the feeding regime. Other animals receive a similar meal schedule and are allowed to run voluntarily on an activity wheel—these animals die of starvation. Wheel running becomes excessive when food is restricted, and animals may run more than 10 Km/d. Interestingly, excessive physical activity interferes with food consumption even though the experimental procedures ensure that eating and running cannot occur at the same time (Epling et al., 1983). This chapter describes the animal model and the biobehavioral theory of activity anorexia and extends the theory to human self-starvation.

2. The Animal Model of Activity Anorexia

Our knowledge of activity anorexia in animals is based on research reported over the last few decades. Our experiments are based on procedures described in the 1960s by Routtenberg and Kuznesof (1967). In our laboratory adolescent male rats (approx 50 d old) are placed in a cage that is attached to a 1.1-m running wheel. Figures 1 show this apparatus; the wheel and side cage can be separated by closing the sliding door. During the first five days of an experiment, the door that separates the side cage from the wheel is closed. Food is freely available in the

From: *Neuromethods, Vol. 18: Animal Models in Psychiatry I*
Eds: A. Boulton, G. Baker, and M. Martin-Iverson

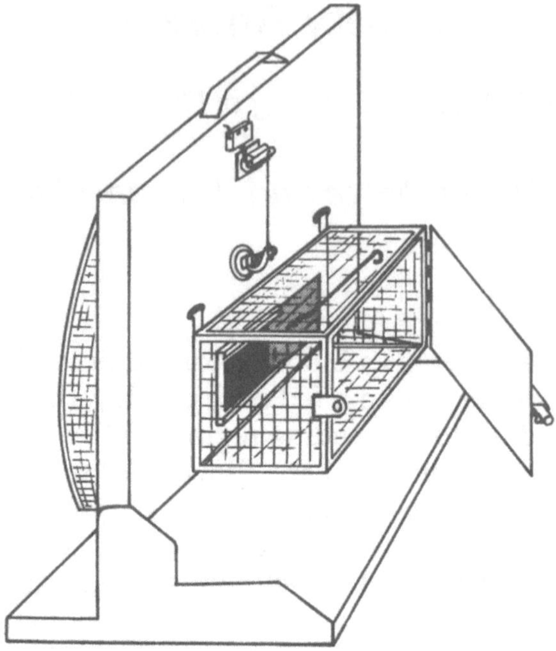

Fig. 1. A standard 1.1-m Wahmann running wheel with an attached side cage. A sliding door prevents or permits access between cage and wheel. In the experimental condition, the wheel remains open except during the 90-min feeding, ensuring that running does not compete with eating.

cage and each animal can eat as much as it wants. The amount eaten is measured daily and food spillage is taken into account. Also, animals are weighed on a daily basis. The food and weight measures provide baseline points for the experimental interventions. The experimental procedures are designed to combine food restriction and physical activity. Experimental and control animals are restricted to a single 60- or 90-min meal daily, depending on the experiment. We have found that the 90-min feeding is preferable, because control animals easily adjust to this schedule and survive. Following the meal period, the doors to the wheels are opened and experimental animals are allowed to run. Control animals also can enter the wheels, but the wheels are locked.

Several procedural points are noteworthy. The animals are given access to the wheels except during the feeding period. In

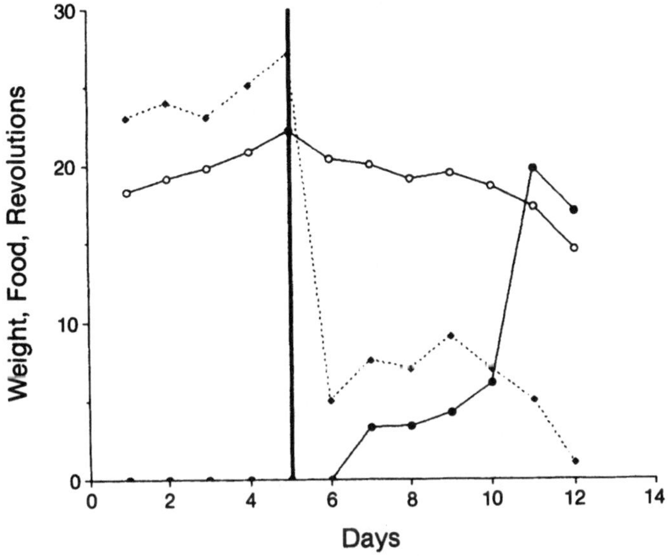

Fig. 2. Activity anorexia in a rat. The figure shows food intake (g; ●···●), body weight (g × 10; ○—○), and wheel revolutions (× 1000; ●—●) over days. The solid line is the point at which the animal was changed from free food to a single 90-min daily meal and given the opportunity to run on a wheel. Note the substantial drop in food intake and the rapid increase in wheel running. Republished with permission from Epling W. F., and Pierce W. D. (1989) Excessive activity and anorexia in rats, in *The Menstrual Cycle and its Disorders: Influences of Nutrition, Exercise and Neurotransmitters* (Pirke K. M., Wuttke W., and Schweiger I. U., eds.), Springer-Verlag, NY.

this way running does not compete with eating. When wheels are available, there are no requirements for animals to run. They can stay in their cages, sit in the running wheel, walk rather than run on the wheels, or emit any other available response. Finally, it is important to note that control animals typically adapt to the feeding schedule within seven days and remain healthy.

The initial effect of placing animals on one meal per day is a large drop in food consumption (*see* Fig. 2). This is not surprising because the animals have not experienced rapid change in food supply and are not adapted to the new feeding schedule. When food restriction and running occur together a number of interesting effects result. Experimental animals begin to run on the wheels. They increase running over days, even though there is no requirement to do so. This is an unusual response by these animals, since energy expenditure is increasing at a time when food intake is limited. Within a week, running increases from several hundred to several thousand revolutions/d. This is a massive amount of energy expenditure, since an animal that weighs <0.5 kg may run the equivalent of 10 km or more.

While running is accelerating to excessive levels, there is an even more suprising effect—animals stop eating. At first the rats increase food intake. However, the intense running begins to interfere with eating. Food intake is suppressed and eventually begins to decline. If this process of increasing activity and declining food intake is allowed to continue, body weight drops sharply and animals die of self-starvation.

Anorexia does not occur in the control animals. These animals have the same food schedule and living conditions, but they cannot run. Therefore, running or physical activity is an important factor in the development of anorexia. We have conducted other experiments with additional control groups. For example, in one control group, animals have access to an activity wheel and food is continuously available. These animals do not run excessively, and after a few days they eat normally.

Overall, the research on activity anorexia in rat has clarified the determinants of this process. The critical conditions for activity anorexia are the simultaneous occurrence of dietary restriction and excessive physical activity. These two factors combine in a multiplicative fashion to produce starvation and death.

2.1. Age, Sex, and Anorexia in Mice

Activity anorexia occurs in several strains of rats, including Sprague-Dawley and JCR:LA/N—corpulent (lean and fat). In addition, we have found this effect in Alas-strain mice (Epling et al., 1981). The procedures are similar to those used with rats, but the wheels are much smaller (15 cm in diameter) and the single feeding is 3 h in length.

In our laboratory, Doris Milke (1982) investigated the effects of age and sex on the development of activity anorexia in Alas mice. All young mice (30–35 d old) met a starvation criterion set at 70% of free-feeding body weight. Older mice (over 200 d of age) were less likely to starve, and only four of 16 animals met starvation criteria. Regardless of age, animals that starved ran excessively on the activity wheel, and those that survived did not. There was no effect of sex in the younger animals, but, in older mice, females (37%) starved more than males (12%).

Figure 3 shows the anorexic effect in two female mice—one 37 and the other 210 d old at the start of the experiment. In both mice, wheel running increases exponentially over days, but peaks more rapidly for the younger animal. At the start of the experiment, food intake is low, increases, and then declines as activity becomes excessive. Body weight progressively decreases for both animals, and, for ethical reasons, the experiment was stopped when they reached 70% of baseline weight. Generally older animals take longer to develop the anorexic effect and are less susceptible to activity anorexia.

The research with mice shows that activity anorexia occurs for more than one species of rodent. The importance of this finding is that the relationship between running and food intake is not species-specific, since rats and mice are genetically distinct. Also, it is interesting that younger animals are more likely to develop anorexia than older ones. There is also tentative evidence that females may be more susceptible than males. Such differences by age and sex may reflect baseline body weight (e.g., females weigh less than males) and levels of physical activity (e.g., females are more active, especially during estrus).

2.2. Level of Activity and Anorexia

The process of activity anorexia occurs when physical activity is combined with dietary restriction. When food is restricted, the amount of activity will determine the chances of an animal becoming anorexic. This is in fact what happened when we controlled the amount of time that animals could run on their wheels (Epling and Pierce, 1984).

In this study 42 adolescent male rats were randomly assigned to a control group (0 h activity) and to one of five activity conditions. The amount of time that animals could run on the wheels was 2, 6, 12, 18, or 22 h/d. The only difference among the

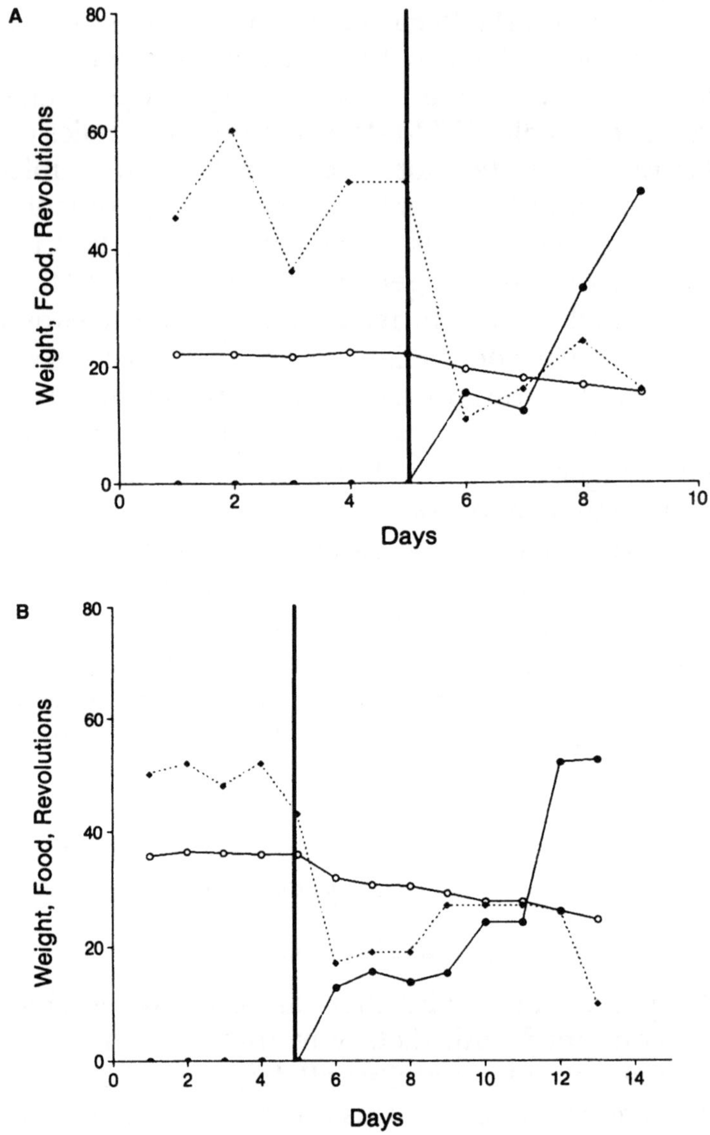

Fig. 3. Activity anorexia in a young female mouse (A) compared with that in a mature female mouse (B). The figure shows food intake (g/10; ●···●), body weight (g; ○—○), and wheel revolutions (× 500; ●—●) over days. The solid line is the point at which each animal was changed from free food to a single, 180-min meal daily and given the opportunity to run on a wheel. Activity anorexia took longer to develop and was less severe in the older mouse.

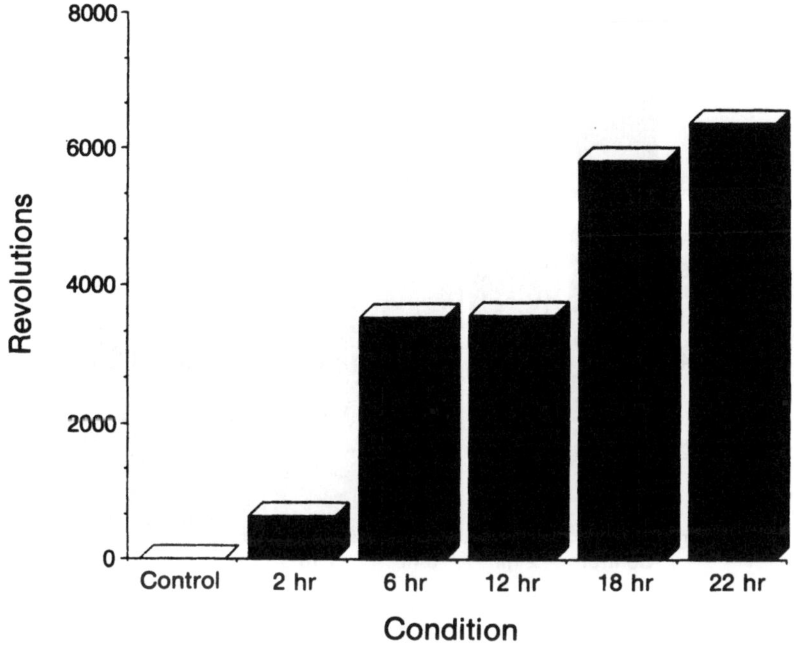

Fig. 4. Mean number of wheel revolutions as a function of opportunity to run (hours access to free-turning wheel). Results indicate that the more opportunity to run, the greater the number of wheel turns. Republished with permission from Epling W. F., and Pierce W. D. (1989) Excessive activity and anorexia in rats, in *The Menstrual Cycle and its Disorders: Influences of Nutrition, Exercise and Neurotransmitters* (Pirke K. M., Wuttke W., and Schweiger I. U., eds.), Springer-Verlag, NY.

groups was the opportunity to run. As in the previous experiments, rats were freely fed for five d and then changed to one 90-min feeding daily.

Opportunity to run influenced the amount of activity of the animals. Figure 4 shows the relationship between duration of access to the running wheels and average daily wheel turns. Generally, the more time the animals had to run, the higher their daily activity. Other results showed that food intake was affected by the amount of running; this is portrayed in Fig. 5. Food intake declined with more opportunity to run, and this decline was a function of the daily rate of change of wheel running.

We were able to classify each animal on a continuum of anorexia. In order to classify animals in terms of anorexia, the

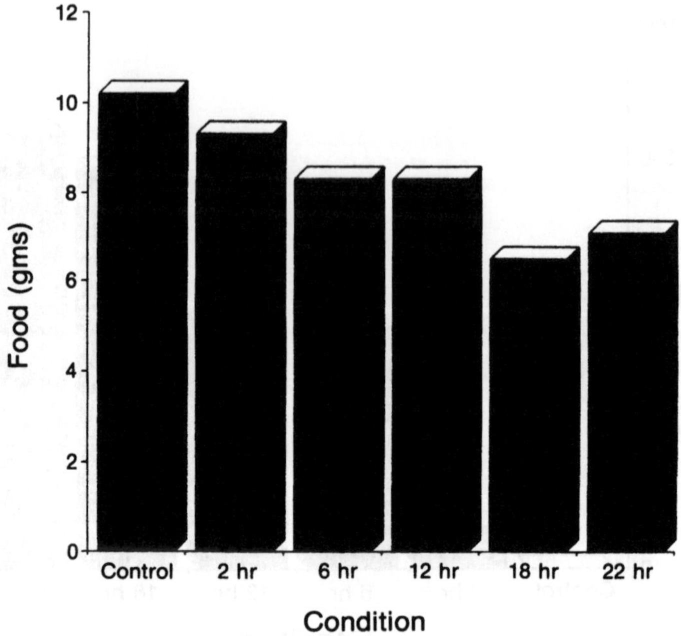

Fig. 5. Mean food intake (g) per day as a function of opportunity to run (hours access to free-turning wheel). Results indicate that food intake declines with opportunity to run. Republished with permission from Epling W. F., and Pierce W. D. (1989) Excessive activity and anorexia in rats, in *The Menstrual Cycle and its Disorders: Influences of Nutrition, Exercise and Neurotransmitters* (Pirke K. M., Wuttke W., and Schweiger I. U., eds.), Springer-Verlag, NY.

individual functions of wheel running, food intake, and body weight were examined. It was possible to define four patterns on the basis of food intake and body weight. Figure 6 presents these patterns as idealized curves. The "strong-anorexia" effect is depicted in Fig. 6a. Here body weight is declining, with some downward acceleration occurring during the last few days. Food intake is low upon entry into the experimental phase, but tends to increase over the first few days. This is followed by a leveling off of intake and then a final period of decline. Strongly anorexic animals showed clear suppression of food intake and sharply declining body weight.

Some animals showed a pattern that we classified as weak anorexia (*see* Fig. 6b). In these cases, food intake initially increases

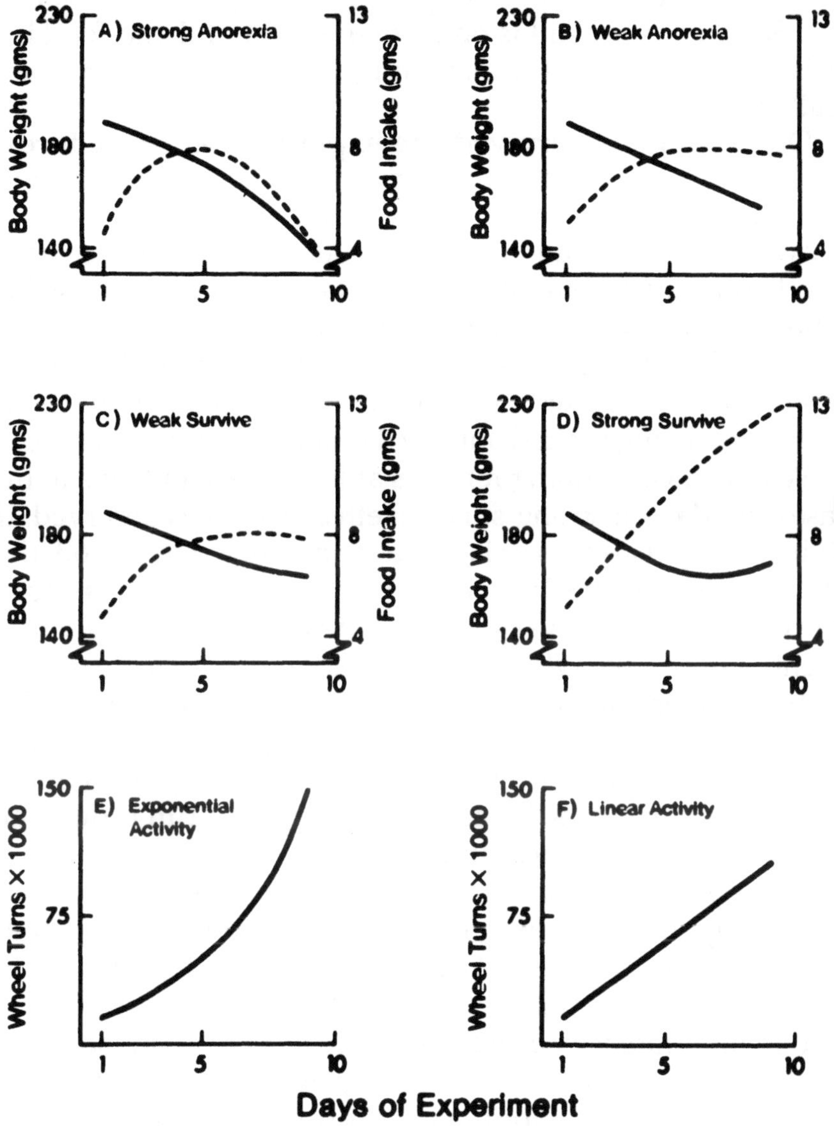

Fig. 6. Idealized functions for body weight (——) and food intake (- - - -) that allowed for classification of four anorexic or survival effects. The figure also shows wheel-running curves associated with strong anorexia (exponential) and survival or weak anorexia (linear). Republished with permission from Epling W. F. and Pierce W. D. (1984) Activity-based anorexia in rats as a function of opportunity to run on an activity wheel. *Nutr. Behav.* **2**, 37–49. Copyright by Alan R. Liss Inc., 1984.

and levels off, but remains suppressed throughout the experiment. Since body weight is on a downward trend and food intake is still not sufficient to maintain weight, these subjects may be called anorexic.

Other animals survived the experimental procedures. "Weak survival" and "strong survival" behavior patterns were identified. Figure 6c depicts the weak-survival effect, in which body weight declines and then levels off. Food intake increases to moderate values, but also levels. Intake is adequate to stabilize body weight at low values. Figure 6d represents the strong-survival effect. Body weight falls, levels, and then begins to increase. Food consumption shows a linear increase over days, with this rise being limited only by the single-meal presentation.

When wheel running remains at low levels and fails to increase over days, a strong-survival effect is always observed. If activity slowly increases or is linear, as portrayed in Fig. 6d, the weak-anorexia or weak-survival pattern occurs. The strong-anorexia pattern is always associated with the activity curve presented in Fig. 6e. Wheel running exponentially increases on a daily basis for animals who are strongly anorexic. Generally, our results show the rate of change in daily physical activity determines food intake. The greater the rate of change in activity, the less an animal eats. Finally, opportunity to run orders the rate of change in daily activity.

Our results are in accord with several studies demonstrating that high-intensity exercise produces decreased food consumption in rats (Athrens et al., 1972; Crews et al., 1969; Oscai and Holloszy, 1969). Katch et al. (1979) found that male rats exposed to high-intensity exercise showed a decrease in food consumption when compared with a group exposed to low-intensity exercise, matched for caloric expenditure. Both exercise groups ingested less food than no-exercise controls.

Apparently, the greater the change in intensity of exercise, the less food consumed. The available evidence suggests that activity operates on the mechanisms that control feeding. Opportunity to run orders the intensity of exercise and thereby orders food consumption.

2.3. Appetite and Satiety

We have argued that a cycle of increasing activity and decreasing food intake is the basis of activity anorexia. There are several ways that physical activity could decrease body weight. A common sense interpretation is that exercise burns calories. This account is not satisfactory, because it fails to explain the effects of physical activity on food intake. Another possibility is that exercise alters the tendency to eat meals when they are available (i.e., appetite). On the other hand, exercise may affect the animals' tendency to stop eating during a meal (i.e., satiety).

Kanarek and Collier (1983) have suggested that satiety is the major reason for activity anorexia. They argue that physical activity interferes with the amount eaten during a single daily meal. In other words, exercise makes the animal stop eating even though it has sufficient time to consume more food. Briefly, they found that animals exposed to the activity-anorexia procedures would eat a sufficient amount of food if they were given four 15-min meals rather than a single 60-min feeding.

According to Kanarek and Collier, animals give up eating after a short period of time even though they have an hour or more to feed. Survival depends on the animals adapting to the meal schedule by increasing the amount eaten during the feeding period. Wheel running is viewed as one of several factors that interfere with the animals ability to adjust to the single daily feeding. When meals are scheduled more frequently, the animal eats at each feeding and maintains adequate food intake and body weight.

This explanation of activity anorexia rests on the assumption that physical activity interferes with food intake only when the feeding schedule is severely restrictive. These are highly specific circumstances and appear to limit the generality of activity anorexia. The implication is that other feeding schedules would permit the animal to adapt, even though physical activity remained excessive.

Unfortunately, in their 1983 experiment, Kanarek and Collier failed to note that the change in feeding schedule controlled the amount of activity that animals exhibited. To illustrate, on the eighth day of their experiment, the animals given one meal ran approx 3000 revolution; those fed four times completed only about 1000 wheel turns. Thus, the increased food intake of the animals with the four brief meals may reflect the reduction in exercise, rather than the change in feeding schedule. In other words, more frequent feedings reduced physical activity, and this may have allowed food intake to recover.

Another line of evidence opposes a satiety account of activity anorexia. According to the satiety hypothesis, physical activity prevents an animal from adjusting to a restricted feeding schedule. This assumption means that exercise prior to a meal somehow regulates eating during the meal. However, Levitsky (1974) has shown that wheel running does not affect the duration or size of a meal (i.e., satiety) in free-feeding animals. The major effect of physical activity is to decrease the initiation or frequency of meals (i.e., appetite). The satiety hypothesis does not explain how activity affects appetite in free-food situations and satiety in the restricted-food condition.

For these reasons, we think it is premature to conclude that activity contributes to anorexia only by interfering with the animals ability to continue eating during a meal. We found that physical activity directly reduces food intake by altering the reinforcement value of eating. The motivational interaction of eating and exercise provides an alternative account of the effect of physical activity on appetite and satiety.

2.4. The Motivational Interaction of Eating and Exercise

One way to understand the activity–anorexia cycle is to consider how deprivation of food affects the reinforcing value of physical activity and how engaging in exercise changes the reinforcing value of eating. Our analysis of this problem is that eating and running are motivationally interrelated. Depriving an animal of food should increase the reinforcing value of exercise. Rats who are required to press a lever in order to run on a wheel should press the bar more when they are deprived of food. Additionally, running on a wheel should reduce the reinforcing value of food. Rats who are required to press a lever for food pellets should not work as hard following a day of exercise. We designed two experiments to test these ideas (Pierce et al., 1986).

We asked whether food deprivation increased the reinforcing effectiveness of wheel running. Although it is well known that food restriction increases physical activity (Finger, 1951; Hall and Hanford, 1954; Reid and Finger, 1955; Cornish and Mrosovsky, 1965; Duda and Bolles, 1963), the reason for this increase is not known. One hypothesis is that food deprivation produces general arousal that results in spontaneous activity (Teghtsoonian and Campbell, 1960). General arousal is not, however, a satisfactory account, because it fails to explain the motivational interrelations between eating and running.

A reinforcement analysis suggests that depriving an animal of one reinforcer (i.e., food) changes the reinforcing value of a different reinforcer (i.e., running). This is an interesting implication, because increased reinforcing effectiveness is usually achieved by withholding the reinforcing event. For example, in order to increase the reinforcement value of wheel running, access to the wheel is prevented. Surprisingly, our analysis predicts that the reinforcement value of wheel running will increase by withholding food rather than running.

We used nine young rats of both sexes to test the reinforcement effectiveness of wheel running as food deprivation changed. The animals were trained to press a lever to obtain 60 s of wheel running. Figure 7 shows the apparatus used for this research. The apparatus is a modified Wahmann running wheel. The wheel was equipped with a solenoid-operated brake. When the solinoid was activated, a rubber-tipped metal shaft contacted the wheel, causing the wheel to stop gradually. A retractable lever (Lehigh Valley Electronics) was mounted on a metal plate that fit over the entrance to the wheel. When the animal pressed the lever, a brake was removed and the running wheel was free to turn. After 60 seconds the brake was again activated and the rat had to press the lever to obtain more time to run.

Once lever pressing for wheel running was consistent, each animal was tested when it was at 75 and 100% of *ad libitum* weight. In order to measure the reinforcement effectiveness of wheel

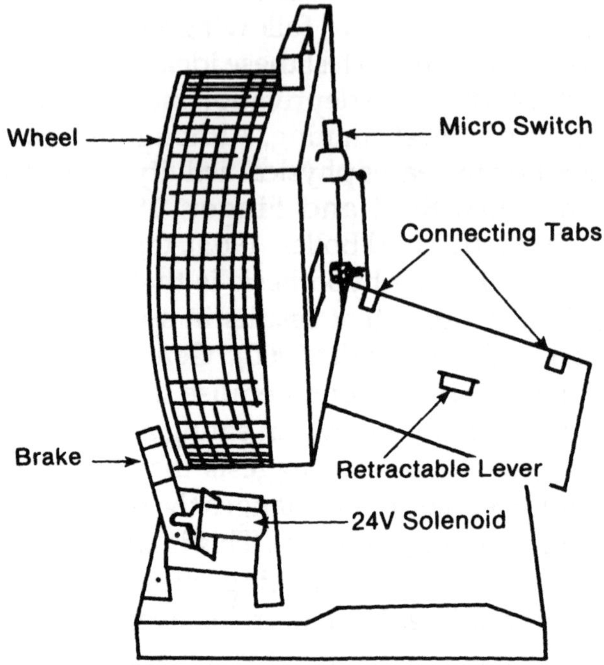

Fig. 7. Modified Wahmann running wheel. Apparatus was used to study lever pressing (responses) for 60 seconds of wheel running as a function of percentage body weight. The retractable lever was placed in the entrance to the wheel. When the rat met the response requirement, the lever retracted and the wheel-brake disengaged for 60 seconds. Republished from Pierce, Epling, and Boer (1986) Deprivation and satiation: The interrelations between food and wheel running. *J. Exper. Anal. Behav.* **46,** 199–210. Copyright 1986 by the Society for the Experimental Analysis of Behavior, Inc.

running, the animals were required to press the lever for an opportunity to run on an incrementing fixed-ratio schedule. Briefly, the rats were required to press five times to obtain 60 seconds of wheel running, then 10 times, 15, 20, 25, and so on. The point at which they gave up pressing for an opportunity to run was used as an index of the reinforcing effectiveness of exercise.

Inspection of Fig. 8 indicates that all animals pressed more for the opportunity to run when they were deprived food (i.e., at 75% body weight). The results also reveal a strong sex difference in the reinforcement effectiveness of wheel running. At both levels of deprivation, female animals emitted more responses for an opportunity to run than males. This may simply reflect

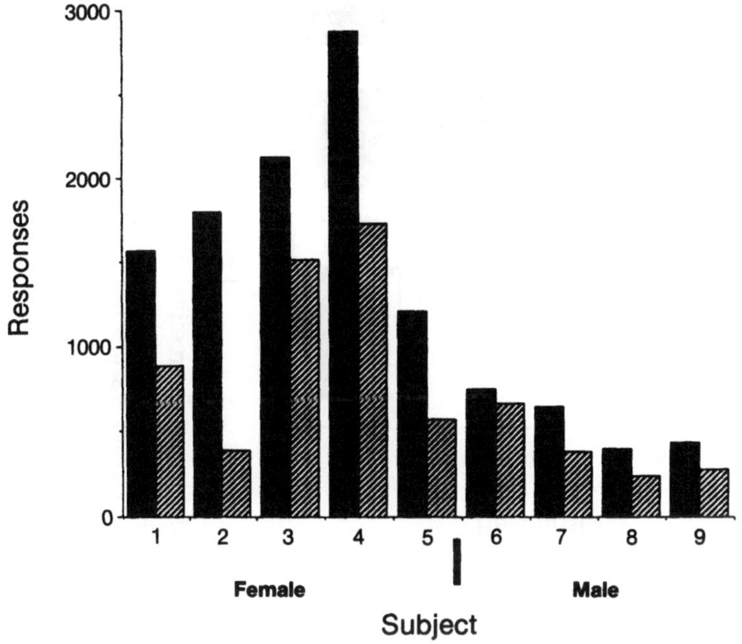

Fig. 8. Total number of bar presses (response) emitted on a progressive-ratio schedule of wheel running at 75 (■) and 100% (▨) of free-feeding body weight for five females and four male rats. Results indicate that food deprivation increases wheel running and that the effect is stronger for females. Republished with permission from Epling W. F., and Pierce W. D. (1989) Excessive activity and anorexia in rats, in *The Menstrual Cycle and its Disorders: Influences of Nutrition, Exercise and Neurotransmitters* (Pirke K. M., Wuttke W., and Schweiger I. U., eds.), Springer-Verlag, NY.

the lower body wts of females or it may be a result of some unknown biological or environmental difference between the sexes. This finding may have importance, since most human anorexias occur in women.

We also used a reversal design to investigate the reliability of the change in reinforcing value of wheel running. Subject 1 pressed a bar for 60 seconds of wheel running at 75, 100, and then again at 75% body weight. Subject 2 was first tested at 100, then at 75, and finally at 100% weight. Figure 9 shows the change in reinforcing value of running as a function of body weight. Both animals completed more bar presses at 75 than at 100% weight. Also, the effect is reliable, because the number of presses

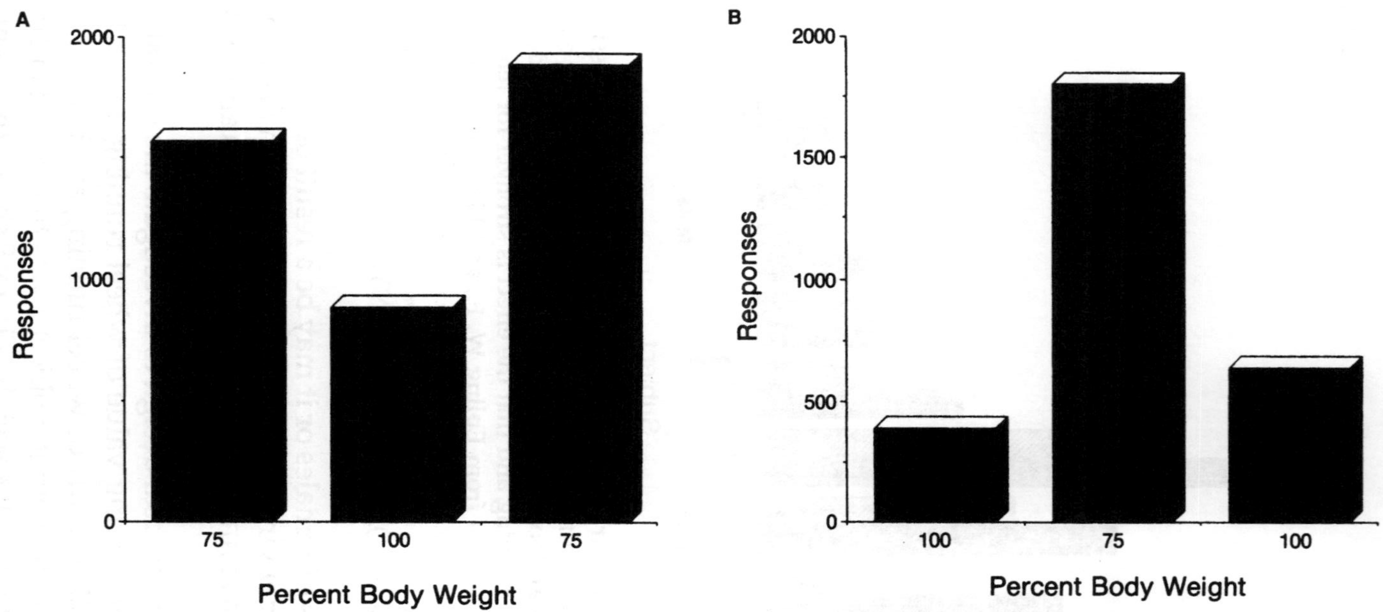

Fig. 9. Total number of bar presses (responses) for wheel running on a progressive-ratio schedule as a function of food deprivation (percentage body weight) for two rats. Subject one (A) was tested at 75, 100, and then again at 75% free-feeding weight; subject two (B) was tested at 100, 75, and then at 100% body weight. Results indicate that food deprivation increases responses for wheel running regardless of order of condition. Based on Pierce, Epling, and Boer (1986) Deprivation and satiation: The interrelations between food and wheel running. *J. Exper. Anal. Behav.* **46**, 199–210. Copyright 1986 by the society for the Experimental Analysis of Behavior, Inc.

is in the expected direction and reverses with changes in food deprivation.

In the second experiment of this study we investigated the effects of wheel running on the reinforcing effectiveness of food (Pierce et al., 1986). Four male rats were trained to press a lever for food pellets in a standard operant-conditioning chamber. When lever pressing for food reliably occurred, we tested for the effects of running on the reinforcing value of food. In this case, we expected that a day of wheel running would decrease the reinforcement effectiveness of food.

Test days were arranged to measure the value of food reinforcement. One day before each test, animals were placed in their wheels without food (20 h food deprivation). On some of the days before a test the wheel was free to turn; and on other days, it was not. When the wheel was free to turn, animals could choose to exercise or to remain in an attached cage. Three of the four rats ran moderately (<1000 revolutions) in their activity wheels on the day of exercise. One rat did not run (subject 40) when given the opportunity. This animal was subsequently forced to exercise on a motor-driven wheel. All animals were given rest periods in their home cages before the food-reinforcement test. This procedure was designed to reduce the effects of fatigue.

The reinforcement effectiveness of food was assessed by counting the number of lever presses for food as food became more and more difficult to obtain. To illustrate, an animal had to press five times for the first food pellet, 10 times for the next, then 15, 20, 25, and so on. As in the first experiment, the giving up point was used as an index of reinforcement effectiveness. Presumably, the more effective or valuable the reinforcer (i.e., food), the harder the animal would work for it.

When test days were preceded by a day of exercise, the reinforcing effectiveness of food decreased sharply. This effect can be seen in Fig. 10. Animals pressed the lever for food more than 200 times when the wheel was locked (i.e., no running). When running preceded test sessions (i.e., open wheel), three of the animals made <38 bar presses for food. Subject 40 did not run during the 20 h of open-wheel time, and lever pressing for food was not affected. However, this animal was forced to run on a motor-

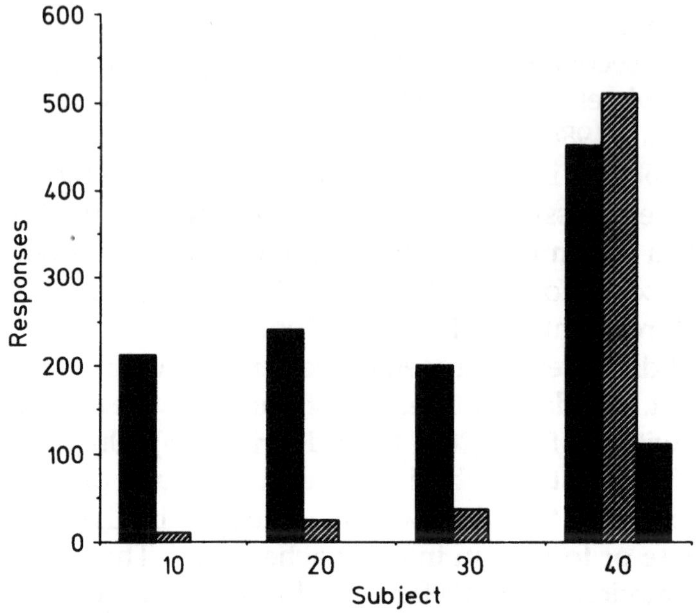

Fig. 10. Total number of lever presses (responses) on a progressive-ratio schedule of food reinforcement for locked-wheel (■) or open-wheel (▤) conditions. Subject 40 did not run when the wheel was open and exposed to a forced-running procedure The effect of forced running is shown by the third bar for this subject. Results show that wheel running, whether forced or voluntary, reduces responses for food. Based on Pierce, Epling, and Boer (1986) Deprivation and satiation: The interrelations between food and wheel running. *J. Exper. Anal. Behav.* **46,** 199–210. Copyright 1986 by the society for the Experimental Analysis of Behavior, Inc.

ized wheel. At this point the animal had gone 48 h without food—nonetheless lever pressing for food substantially declined.

Additional evidence from this experiment suggested that the effects of exercise were similar to those of feeding the animal. The same four subjects were returned to their home cages and given free access to food for 16 days. Following this, animals were deprived of food for 20 hours and exposed to a variable-interval (VI) food-reinforcement schedule. Subjects were repeatedly deprived of food and placed on the VI schedule until response rates stabilized (condition 1).

Response rates were also obtained when animals had free access to food in their home cages prior to a VI-test session (con-

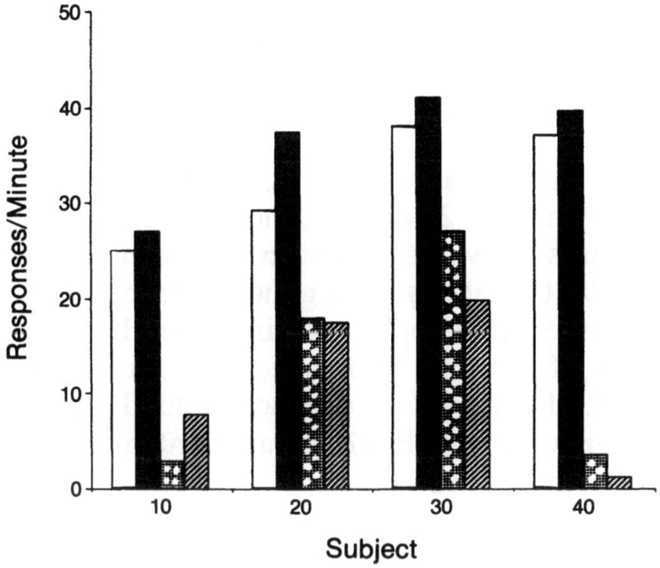

Fig. 11. Bar presses for food per minute (rate) on a variable-interval 30-s schedule. Rats were tested after 20-h food deprivation in the home cage (Condition 1; □), 20-h food deprivation in a locked wheel (Condition 2; ■), 20-h food deprivation in an open free-turning wheel (Condition 3; ▣), and after free-feeding in the home cage (Condition 4; ▨). Results indicate that wheel running lowers rate of response even when animals are food deprived. The effect of wheel running on food-related behavior is similar to free-feeding the animals. Based on a reanalysis of Table 5 from Pierce et al. (1986) Deprivation and satiation: The iterrelations between food and wheel running. *J. Exper. Anal. Behav.* **46,** 199–210. Copyright 1986 by the society for the Experimental Analysis of Behavior, Inc.

dition 2). The experiment compared bar presses per min on the food reinforcement schedule when animals did (condition 3) or did not (condition 4) run the day before a test session. In both of these conditions (3, 4), the rats were deprived of food for 20 hours. Subject 40 did not run during the day of exercise and was again forced to run on a motorized wheel.

The results of this experiment are shown in Fig. 11. When animals were deprived of food for 20 hours and placed in a wheel that could not turn (condition 4), rates of responding were similar to those obtained after 20 hours of food deprivation in the home cage (condition 1). This shows that mere placement in a running wheel does not affect the reinforcement value of food. In contrast, after one day of running (condition 3) response rates are substantially less, indicating that the exercise reduced the reinforcing effectiveness of food. Interestingly, the rate of bar pressing for food following a day of running (condition 3) is similar to the rates obtained when animals ate freely before the test session (condition 2). Overall, these results suggest that exercise decreases the value of food reinforcement and that this effect is similar to that of satiating the animal on food.

In summary, food deprivation increases the reinforcement effectiveness of running. Spontaneous or forced running decreases the reinforcement effectiveness of food. These two relationships provide a behavioral account of the processes underlying activity anorexia.

3. Biological Basis of Activity Anorexia

Activity anorexia results from the interrelationships of (a) deprivation and food schedule on physical activity and (b) activity on food consumption. Briefly stated, strenuous locomotor activity works to suppress appetite. This decrease in the value of food reinforcement serves to affect food schedule and/or deprivation, which further increases activity. These factors—excessive physical activity and food restriction—appear to be a set of conditions sufficient for producing activity anorexia.

3.1. Evolution and Activity Anorexia

An evolutionary account of eating and activity points to the survival value of such behavior. During times of food scarcity, organisms can either stay and conserve energy or become mobile and travel to another location. The particular strategy adopted by a species depends on natural selection. Thus, if travel led to reinstatement of food supply and remaining resulted in starvation then those animals that traveled would gain reproductive advantage. This means that locomotion induced by changes in food supply became a major evolutionary adaptation of some species.

Many species (e.g., rats, mice, gerbils, guinea pigs, and so on) increase locomotion when deprived of food (Collier, 1969; Cornish and Mrosovky, 1965; Epling et al., 1981; Vincent and Pare, 1976). In terms of primates, Loy (1970) observed a colony of free-ranging rhesus monkeys who inhabited an island off the southeast corner of Puerto Rico. During early July, 1968, a severe food shortage occurred. The island was covered with heavy vegetation and the monkeys supplemented their diet by eating plants and an occasional insect. However, their main source of food was commercial monkey chow that was distributed at several feeding stations around the island. Unfortuntely, the June food shipment was delayed and the animals experienced a severe famine.

These monkeys showed a marked reduction in social behavior. For example, monkeys spend a considerable amount of time grooming one another. This behavior and sexual activity substantially decreased during the famine. Most important, Loy reported that there was a dramatic increase in food-directed physical activity. He stated,

> There was a great increase in the amount of time spent foraging from the vegetation cover. Some monkeys, especially the young, appeared to lose weight. A few of the monkeys that subsequently died appeared lethargic immediately before they were missed from their group. These same monkeys foraged *constantly* up to their death or disappearance (p. 265; emphasis ours).

The increase in foraging is interesting and consistent with the physical activity observed in laboratory animals exposed to food shortage. Like foraging, the apparently "spontaneous" running of laboratory animals is controlled by food allocation, and this suggests that wheel running is displaced food-seeking behavior.

There are other primates who show increased locomotion when their food supply decreases. Devore and Hall (1965) have studied the ecology of baboons living in different regions of the African continent. They have studied the movement and range of baboon troops as related to food supply. These researchers noted,

> Studies in southern Africa, Kenya, and on hamadryas in Ethiopia all indicate the average distance travelled by baboons during

> a day is three miles. This measures the shortest distance from the sleeping place in the morning, along the route travelled during the day, and back to a sleeping place in the evening. The distance the group or any individual baboon walks is much greater, since feeding activity is meandering. There is also considerable latitude in the distance travelled on any particular day. This varies from only a few yards (when a Kenya group sleeps in a fig tree and feeds in and under the tree throughout the following day) to a maximum distance of 12 miles (observed once for a group of 65 in South-West Africa). The contrast in available food between a heavily laden fig tree and the sparse vegetation of the study in South-West Africa suggests that *available food is the single most important factor affecting length of day range*. This is supported by observations at different seasons, which show that during the seasons when suitable vegetable foods are most plentiful average day ranges are shorter... A second reason for the longer average day range in the dry season (Kenya) or winter (Cape) is that a group is more likely to shift to a *new core area*...during these seasons. It is likely that this shifting is also related to the available food supply, representing movement to a new locus of foraging activity after reduction of the available food in the former locus (pp. 31,32; emphasis ours).

These observations of baboon behavior make it clear that there are two aspects to food-seeking behavior. Within a foraging area, these animals will travel up to 12 miles a day to obtain food. However, when the regional food supply drops below some optimal level, the baboons may travel large distances to another foraging area. Again, these findings are consistent with the excessive increase in wheel running of laboratory animals deprived of food.

Apparently, many species have adapted to food shortages by increasing physical activity. When food supply decreases, travel to another location may be necessary. Travel may be set off by decreasing contact with food, reduction in the caloric value of food items, and by internal cues associated with loss of body weight. Current evidence suggests that internal cues associated with weight loss play a major role in initiating locomotor activity (Kanarek and Collier, 1983).

In our laboratory, animals show large increases in wheel running at several depletion points (typically between 85 and

75% body weight). Interestingly, approx 10% of the animals do not display excessive activity. This finding suggests that the tendency to travel when food is depleted is distributed in the population. Some animals leave when weight loss is minimal, some at more severe levels, and some stay and do not travel. These differences in behavioral characteristics make evolutionary sense.

When food is becoming scarce, natural variation ensures that some animals leave the location and travel to another area. This same variation guarantees that other members of the species remain in the same location. Those individuals who travel may, or may not, find a new supply of food. If food is contacted, then these individuals are more likely to survive and reproduce. Thus, the tendency to travel during a famine increases in the population. On the other hand, the best strategy may be to stay, since the food shortage could be temporary. Also, the amount of food in the location may now be sufficient for those few animals that remain. If staying leads to survival and reproductive success, then this tendency is selected. With both kinds of selection operating, it is likely that a range of variation in the tendency to travel will be maintained in the population.

A major problem for an evolutionary analysis of activity anorexia is to account for the decreased appetite of animals who are traveling to a new food patch. The fact that increasing energy expenditure is accompanied by decreasing caloric intake seems to violate common sense. From a homeostatic (i.e., energy balance) perspective, food intake and energy expenditure should be positively related.

In fact, this is the case if an animal has the time to adjust to a new level of activity and the food supply is not greatly reduced (Tokuyama et al., 1982). However, during a periodic depletion of food sources, travel should not stop when food is infrequently contacted or is difficult to acquire, since stopping to eat would be negatively balanced against reaching a more abundant food patch. Increasing contact with food would signal a replenished food supply and reduce the tendency to travel. Our research indicates that the distance covered in a day is regulated by the amount of food consumed, rather than by improvement in body weight (Russell et al., 1987). Apparently, the decision to travel is related to weight loss and the decision to stop is regulated by food consumption.

Daily food consumption depends on the density and size of the food items contacted. In this regard, it is important to recall that the reinforcing value of food is low when physical activity is high. This process had selective advantage, since animals that found food less attractive would travel more quickly to a place where food was more plentiful.

A decline in the reinforcing value of food means that animals will not work hard for nourishment. When food is scarce, considerable effort may be required to obtain it. For this reason animals ignore food and continue to travel. However, as food becomes more plentiful and the effort to acquire it decreases, the organism begins to eat. Food consumption and increasing body weight lower the reinforcement value of physical activity, and travel stops. On this basis, animals who expend large amounts of energy on a trek or migration become anorexic.

The evolutionary processes that produced anorexia during a famine-induced travel were probably important to early humans. It is difficult to be sure, but human tribes of hunters and gatherers likely reacted to sudden food shortages in the same way as other primates. The primate evidence suggests that natural selection favored travel in times of food scarcity *(see above)*.

Modern humans possess the genetic complement that favors physical activity in times of food shortages. Most humans in Western culture do not experience natural famines. However, a culture that encourages dieting and exercise may inadvertently trigger the activity–anorexia cycle. That is, in humans, sociocultural factors can produce and maintain conditions of food depletion even though food is readily available. Thus, anorectics may not experience increased contact with food, activity continues to spiral upward, and food consumption remains low or declines.

An evolutionary analysis provides one type of understanding of activity anorexia. However, this explanation is based on "ultimate" causation, and the genetic history is indirectly inferred. Another kind of biological explanation is based on "proximate" causation, involving specific physiological mechanisms that link eating and physical activity.

3.2. Physiology and Activity Anorexia

Recent evidence points to the role of endogenous opiates as mediators of the relationship between eating and physical activity. In a review article, Dum and Herz (1987) suggested a link between the endogenous opiates and motivation for food. Also, Marrazzi and Luby (1986) proposed that anorexia in humans resulted from an addiction to endogenous opiates. They stated,

> Opioid systems in the brain are assumed to play a fundamental role in adaptation to starvation and the down–regulation of metabolic set points. There is now substantial evidence that opioids are mobilized in states of prolonged food deprivation, and we are hypothesizing that they are the substrate for an auto-addictive process responsible for the relentlessness of chronic anorexia nervosa (p. 191).

In terms of activity anorexia, the endorphins are the most interesting of these opiates.

We suggest that β-endorphin, and perhaps other brain opiates, mediate(s) the relationship between increasing physical activity and decreasing food intake. Generally, declining body weight stimulates an increase in physical activity. Physical activity increases production of β-endorphin and this increase reinforces exercise (i.e., runner's high; Shainberg, 1977).

There is evidence that β-endorphin and other brain opiates have opposite effects on appetite, depending on body weight. At normal or obese weight a variety of animals will decrease eating when the opiate-blocking agent naloxone is injected (Sanger, 1981). Since naloxone blocks endogenous opiates, this research suggests that the opiates stimulate eating when animals are at normal weight or higher. On the other hand research indicates that opiates *suppress* appetite when exercise is intense or when body weight is low.

In terms of the suppressive effects of endogenous opiates, Davis and his associates (Davis et al., 1985) found that exercise-trained rats ate less when given an injection of 2-deoxy-D-glucose, which indirectly induces release of endogenous opioids. In addition, Sanger and McCarthy (1980) found that rats who were deprived of food for 24 hours ate less when injected with morphine. Since morphine has effects similar to those of the endogenous opiates, this suggests that opiates reduce appetite *under conditions of food deprivation* (Epling and Pierce, 1988). A similar effect seems to occur in humans who are starving. A recent study found that 2-deoxy-D-glucose increased judgments of hunger in normal-weight people, but decreased hunger in anorexic patients (Nakai et al., 1987).

A reduction of food intake by anorexic patients may result from the intense physical activity that these individuals exhibit (Katz et al., 1986; Kron et al., 1978). There is evidence that intense physical exercise increases endorphin levels. Research has shown that intense exercise by humans resulted in elevated levels of plasma β-endorphin (Colt et al., 1981; Farrell et al., 1982).

Other researchers have found that runners had a higher concentration of endorphin after completing a marathon (Appenzeller et al., 1980). Endorphin levels remained elevated for up to two hours after the race. McMurray and his coworkers investigated the effects of exercise intensity on release of β-endorphin (McMurray et al., 1987). Twenty men and women in good health volunteered to ride a stationary bicycle. The workload on the bicycle was increased from low to high intensity. Blood samples were obtained before and after each exercise session. β-endorphin increased when exercise levels were at 80% of the individual's maximum output. Generally the evidence is clear that intense exercise produces increases in plasma β-endorphin.

According to the activity-anorexia hypothesis, physical activity suppresses food intake. We have shown that, under starvation conditions, endorphins reduce food intake in animals and decrease reports of hunger in humans. In addition, strenuous physical activity increases blood concentrations of β-endorphin. If activity anorexia is mediated by endorphins, then anorectics should show increased levels of endogenous opiates. In fact, elevated opiate levels have been measured in diagnosed anorectics.

Kaye and associates found higher than normal levels of endogenous opiates in the cerebrospinal fluid of anorexia nervosa patients (Kaye et al., 1982). These elevated levels were found in patients who were severely underweight, but not in patients who were close to normal body weight. This finding is somewhat controversial, since Gerner and Sharp (1982) carried out a similar study and found that β-endorphin levels were normal for anorectic patients. Unfortunately, the weight status of the patients in this study was not reported. The discrepancy between the two studies may therefore be a result of the difference in patients' weights.

There is some evidence that endogenous opiates function to suppress appetite in anorexic patients. A study by Moore and colleagues found that weekly weight gain was substantially increased when anorexia patients were given a constant intravenous infusion of the opiate blocker naloxone (Moore et al., 1981). Although food intake was not measured in this study, it is difficult to see how a ten-fold improvement in weight gain could be accomplished without a substantial increase in eating. Nonetheless, a possibility is that endorphins affect the conservation of nutrients and lower energy expenditure (Margules, 1979). An opiate blocker would therefore contribute to weight gain in the absence of increased food intake. However, Reid and Wideman (1982) failed to confirm the conservation hypothesis in starving rats. This suggests that the opiate blocker resulted in more eating by anorexic patients. Finally, recall that anorectics reported less hunger when given an opiate stimulant. Overall, these findings are consistent with an opioid-induced suppression of eating in anorexic patients.

We have previously noted that the release of endogenous opiates may function as reinforcement for physical activity. The reinforcement hypothesis suggests that injection of an opiate blocker will decrease the intense wheel running of anorexic animals. This is because the euphoric effects of opiates are diminished by blocking the receptors. In our laboratory, research by Douglas Boer has explored the effects of naloxone on the anorexic running of six male rats (Boer et al., 1988). In this study, animals were made anorexic by feeding them for 90 min/d and providing a running wheel. Once wheel running exceeded 5000 revolutions/d (5 km), each animal was given a dose of naloxone (0.5 mL) or saline (0.5 mL) on alternate days. The average number of wheel revolutions was approx 5800 on saline days, and 4800 on days when the animals were injected with the opiate blocker (naloxone). Importantly, each animal showed less running on the days when naloxone was given. These findings provide preliminary evidence for the hypothesis that anorexic running is strengthened and maintained by opiate release.

At this point, it is useful to summarize the evidence for activity anorexia mediated by endogenous opiates. Starvation and loss of weight increase physical activity. Physical activity produces an increase in endogenous opiates. One important property of opiates is their "euphoric" or reinforcing effects. Because of these effects, physical activity increases and this produces a further elevation in endogenous opiates. Importantly, when body weight is reducing as a result of food restriction, high levels of opiates decrease appetite. Notably, anorexic patients are hyperactive and show high levels of endogenous opiates. These patients also gain weight when endogenous opiates are blocked and do not feel hungry when opiates are stimulated.

3.3. Anorexia, Exercise, and Fertility

Anorexic women and female athletes often experience problems with menstrual cycles. There appear to be two physiological processes that regulate female fertility. One process involves estrogen changes as a result of declining body fat. Frisch (1988) has accumulated evidence that menstrual cycle is regulated by the relative amount of fat or adipose tissue. From this perspective, physical activity affects menstrual cycle because exercise contributes to lean-body mass. We are essentially in agreement with Frisch's account of female fertility; however, there is evidence that a second process contributes to menstrual problems of anorectics and women athletes. This process involves the effects of exercise on the release of endogenous opiates that, in turn, affect menstrual cycle. Problems of menstruation during activity anorexia are apparently attributable to both of these processes.

The most prominent feature of anorexia is starvation. According to Frisch, problems of menstruation usually occur when a woman becomes too lean. Evolution has ensured that reproduction will not occur at times when food is scarce. Frisch states,

> It is not suprising that the reproductive function...falters when a woman becomes too lean. Such a response would have given our female ancestors a selective advantage by ensuring that they conceived only when they could complete a pregnancy successfully. Reproduction, after all, requires energy, or calories: some 50,000 to 80,000 calories to produce a viable infant and then from 500 to 1000 calories a day for lactation... In ancient times, when the food supply was scarce or fluctuated seasonally and when breast milk was a newborn's only food, a woman who became pregnant when she lacked an adequate store of body fat—the most readily mobilized fuel in her body—could have endangered both her own life and that of her developing fetus and newborn infant (p. 88).

The central idea is that women must attain a sufficient lean-to-fat ratio in order to menstruate. The hypothalamus releases gonadotropin-releasing hormone (GnRH) that starts the process of menstruation and ovulation. At the present time, it is not clear how the lean-to-fat ratio affects the release of GnRH. Loss of body fat may lower body temperature or change metabolism and this may signal the hypothalamus. Frisch favors an account based on changes in estrogen level resulting from loss of body fat. When the amount of fat declines relative to leanness, a relatively inactive form of estrogen increases. Presumably, these changes in estrogen signal the hypothalamus to stop the release of GnRH. Since GnRH regulates the onset of menstrual cycle, a disruption produces irregularity, delay, or cessation of the monthly cycle.

In terms of activity anorexia, Frisch's model of fertility suggests that physical activity contributes to fat reduction, which, in turn, leads to menstrual problems. In this regard, Frisch indicated that

> several recent studies, including our own, have shown that dieting is not the only way women become lean enough to impair their hypothalamic function and disrupt menstruation. Well-trained athletes of all kinds, such as runners, swimmers and ballet dancers, have a high incidence of delayed menarche, irregular cycles, and amenorrhea. This pattern implies that exercise could be the cause—presumably by building muscles and reducing fat, thus raising the ratio of lean mass to fat (p. 93).

From this perspective the hyperactivity of anorectics may contribute to problems of menstruation by altering the lean-to-fat ratio. However, there is evidence that a second process may be involved. This process occurs when physical activity increases the release of endogenous opiates.

Physical activity can produce problems of menstruation for women athletes. Olympic and college track and field athletes often experience delayed onset of menstrual cycle (Cumming and Rebar, 1983; Malina et al., 1973). Ballet dancers who are very active may also have delays of menstruation (Frisch et al., 1980). The earlier young women begin training, the greater the chance of delayed menarche (Frisch et al., 1981). Importantly, research findings show that almost 50% of exercising women with amenorrhea are not excessively lean (Lutter and Cushman, 1982). Taken together, the evidence suggests that physical activity, *independent* of the lean-to-fat ratio, can affect menstrual cycle.

We have already documented the increase in endogenous opiates that occurs during intense exercise. Cumming and Rebar (1985) suggest that opiate release could be involved in exercise-associated problems of menstruation. Opiate levels are known to affect the release of luteinizing hormone through their action on the hypothalamus and GnRH (Quigley et al., 1980; Ropert et al., 1981). Luteinizing hormone (LH) plays a major role in the control of menstrual cycle. Opiates seem to decrease the release of LH in women who exercise excessively (Warren, 1983).

The direct effect of exercise on LH release has been investigated by Cumming and his associates (Cumming et al., 1985). These researchers compared LH levels of female long-distance runners to those of sedentary women of similar height and weight. All women were experiencing normal menstrual cycles at the time of the study. Blood samples were collected from an intravenous catheter every 15 minutes for six hours. The number of LH pulses and the total amount of LH were lower in the runners. Since none of the women had exercised for 24 hours before the study, training seems to have a long-lasting effect on LH release.

The opiate hypothesis suggests that LH suppression is a result of increased levels of endogenous opiates generated by exercise. One test of this idea is to give the opiate blocker naloxone to exercising women. An opiate blocker should increase LH levels in these women. In fact, Cumming and Rebar (1985) conducted such a study and found that LH levels did not improve in women with exercise-induced amenorrhea. In contrast, McArthur and coworkers found that naloxone produced the expected increase in LH levels (McArthur et al., 1980).

These contrasting findings may have occurred because of the conditional effects of naloxone. Petraglia and associates studied the LH response in 18 healthy women (Petraglia et al., 1986). Twelve young women were given naloxone at different times in their menstrual cycle. The six older women were postmenopausal and were also given the opiate blocker. The effects of naloxone on LH depended on age and cycle phase. Naloxone increased LH levels only in normally menstruating women during the luteal phase of the cycle. Since naloxone produces variable effects in healthy women, it is also likely that the drug has inconsistent effects in women with exercise-associated amenorrhea. Thus, the opiate hypothesis of menstrual problems is difficult to test by administering opiate blockers.

A more direct test of the opiate hypothesis involves the administration of opioid drugs. An opiate agent, such as morphine, binds at the same receptors as the endogenous opiates. For this reason, morphine and natural opiates should produce similar effects on LH levels. In the study by Petraglia and associates, the same women were given the opiate drug morphine. Interestingly, as expected, morphine decreased LH in these women. This decrease in LH did not depend on age or cycle phase. The findings suggest that increases in endogenous opiates produce decreases in LH levels. Unfortunately, there are no studies of the effects of opiate drugs on LH levels of exercising women. Such a study would provide a stronger test of the opiate account of menstrual problems during activity anorexia.

In summary, menstrual problems during activity anorexia can be attributed to three major processes. The most obvious process is that starvation produces lean-body mass. When women are excessively thin, estrogen levels change, and this affects menstrual cycle. A second process is based on physical activity affecting the lean-to-fat ratio. As women become more fit as a result of exercise, the proportional amount of body fat decreases, and this also affects estrogen level and menstrual cycle. Finally, exercise increases the release of endogenous opiates. These natural opiates are known to decrease LH, and this hormone is involved in the regulation of menstrual cycle. In this case, physical activity directly influences menstruation and does not depend on changes in body weight.

We have shown that activity anorexia is a natural response to an unexpected food shortage. When food is depleted, natural selection has ensured that most individuals leave their territory and travel to another location. This same selection pressure guarantees that some individuals stay behind. When traveling is the dominant tendency, research suggests that the "decision to go" is a response to significant loss of weight. Once travel has begun, animals refuse to eat—they become anorexic. This refusal is apparently a result of the declining value of food reinforcement. We have shown that this decline ensures that travel continues until the individual contacts an abundant food supply. As food density increases, the animal begins to eat and the reinforcing value of travel declines. Many species, including humans, seem to have evolved this kind of response to unpredictable food shortages.

These evolutionary tendencies are regulated by specific physiological processes. Starvation and loss of weight increase physical activity, and physical activity stimulates the release of brain opiates. Brain opiates reinforce the continuation of physical activity. Because of these effects people show increases in exercise, sports, and fitness that further increase brain opiates. When body weight is low, opiates decrease appetite. Notably, anorexic patients are hyperactive and show high levels of brain opiates. Drugs that block opiates cause anorectics to gain weight. These patients also indicate that they do not feel hungry when opiates are stimulated.

The relationship between exercise and brain opiates appears to account for the loss of normal menstrual function in female anorectics. Exercise contributes to increased opiate release that, in turn, affects hormones that regulate menstrual cycle. Thus, women who are diagnosed as having anorexia nervosa as a result of extreme thinness and menstrual problems are apparently exhibiting activity-based anorexia.

4. Extending Activity Anorexia to Humans

The behavioral and biological relationships that contribute to activity anorexia in animals may also account for the increasing incidence of anorexia in Western culture. In order to generalize the animal model to instances of human self-starvation, it is necessary to show that in humans

1. Physical activity is associated with anorexia;
2. Physical activity decreases food intake;
3. Lower food consumption increases physical activity;
4. The onset of anorexia develops in a manner that is similar to the pattern observed in laboratory animals;
5. The psychological and physical symptoms follow, rather than precede, starvation, and
6. Social reinforcement based on cultural practices encourages people to combine diet with exercise.

4.1. Physical Activity and Anorexia Nervosa

Excessive physical activity has been frequently noted by researchers of anorexia nervosa (Blitzer et al., 1961; Crisp, 1965; Halmi, 1974; King, 1963; Kron et al., 1978; Slade, 1973; Thoma, 1967). Kron and associates (1978) conducted a retrospective study of hospitalized anorectics and concluded that hyperactivity is a central feature of anorexia nervosa. Their summary of their findings makes it clear that excessive physical activity is critical to the onset and process of anorexia:

> We reviewed the charts of 33 patients hospitalized with this illness during the past 10 years. ...In 25 of the 33 charts, the presence of hyperactivity was recorded; only one patient was specifically noted to show unremarkable physical activity. ...These preliminary findings suggest that "hyperactivity" is an early and enduring clinical feature of anorexia nervosa and not merely secondary to either a conscious attempt to lose weight or weight loss *per se* (p. 439).

Generally, there is strong evidence that excessive physical activity is associated with human anorexia.

4.2. Exercise and Food Intake

Humans reduce food intake when physical activity becomes excessive. Studies with humans have shown that exercise is related to a decrease in caloric intake (Edholm et al., 1955; Epstein et al., 1978; Johnson et al., 1972; Mayer et al., 1956; Watt et al., 1976). Edholm et al. (1955) reported that military cadets ingest less food on drilling days than they do on days of lower activity. Caloric intake on days of greater activity was significantly depressed in comparison with other days. Epstein et al. (1978) found that obese school children would voluntarily decrease food intake following a prelunch exercise period.

The effects of physical activity on food intake can also be observed when exercise is reduced. Mavissakalian (1982) reports the successful treatment of two 17-year-old anorexic females both of whom were admitted to a hospital following excessive weight loss. A predominant symptom was compulsive exercising, and this behavior was present for both patients. A behavior-modification treatment package required increased food consumption and a one-hour period of rest following each meal. The rest component was added to specifically prevent the compulsive exercising which the patients manifested. At discharge, both patients' families were instructed in procedures required to continue treatment. With continuing treatment, both women became sedentary and overweight. This suggests that the reduction of physical activity caused an increase in food intake. Generally, increased physical activity reduces food intake, and decreased exercise increases it.

4.3. Food Intake and Physical Activity

Studies suggest that increased body weight and food consumption are associated with inactivity (Bloom and Eidex, 1967; Mayer, 1965). Additionally, increased mobility and arousal have been described for starving populations exposed to economic hardship (Howard, 1839), malnourished German school children during World War One (Blanton, 1919), and starving people in Germany after World War Two (Russell-Davis, 1951). Of course, when starvation becomes extreme, activity decreases. The survivors of Belsen concentration camp were close to death and very inactive when liberated (Lipscomb, 1945). This observation parallels the sudden decline in wheel running by laboratory animals who are near death as a result of activity anorexia.

This same inactivity was noted by Keys and colleagues (Keys et al., 1950) when 36 conscientious objectors were required to undergo six months of semistarvation. Although Keys emphasized the inactivity of the men, his procedure may have masked the expected increase in physical activity when food is restricted. In this study, the men were required to engage in moderately strenuous exercise. They had to participate in a regular physical-activity program, hike 22 miles a week, and walk 2–3 miles a day to the mess hall. Each man was also required to do a weekly 30-min test on a motor-driven treadmill at 3.5 mph on a 10% grade. Nonetheless, there is evidence that food deprivation did engender excessive physical activity. Keys and colleagues reported that

> ...some men exercised deliberately at times. Some of them attempted to lose weight by driving themselves through periods of excessive expenditure of energy... (p. 828).

Overall, the study by Keys et al. and other evidence suggests that humans increase physical activity when food intake declines and become inactive when starvation is severe.

4.4. Temporal Sequence of Anorexic Symptoms

Another way of generalizing the theory of activity anorexia is to show that the onset of anorexia in humans is consistent with the pattern observed in animals. In the laboratory, food restriction generates excessive physical activity that interferes with eating. A similar pattern has been reported for anorexic patients (Beumont et al., 1983). Beumont and associates asked 25 anorexic patients to identify their symptoms and the order of occurrence. Of the 28 reported symptoms, only "manipulating food servings" and "increased sport activity" were present in all patients. Generally, the ordering of the symptoms indicated that behaviors involving dieting and food restriction occur early in the sequences. These changes in food allocation were followed by increased sport activity and "exercising alone."

In humans, anorexia may also be initiated by sudden increases in physical exercise that, in turn, alters dietary intake. Katz (1986) has described two patients "...in whom participation in long distance running clearly preceded the appearance of anorexia nervosa and appeared to play a role in its onset. ..." In one case, a 32-year-old physician began to jog in order to increase his fitness. When the running reached 35 miles per week, he began to lose weight (from 175 to 135 pounds in five months). He became obsessed with his weight and food intake, preoccupied with preparing gourmet food, and gave up eating meat. Eventually, the man's weight dropped to 125 pounds and he was running 50 m/wk. For the first time, he started to binge eat and vomit after the meal. Now he was subsisting on salads, coffee, or diet soda, and his weight went to 115 pounds. Although extremely thin, the man said he felt fat and frequently asked his wife for reassurance that he was not overweight. He became depressed, more obsessed with his weight and diet, and lost interest in sexual intercourse. A similar history was reported for the second patient, who also developed anorexia following excessive physical activity.

4.5. Anorexic Symptoms and Starvation

The Katz study is also important because many of the psychological and physical symptoms of anorexia nervosa followed, rather than preceded, activity-induced starvation. In other words, preoccupation with food, bingeing, vomiting, distortion of body image, loss of libido, and depression came after the onset of excessive exercise and food restriction. This suggests that these "symptoms" are not causes of anorexia.

Further evidence that symptoms are not causes is provided by the experimental study of semistarvation in conscientious objectors (Keys et al., 1950). The men were young, healthy, and psychologically normal when they began the study. The men ate normally for the first three months of the study, and various measures of personality, eating, and behavior were gathered. Over the next six months, the food ration was cut in half and the men were reduced to approx 75% of their original weight. Finally, the men were returned to a normal diet over the last three months of the experiment.

Starvation resulted in dramatic changes of behavior, personality, and physical well-being. The behavioral and personality changes were, in many instances, consistent with the psychological symptoms of anorexia nervosa. The men became preoccupied with food. Their conversation was dominated by food-related issues; their reading material focused on cooking; some began collecting cooking utensils. Eating habits began to change. The men spent the day planning how to eat the small, daily food ration. On some occasions, the men would cut their food into tiny bites so that each morsel could be appreciated.

Some of the men became bulimic. They would lose control, gorge themselves, and then feel guilty about breaking the diet. In some instances, these bulimic episodes were followed by self-induced vomiting. There were other unusual changes in their emotional well-being. For example, some men reported long periods of depression, irritability, anxiety, and outbursts of anger. All of these effects have been reported as symptoms of anorexia nervosa.

The men were given the Minnesota Multiphasic Personality Inventory (MMPI). Test scores significantly changed with starvation. During the baseline phase, the men scored in the normal range on all personality dimensions. When subjected to starvation many of the men showed serious changes in personality. According to test scores, the men became neurotic, and two of the volunteers showed signs of severe psychosis. These emotional and personality changes did not immediately recover when the men were returned to normal eating conditions. Many professionals have suggested that the personality and emotional disposition of the anorexic causes self-starvation. The Keys et al. (1950) study makes it clear that personality and behavioral changes are the result of starvation, rather than the cause.

4.6. Social Reinforcement and Anorexia

If the symptoms of anorexia are not causes of self-starvation then other factors must determine the onset and maintenance of anorexia. We have suggested that the interrelations between food restriction and excessive physical activity are the critical determinants. In the laboratory, the experimenter imposes the food restriction on the animals, and this initiates the activity–anorexia cycle. In order to generalize the theory of activity anorexia it is necessary to specify the conditions that "impose" food restriction or exercise on humans. Our contention is that contingencies of reinforcement set by Western culture encourage people to diet and exercise, thereby increasing the chances that an individual will be trapped by the biobehavioral processes of activity anorexia.

Our culture currently values a thin, trim appearance in women (Lakoff and Scherr, 1984; Mazur, 1986) and physical fitness in both sexes (Beck et al., 1976; Garner et al., 1985). Several researchers have noted that, in Western culture, the mass media conveys these values (Bruch, 1978; Kurman, 1978; Wooley and Wooley, 1980), and people learn by observation (Bandura, 1977) to uphold and promote the beauty standards. Acceptance of the thinness and fitness values means that people provide social approval, economic advantages, and privileges to individuals who attain the cultural beauty standards (Brigham, 1980; Green et al., 1984; Jones et al., 1978; Umberson and Hughes, 1984; Unger et al., 1982). Since dieting and exercising are ways of achieving these standards, this behavior is reinforced (Garner et al., 1985). Importantly, social conditioning ensures that some people will inadvertently combine dieting with physical activity in a way that produces activity anorexia (Davies and Furnham, 1986; Dwyer et al., 1970; Jakobovits et al., 1977; Miller et al., 1980).

5. Conclusion

The animal model of activity anorexia specifies the major determinants of a biologically based form of self-starvation. Food restriction induces high rates of physical activity. As activity increases relative to baseline levels, food intake is suppressed. This results in a cycle of exponentially increasing physical activity that further reduces food intake. Research indicates that the reinforcing value of food declines with exercise. Also the reinforcement effectiveness of physical activity increases with food deprivation. Together, these relationships provide a behavioral account of the activity–anorexia cycle.

The behavioral relationships reflect a particular evolutionary history. An evolutionary account points to the survival value of anorexia during times of food scarcity. Reduction of food supply induces travel because animals that became mobile found food, survived, and reproduced. During a migration, stopping to eat was negatively balanced against reaching another food patch. Thus, natural selection resulted in a reciprocal relation between eating and physical activity. This relationship is based on specific physiological mechanisms.

Exercise increases endorphins, and endorphins decrease appetite when food is restricted. The release of brain opiates automatically reinforces exercise and increases it to excessive levels. Anorexia patients and women athletes show elevated opiate levels. Both women athletes and anorexics have a high incidence of menstrual problems. Elevated opiate levels reduce LH, and this hormone is associated with the control of menstrual cycle. The survival value of decreased fertility during famine and migration is apparent.

Food restriction and physical exercise are common practices of individuals in the Western culture. People in this culture accept and promote the values of thinness and fitness. Since many people diet, exercise, or both, it is likely that some will initiate the biobehavioral processes of activity anorexia. Although these anorexias are diagnosed as "nervosa," evidence suggests that the psychological and physical symptoms do not cause the starvation. Our theory suggests that most cases of anorexia nervosa are in fact instances of activity anorexia.

Acknowledgments

We would like to thank Alberta Mental Health Research Fund who founded several of the experiments reported here.

Reprint requests should be addressed to W. F. Epling or W. D. Pierce, Center for Experimental Sociology, The University of Alberta, Edmonton, Alberta, Canada, T6G 2H4. This paper is based on chapters of our book *Solving the Anorexia Puzzle: A Scientific Approach* (Hogrefe and Huber, Toronto, in press).

References

Ahrens R. A., Bishop C. L., and Berdanier C. D. (1972) Effect of age and dietary carbohydrate source on the response of rats to forced exercise. *J. Nutr.* **102,** 241–248.

Appenzeller O., Standefer J., Appenzeller J., and Atkinson R. (1980) Neurology of endurance training 5: Endorphins. *Neurology* **30,** 418–419.

Bandura A. (1977) *Social Learning Theory* (Prentice-Hall, Englewood Cliffs, NJ).

Beck S. B., Ward-Hull C. I., and Mclear P. M. (1976) Variables related to womens somatic preferences of the male and female body. *J. Per. Soc. Psychol.* **4,** 1200–1210.

Beumont P. J. V., Booth L., Abraham S. F., Griffiths D. A., and Turner T. R. (1983) A temporal sequence of symptoms in patients with anorexia nervosa: Preliminary report, in *Anorexia Nervosa: Recent Developments in Research* (Darby P. L., Garfinkel P. E., Garner D. M., and Coscina D. V., eds.) Alan R. Liss, New York, pp. 129–136.

Blanton S. (1919) Mental and nervous changes in children of the Volkschulen of Trier, Germany, caused by malnutrition. *Ment. Hyg.* **3,** 343–386.

Blitzer J. P., Rollins N., and Blackwell A. (1961) Children who starve themselves. *Psychom. Med.* **23,** 69–38.

Bloom W. L. and Eidex M. F. (1967) Inactivity as a major factor in adult obesity. *Metabolism* **16,** 679–684.

Boer D. P., Epling W. F., Pierce W. D., Russell J. C., and Soetart L. (1988) Activity-based anorexia: Blocking of anorexic activity by naloxone, an endogenous opiate antagonist. Paper presented to the annual meeting of The Association for Behavior Analysis, Philadephia, PA.

Brigham J. C. (1980) Limiting conditions of the "physical attractivness stereotype": Attributions about divorce. *J. Res. Per.* **14,** 365–375.

Brusch H. (1978) *The Golden Cage* (Harvard University Press, Cambridge, MA).

Collier G. (1969) Body weight loss as a measure of motivation in hunger and thirst. *Ann. NY Acad. Sci.* **157,** 594–609.

Colt E. W. D., Wardlaw S. L., and Frantz A. G. (1981) The effect of running on plasma endorphin. *Life Sci.* **28,** 1637–1640.

Cornish E. R. and Mrosovsky N. (1965) Activity during food deprivation and satiation of six species of rodent. *Anim. Behav.* **13,** 242–248.

Crews E. L., Fuge K. W., Oscai L. B., Holloszy J. O., and Shank R. E. (1969) Weight, food intake, and body composition: Effects of exercise and of protein deficiency. *Am. J. Physiol.* **216,** 275–287.

Crisp A. H. (1965) Clinical and therapeutic aspects of anorexia nervosa: Study of 30 cases. *J. Psychosom. Res.***9,** 67–78.

Cumming D. C. and Rebar R. W. (1983) Exercise and reproductive function in women. *Am. J. Industr. Med.* **4,** 113–125.

Cumming D. C. and Rebar R. W. (1985) Hormonal changes with acute exercise and with training in women. *Sem. Reprod. Endocrinol.* **3,** 55–64.

Cumming D. C., Vickovic M. M., Wall S. R., Fluker M. R., and Belcastro A. N. (1985) The effect of acute exercise on pulsatile release of lutienizing hormone in women runners. *Am J. Obstet. Gynecol.* **141,** 482–485.

Davies E. and Furnham A. (1986) The dieting and body shape concerns of adolescent females. *J. Child Psychol. Psychiat.* **27,** 417–428.

Davis J. M., Lamb D. R., Yim G. K., and Malvern P. V. (1985) Opioid modulation of feeding behavior following repeated exposure to forced swimming exercise in male rats. *Pharmacol. Biochem. Behav.* **23,** 709–714.

Devore I. and Hall K. R. L. (1965) Baboon ecology, in *Primate Behavior: Field Studies of Monkeys and Apes* (Devore I., ed.) Holt, Rinehart, and Winston, New York.

Duda J. J. and Bolles R. C. (1963) Effects of prior deprivation, current deprivation, and weight loss on the activity of the hungry rat. *J. Comp. Physiol. Psychol.* **56,** 569–571.

Dum J. and Hertz A. (1987) Opiods and motivation. *Interdiscip. Sci. Rev.* **17,** 180–189.

Dwyer J. T. Feldman J. J., and Mayer J. (1970) The social psychology of dieting. *J. Health Soc. Behav.* **11,** 269–287.

Edholm O. G., Fletcher J. G., Widdowson E. M., and McCance R. (1955) The energy expenditure and food intake of individual men. *Br. J. Nutr.* **9,** 286–300.

Epling W. F. and Pierce W. D. (1984) Activity-based anorexia in rats as a function of opportunity to run on an activity wheel. *Nutr. Behav.* **2,** 37–49.

Epling W. F. and Pierce W. D. (1988) Activity-based anorexia: Biobehavioral perspective. *Int. J. Eat. Disord.* **7,** 475–485.

Epling W. F. and Pierce W. D. (1989) Excessive activity and anorexia in rats, in *The Menstrual Cycle and Its Disorders: Influences of Nutrition, Exercise, and Neurotransmitters* (Pirke K. M., Wuttke W., and Schweiger I. U., eds.), Springer-Verlag, New York, pp. 79–87.

Epling W. F., Pierce W. D., and Stefan L. (1983) Schedule induced self-starvation, in *Quantification of Steady-State Operant Behaviour* (Bradshaw C. M., Szabadi E., and Lowe C. F., eds.), Elsevier/North Holland, Amsterdam, pp. 393–396.

Epling W. F., Pierce W. D., and Stefan, L. (1981) A theory of activity-based anorexia. *Int. J. Eat. Disord.* **3,** 27–46.

Epstein L. H., Masek B. J., and Marshall W. R. A. (1978) A nutritionally based school program for control of eating in obese children. *Behav. Ther.* **9,** 766–778.

Farrell P. A., Gates W. K., Muksud M. G., and Morgan W. P. (1982) Increases in plasma β-endorphin / β-lipotropin immunoreactivity after treadmill running in humans. *J. Appl. Physiol.* **52,** 1245–1249.

Finger F. W. (1951) The effect of food deprivation and subsequent satiation upon general activity in the rat. *J. Comp. Physiol. Psychol.* **44,** 557–564.

Frisch R. E. (1988) Fatness and fertility. *Sci. Am.* **258,** 88–95.

Frisch R. E., Wyshank G., and Vincent L. (1980) Delayed menarche and amenorrhea in ballet dancers. *New Eng. J. Med.* **303,** 17–19.

Frisch R. E., Gotz-Welbergen A. V., McArthur J. W., Albright T., Witschi J., Bullen B., Birnholtz J., Reed R. B., and Herman H. (1981) Delayed menarche and amenorrhea of college athletes in relation to age of onset of training. *J. Am. Med. Assoc.* **246,** 1559–1563.

Garner D. M., Rockert W., Olmstead M. P., Johnson C., and Cosina D. V. (1985) Psychoeducational principles in the treatment of bulimia and anorexia nervosa, in *Handbook of Psychotherapy for Anorexia Nervosa and Bulimia* (Garner D. M. and Garfinkel P., eds.), Guilford, New York, pp. 513–572.

Gerner R. H. and Sharp B. (1982) CSF β-endorphin immunoreactivity in normal, schizophrenic, depressed, manic and anorexic patients. *Brain Res.* **7,** 244–247.

Green S. K., Buchanan D. R., and Heuer S. K. (1984) Winners, losers and choosers: A field investigation of dating initiation. *Pers. Soc. Psychol. Bull.* **10,** 502–511.

Hall J. F. and Hanford P. V. (1954) Activity as a function of a restricted feeding schedule. *J. Comp. Physiol. Psychol.* **47,** 362,363.

Halmi, K. (1974) Anorexia nervosa: Demographic and clinical features. *Psychosom. Med.* **36,** 18–26.

Howard R. B. (1839) *An Inquiry into the Morbid Effects of Deficiency of Food Chiefly With Reference to Their Occurance Amongst the Destitute Poor* (Simpkin, Marshall, London).

Jakobovits C., Halstead, P., Kelley L., Roe D. A., and Young C. M. (1977) Eating habits and nutrient intakes of college women over a thirty-year period. *J Am. Diet. Assoc.* **71,** 405–411.

Johnson R. E., Mastropaolo J. A., and Wharton M. A. (197) Exercise, dietary intake, and body composition. *J Am. Diet. Assoc.* **61,** 399–403.

Jones W. H., Hannson R., and Philips A. L. (1978) Physical attractiveness and judgments of psychotherapy. *J. Soc. Psychol.* **105,** 79–84.

Kanarek R. B. and Collier G. H. (1983) Self-starvation: A problem of overriding the satiety signal? *Physiol. Behav.* **30,** 307–311.

Katch F. I. Martin R., and Martin J. (1979) Effects of exercise intensity on food consumption in the male rat. *Am. J. Clin. Nutr.* **32,** 1401–1407.

Katz J. L. (1986) Long distance running, anorexia nervosa, and bulimia: A report of two cases. *Comp. Psychiat.* **27**, 74–78.

Kaye W. H., Picker D. M., Naber D., and Ebert M. H. (1982) Cerebrospinal fluid opioid activity in anorexia nervosa. *Am. J. Psychiat.* **139**, 43–45.

Keys A., Brozek, J., Henschel A., Mickelson O., and Taylor H. L. (1950) *The Biology of Human Starvation* (University of Minnesota Press, Minneapolis).

King A. (1963) Primary and secondary anorexia nervosa syndromes. *Br. J. Psychiat.* **109**, 470–479.

Kron L., Katz J. L., Gorzynski G., and Weiner H. (1978) Hyperactivity and anorexia nervosa: A fundamental clinical feature. *Comp. Psychiat.* **19**, 433–440.

Kurman L. (1978) An analysis of messages concerning food, eating behaviors and ideal body image on prime-time American network television. *Dissertat. Abstr.* **39a**, 1907,1908.

Lakoff R. T. and Scherr R. L. (1984) *Face Value: The Politics of Beauty* (Routledge and Kegan Paul, Boston).

Levitsky D. (1974) Feeding conditions and intermeal relationships. *Physiol. Behav.* **12**, 779–787.

Lipscomb F. M. (1945) Medical aspects of Belsen concentration camp. *Lancet* **2**, 313–315.

Loy J. (1970) Behavioral response of free-ranging rhesus monkeys to food shortage. *Am. Phys. Anthrop.* **3**, 263–272.

Lutter J. M. and Cushman S. (1982) Menstrual patterns in female runners. *Physiol. Sports Med.* **10**, 60–72.

McArthur J. W., Bullen B. A., Beitins I. Z., Pagano M., Badger T. M., and Klibanski A. (1990) Hypothalamic amenorrhea in runners of normal body composition. *Endocr. Res. Commun.* **7**, 1–25.

McMurray R. G., Forsythe W., Mar M. H., and Hardy C. J. (1987) Exercise intensity-related responses of β-endorphin and catecholamines. *Med. Sci. Sports Exer.* **19**, 570–574.

Malina P. M., Harper A. B., Avent H. H., and Campbell B. E. (1973) Age at menarche in athletes and nonathletes. *Med. Sci. Sports Exer.* **5**, 11–13.

Margules D. L. (1979) β-endorphin and endoloxone: Hormones of the autonomic nervous system for the conervation or expenditure of bodily resources and energy in anticipation of famine or feast. *Neurosci. Biobehav. Rev.* **3**, 155–162.

Marrazzi M. A. and Luby E. D. (1986) An auto-addiction opiod model of chronic anorexia nervosa. *Int. J. Eat. Disord.* **5**, 191–208.

Mavissakalian M. (1982) Anorexia nervosa treated with response prevention and prolonged exposure. *Behav. Res. Ther.* **20**, 27–31.

Mayer J. (1965) Inactivity as a major factor in adolescent obesity. *Ann. NY Acad. Sci.* **131**, 502–506.

Mayer J., Roy P., and Mitra K. P. (1956) Relation between caloric intake, body weight and physical work: Studies in an industrial male population in West Bengal. *Am. J. Clin. Nutr.* **4**, 169–175.

Mazur A. (1986) U. S. trends in feminine beauty and overadaptation. *J. Sex Res.* **22,** 281–303.

Milke D. (1982) *Running Anorexia in Mice as a Function of Age and Sex*. Honors thesis, Department of Psychology, The University of Alberta, Edmonton, Alberta, Canada.

Miller T. M., Coffman J. G., and Link R. A. (1980) Survey on body image weight and diet of college students. *J. Am. Diet. Assoc.* **77,** 51–56b.

Moore R., Mills I. H., and Forester A. (1981) Naloxone in the treatment of anorexia nervosa: Effect of weight gain and lipolysis. *J. Royal Soc. Med.* **74,** 129–131.

Nakai Y., Kinoshita F., Koh T., Tsujii S., and Tsukada T. (1987) Perception of hunger and satiety induced by 2-deoxy-D-glucose in anorexia nervosa and bulimia nervosa. *Int. J. Eat. Disord.* **6,** 49–57

Oscai I. B. and Holloszy J. O. (1969) Effects of weight changes produced by exercise, food restiction, or over-eating in body consumption. *J. Clli. Invest.* **48,** 2124–2128.

Petraglia F., Porro C., Facchinetti F., Cicoli C., Bertellini E., Volpe A., Barbieri G. C., and Genazzani A. R. (1986) Opioid control of LH secretion in humans: Menstrual cycle, menopause and aging reduce effect of naloxone but not of morphine. *Life Sci.* **8,** 2103–2110.

Pierce W. D., Epling W. F., and Boer D. P. (1986) Deprivation and satiation: The interrelations between food and wheel running. *J. Exp. Anal. Behav.* **46,** 199–210.

Quigley M. E., Sheehan K. L., Casper R. F., and Yen S. S. C. (1980) Evidence for increased dopaminergic and opioid activity in patients with hypothalamic hypogonadotropic amenorrhea. *J. Clin. Endocr. Metab.* **50,** 949.

Reid L. D. and Wideman J. (1982) Naltrexone has no effects on body weights of starving rats. *Bull. Psychon. Soc.* **19,** 298–300.

Reid L. S. and Finger F. W. (1955) The rats adjustment to 23-hour food-deprivation cycles. *J. Comp. Physiol. Psychol.* **48,** 110–113.

Ropert J. F., Quigley M. E., and Yen S. S. C. (1981) Endogenous opiates modulate pulsatile luteiniing hormone release in humans. *J. Clin. Endcr. Metab.* **52,** 583–585.

Routtenberg A. and Kuznesof A. W. (1967) Self-starvation of rats living in activity wheels on a restricted feeding schedule. *J. Comp. Physiol. Psychol.* **64,** 414–421.

Russell J. C., Epling W. F., Pierce D., Amy R. M., and Boer D. P. (1987) Induction of voluntary prolonged running by rats. *J. Appl. Physiol.* **63,** 2549–2553.

Russell-Davis D. (1951) *Studies in Malnutrition,* MRC, special report, series number 75 (HM Stationary Office, London).

Sanger D. J. (1981) Endorphinergic mechanisms in the control of food and water intake, appetite. *J. Intake Res.* **2,** 193–208.

Sanger D. J. and McCarthy P. S. (1980) Differential effects of morphine on food and water intake in food deprived and freely-feeding rats. *Psychopharma* **72,** 103–106.

Shainberg D. (1977) Long distance running as mediation. *Ann. NY Acad. Sci.* **301,** 1002–1009.

Slade P. D. (1973) A short anorectic behavior scale. *Br. J. Psychiat.* **122,** 83–85.

Teghtsoonian R. and Campbell B. A. (1960) Random activity of the rat during food deprivation as a function of environmental conditions. *J. Comp. Physiol. Psychol.* **53,** 242–244.

Thoma H. (1967) *Anorexia Nervosa* (Humber-Klitt, Bern: Stuttgart), International University Press, New York.

Tokuyama K., Saito M., and Okuda H. (1982) Effects of wheel running on food intake and weight gain of male and female rats. *Physiol. Behav.* **23,** 899–903.

Umberson D. and Hughes M. (1984) The impact of physical attractiveness on achievement and psychological well being. Paper presented at the meeting of the American Sociological Association, San Antonio, Texas.

Unger R. K., Hilderbrand M., and Madar T. (1982) Physical attractiveness and asumptions about social deviance: Some sex-by-sex comparisons. *Per. Soc. Psychol. Bull.* **8,** 293–301.

Vincent G. P. and Pare W. P. (1976) Activity-stress ulcer in the rat, hamster, gerbil, and guinea pig. *Physiol. Behav.* **16,** 557–560.

Warren M. P. (1983) Effects of undernutrition on reproductive function in the human. *Endocr. Rev.* **4,** 363–377.

Watt E. W., Wiley J., and Fletcher G. F. (1976) Effect of dietary control and exercise training on daily food intake and serum lipids in post-myocardial infaraction patients. *Am. J. Clin. Nutr.* **29,** 900–904.

Wooley S. C. and Wooley O. W. (1980) Eating disorders: Obesity and anorexia, in *Women and Psychotheraey: Assessment of Research and Practice* (Brodsky A. and Hare-Mustin R., eds.), Guilford Press, New York.

An Animal Model of Attention Deficit

Joram Feldon and Ina Weiner

1. Introduction

Attentional dysfunction has been implicated in a variety of psychopathological conditions (Cutting, 1985). Two major syndromes in which such dysfunction is considered to play a central role are attention-deficit hyperactivity disorder (ADHD) and schizophrenia. It is not clear to what extent, if at all, attentional deficits described in these disorders share a common mechanism. However, both schizophrenia and ADHD have been linked to dopaminergic (DA) dysfunction, and, if this neurotransmitter system subserves attentional processes, at least partial commonality of mechanism may exist. Indeed, in view of the lack of clarity surrounding the diagnostic criteria for ADHD and schizophrenia and the repeated calls for their nosological reappraisal, the continuing focus in both of these syndromes on the two dysfunctions, attentional and dopaminergic, is rather remarkable. It appears, then, that an animal model of attentional (dys)function that is related to DA (dys)function could benefit the understanding of both disorders. This chapter presents such a model.

2. Latent Inhibition

There are several animal models of schizophrenia and ADHD that are based on behavioral changes resulting from drug- or lesion-induced alterations in the DA system (e.g., the animal amphetamine model of schizophrenia and the neonatal DA

From: *Neuromethods, Vol. 18: Animal Models in Psychiatry I*
Eds: A. Boulton, G. Baker, and M. Martin-Iverson ©1991 The Humana Press Inc.

depletion model of ADHD), but none of them focuses *a priori* on attentional dysfunction. Our strategy is different. We begin with choosing a behavioral paradigm in animals that reflects the operation of attentional processes believed to be dysfunctional in the clinical syndrome, and assess the influence of DA manipulations on these processes. The paradigm is that of latent inhibition (LI) (Lubow, 1973).

In its basic form, LI is defined within a two-stage paradigm. In the first stage, animals from each of two groups are placed in an environment that will later serve as the test apparatus. One group, designated the "stimulus-preexposed" group, receives a series of stimuli, such as tones or lights. The other, "nonpreexposed," group is allowed to spend an equivalent amount of time in the apparatus. When the preexposure stage is completed (either immediately or, more typically, 24 h later), all of the animals receive the conditioning-test stage of the experiment; the previously exposed stimulus is paired with a reinforcer over a number of trials. The effectiveness of the conditioned stimulus (CS)–unconditioned stimulus (US) pairings is assessed by examining some behavioral index of the conditioned responding. Latent inhibition consists of the fact that the stimulus preexposed group learns the CS–US association more slowly than the nonpreexposed group.

Latent inhibition is a simple, robust behavioral phenomenon that has received considerable attention in the animal learning literature over the last 20 years (Lubow, 1989). It can be demonstrated in a variety of classical and instrumental conditioning procedures, including avoidance, conditioned suppression, taste aversion, discrimination learning, and others, and in many different mammalian species, including humans.

There is general agreement that nonreinforced stimulus preexposure decreases the attention to, or the associability of, that stimulus without affecting its associative strength. This conclusion is based on findings that the preexposed stimulus does not acquire inhibitory properties and is retarded in both excitatory and inhibitory conditioning (Reiss and Wagner, 1972; Rescorla, 1969,1971; Solomon et al., 1974). Consequently, the development of LI is considered to reflect animals' learning not to attend to, or to ignore, irrelevant stimuli (Lubow et al., 1981; Mackintosh, 1973,1975,1983; Moore, 1979; Moore and Stickney, 1980; Schmajuk and Moore, 1985,1988; Solomon, 1980). The fact that LI is considered by many theorists to be a reflection of attentional processes has become of increasing significance to neuroscientists, who see LI as a convenient tool for measuring the effects of various manipulations, such as drug treatments and lesions, on attention (Weiner, 1990). In addition, since LI is also displayed in humans, it provides one of the few animal paradigms that directly relates to our understanding of normal human attentional processes as well as ones in which attentional deficits are implicated, as in schizophrenia (Lubow, 1989; Lubow et al., 1982) or ADHD.

The relationship between LI and the schizophrenic attention deficit is quite straightforward, since the latter has been most often described as an inability to ignore irrelevant, or unimportant stimuli. Indeed, in a recent article, Anscombe (1987) has argued compellingly that all of the major symptoms of schizophrenia can be derived from this single, underlying deficit. The relationship appears more problematic in the case of ADHD, since in this case the attentional deficit is conventionally described as "inattention," a term that appears to imply absence of attention to significant stimuli, rather than excessive attention to nonsignificant stimuli. However, children with ADHD seek stimulation more than normal children, make high rates of errors on tasks requiring attention or reaction to specific stimuli, are less able to inhibit responses to stimuli, engage in "off-task" behaviors, orient themselves to extraneous stimuli, and often switch from one activity to another (Rutter, 1989; Taylor, 1989). These behaviors suggest that children with ADHD may have difficulties in ignoring stimulation that is ignored by normal children. In other words, they may be inattentive to relevant stimuli because of excessive attention to irrelevant stimuli.

The LI experiments described in this chapter demonstrate that the capacity to ignore irrelevant stimuli is dependent on dopaminergic function, and shed light on the mechanisms by which DA activity modulates such capacity. Increase and decrease in DA activity were produced by the administration of amphetamine and neuroleptics, respectively. LI was tested in either an off-baseline conditioned emotional response or a two-way active-avoidance procedure.

2.1. Conditioned Emotional Response (CER) Procedure

The off-baseline CER procedure is carried out in operant chambers equipped with water bottles and drinkometers that detect licking. It consists of four stages:

1. *Baseline:* On each of several days, rats are individually placed into the experimental chamber and allowed to make 600 licks. The rat is then returned to its home cage and allowed access to water for 30 min.
2. *Preexposure (PE):* With the bottle removed, each animal is placed into the experimental chamber. The preexposed (PE) animals receive a predetermined number of tone presentations (2.8 KHz, 3–5 s duration) with an intertrial interval (ITI) of 50 s. The nonpreexposed (NPE) animals are confined to the chamber for the identical period of time, but do not receive the tones.
3. *Conditioning:* With the bottle removed, each animal is given two tone–shock pairings. Tone parameters are identical to those used in preexposure. The 1-mA, 1-s shock immediately follows tone termination. The first tone–shock pairing is given 5 min after the start of the conditioning session. Five minutes later the second pairing is administered. After the second pairing, animals are left in the experimental chamber for an additional 5 min.
4. *Test:* Each animal is placed into the chamber and allowed to drink from the bottle. When the subject completes 90 licks, the tone is presented. The tone continues until 10 additional licks are completed. If the subject fails to complete the last 10 licks within 300 s, the session is terminated and a score of 300 is recorded. The times between licks 80–90 and 90–100 are recorded. The amount of suppression of licking is indexed using a suppression ratio, $A/A + B$, where A is the time to complete licks 80–90 (tone off) and B is the time to complete licks 90–100 (tone on). A suppression ratio of 0.00

indicates complete suppression (no LI) and a ratio of 0.50 indicates no change in response time from the "tone-off" to the "tone-on" period (LI).

The off-baseline CER procedure offers two important advantages for testing drug effects on behavior. First, in both the preexposure and conditioning stages, the animal is not required to perform any overt response, and the test stage can be separated in time from both stages and conducted without drugs. These features enable the elucidation of drug effects on learning and attentional mechanisms, unconfounded with their motivational or motor effects. This is particularly important when testing such drugs as amphetamine and neuroleptics, which exert powerful motivational and motor effects (Weiner and Feldon, 1987; Weiner et al., 1984).

2.2. Two-Way Active Avoidance Procedure

The avoidance procedure consists of two stages:

1. *Preexposure:* Each animal is placed in the shuttle box with the house lights on and receives 50 tone presentations (5 s each) on a variable-interval (VI) 60-s schedule, ranging from 20 to 100 s. The nonpreexposed (NPE) animals are confined to the shuttle box for an identical period of time, but do not receive the tones. At the end of the preexposure session, animals are returned to their home cages.
2. *Test:* Each animal is placed in the shuttle box and receives 100 avoidance trials, presented on a VI 60-s schedule ranging from 30 to 90 s. Each avoidance trial starts with a 5-s tone followed by a 30-s shock, the tone remaining on with the shock. If the animal crosses the barrier to the opposite compartment during the 5-s tone, the tone is terminated and no shock is delivered (avoidance response). A crossing response during shock terminates the tone and the shock (escape response). If the animal fails to cross during the entire tone–shock trial, the tone and the shock terminate automatically after 35 s. The latencies of the avoidance/escape responses are recorded and analyses are carried out on the percentage of avoidances. LI consists of poorer avoidance acquisition in the PE as compared to the NPE group.

Our choice of this procedure had three aims: First, to test the generality of our findings with the CER procedure; second, to test whether the effects of amphetamine and neuroleptics on LI would also be evident in a procedure that involved motor activity and that is known to be markedly affected by both drugs; and third, to test whether the effects of these drugs on attention could be dissociated from their motor effects.

2.3. Critical Parameters

The development of LI is affected by numerous parameters of the experimental design, such as stimulus intensity and duration, interstimulus interval, number of preexposures, and so on. An exhaustive list of these parameters is presented in Lubow et al. (1981) and Lubow (1989). Here we shall mention only two parameters that were manipulated in our experiments and that were found to modulate the effects of drugs we tested on LI.

2.3.1. Number of Preexposures

LI is a positive function of number of stimulus preexposures. In other words, attention decreases as the number of preexposures is increased, until complete LI is achieved. The values are specific to the particular LI preparation but, usually, at least 15 stimulus preexposures are required to produce LI. Number of preexposures modulates the expression of drug effects: If a given drug abolishes LI, this effect will be most clearly evident following a relatively high number of preexposures that produces a robust LI effect. Conversely, if a drug facilitates LI, this effect can be masked under conditions that lead to a complete or nearly complete LI. Therefore, in this case, it is recommended to use a low number of stimulus preexposures, which leads to a small LI effect.

The orderly function that describes the relationship between number of stimulus preexposures and the size of the LI effect is of interest for an additional reason: It can be used to determine the effects of drugs on attentional processes as a function of number of stimulus preexposures. In principle, a drug may shift the curve either to the right (i.e., to retard the development of LI) or to the left (i.e., to speed up the development of LI).

2.3.2. Preexposure–Conditioning Interval

The LI procedure can be run in one session, with conditioning immediately following preexposure, or the two stages can be separated, typically by 24 h. Theoretically, the preexposure–conditioning interval should not affect the development of LI. However, our results show that drug effects on LI can be critically dependent on such an interval. In addition, a 24-h interval between the two stages enables drug administration to be confined to one of the stages.

2.4. The Locus of Drug Action

2.4.1. Preexposure vs Nonpreexposure

When elucidating drug effects on LI, it is important to keep in mind that this paradigm compares two groups: stimulus-preexposed (PE) and stimulus-nonpreexposed (NPE). LI is defined and can be assessed only as a unidirectional difference between these two groups, i.e., poorer learning in the PE group than in the NPE group. This two-group design has two important advantages: First, since the control (NPE) group undergoes regular conditioning (e.g., CER or avoidance), it allows within a single experimental design the assessment of the effect of the tested drugs on regular conditioning, independently of, and in addition to, their effects on attentional processes. The comparison is, of course, to a no-drug condition.

Second, and more critically, it allows the determination of whether the action of the drug on LI stems from a specific effect on attentional processes. Ideally, in the latter case, the action of the drug should be confined to the PE group, whereas the NPE group should be unaffected. However, a drug may act on both attentional processes and conditioning, or only on conditioning, and these actions critically determine the interpretation of the results. We shall illustrate this point by referring to the CER procedure. As elaborated above, in this procedure LI consists of the fact that the PE group is less suppressed in the test (has a higher suppression ratio) than the NPE group. If a given drug acts exclusively on attentional processes, then it should either decrease or increase suppression in the PE group, but not affect suppression in the NPE group. (In comparison to a no-drug condition, the drug–NPE group remains like the no-drug–NPE group, whereas the drug–PE group is less, or more, suppressed than the no-drug–PE group.) The former action will increase the difference between the PE and the NPE groups, i.e., will produce a facilitation of LI. The latter action will decrease the difference between the PE and the NPE group, i.e., will produce an attenuation or abolition of LI. If a given drug also acts on CER conditioning (for example, if the drug is anxiogenic or anxiolytic, i.e., facilitates or disrupts fear conditioning, respectively), then it will exert an identical effect on both the NPE and the PE groups (i.e., an anxiogenic drug will increase suppression in both, and an anxiolytic drug will decrease suppression in both). In this case the question is, what happens to the *difference* between the PE and the NPE groups?

- If the PE group still shows poorer conditioning than the NPE group, and the difference is similar in magnitude to that seen in the no-drug controls, then we conclude that the drug acts on conditioning and does not affect attentional processes.
- If the difference between drug–PE and drug–NPE groups is significantly larger than that seen in the no-drug condition (i.e., there is a larger LI effect), then we conclude that the drug affects conditioning and facilitates attentional processes.
- If the difference between the drug–PE and drug–NPE groups becomes smaller or disappears (i.e., there is no LI effect), then it is critical to determine whether the attenuation/abolition of LI is caused by both groups being suppressed, i.e., the PE group becomes like the NPE group, or by both groups not being suppressed, ie., the NPE group becomes like the PE group. In the latter case, we conclude that the drug disrupts CER acquisition, but reach no conclusion as to its effects on attentional processes; in the former case, we conclude that the drug affects attentional processes, i.e., attenuates/abolishes LI. It is this outcome, i.e., the PE group showing a degree of suppression similar to that of the NPE group, that constitutes a "true" abolition of LI.

2.4.2. Preexposure vs Conditioning

Conventionally, studies investigating the effects of pharmacological/physiological manipulations on LI treat this phenomenon in terms of its final outcome, i.e., retardation in associative learning as a consequence of preexposure. A pharmacological or physiological manipulation that affects attentional processes should affect this final outcome.

We have taken a different approach. Rather than treating LI in terms of the final outcome, we are concerned with the processes underlying this outcome. Hence, the primary question is not how a given manipulation affects the final retardation of learning, but how it affects the processes responsible for this retardation. This approach asserts that, when testing drug effects on LI, it is important to distinguish between the acquisition of LI (i.e., learning not to attend to a stimulus that predicts no significant consequences), which takes place in preexposure, and the expression of LI (i.e., retardation of the association between this stimulus and reinforcement) in conditioning. It is in the conditioning stage that the control over behavior is actually exerted by the irrelevant stimulus, since in this stage the animal continues to respond to the preexposed stimulus as irrelevant in spite of the fact that it now signals a significant outcome, i.e., reinforcement.

A given drug may affect LI by acting on processes taking place in preexposure, on those occurring in conditioning, or both. The determination of the locus of drug action is critical for understanding its mechanism of action. The significance of this approach is illustrated by the fact that two drugs may yield an identical influence on the final outcome of LI, but that this influence may be exerted via different mechanisms. For example, both chlordiazepoxide and amphetamine abolish LI when given in both the preexposure and the conditioning stages. However, chlordiazepoxide acts in the preexposure stage (Feldon and Weiner, 1989), whereas, as we shall describe, amphetamine acts in a different manner. In order to determine the site of drug action on LI, we use a "drug–no-drug" design: The drug is administered in preexposure only, in conditioning only, in both stages, or in neither.

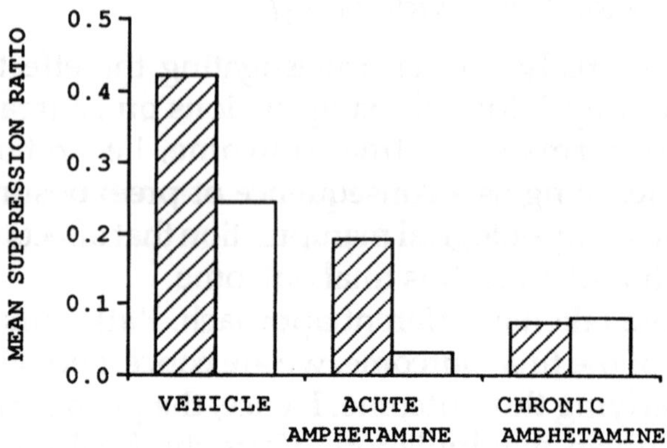

Fig. 1. Mean suppression ratios of the preexposed (PE; ▨) and nonpreexposed (NPE; □) groups under three drug conditions: vehicle, acute amphetamine (1.5 mg/kg *dl*–), and chronic amphetamine (15 d). Preexposure and conditioning were given in one session (from Weiner et al., 1984).

3. The Effects of Amphetamine on LI

In all the experiments described, amphetamine was administered ip 15 min before the behavioral procedure. The suppression test in the CER procedure was conducted without drugs, unless otherwise specified. The number of preexposures varied between 30 and 50.

3.1. The CER Procedure

3.1.1. Number of Injections and the Preexposure–Conditioning Interval

In our initial experiments, preexposure and conditioning stages were given in one session, and the test followed 2 h later (animals were returned to home cages for 2 h). In this procedure, the administration of *dl*-amphetamine (1.5 mg/kg) before the preexposure–conditioning session left LI intact. In contrast, when animals were pretreated with the drug for 14 days before the LI procedure, LI was abolished (Weiner et al., 1984). The results are presented in Fig. 1. It can be seen that acute administration of the drug increased suppression in both the PE and the NPE groups, but spared the difference between them, i.e., left LI intact. In contrast, in the chronic drug condition, the PE group was as suppressed as the NPE group.

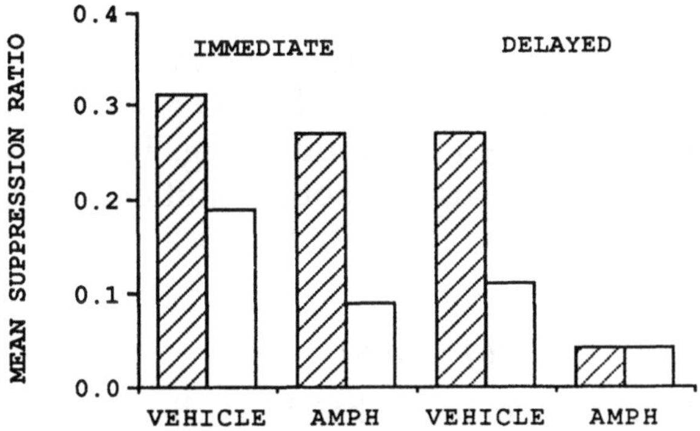

Fig. 2. Mean suppression ratios of the preexposed (PE; ▨) and nonpreexposed (NPE; □) groups at two delays between preexposure and conditioning (0 and 24 h) and two drug conditions in preexposure and conditioning (vehicle and 1.5 mg/kg *dl*-amphetamine) (from Weiner et al., 1988).

In order to determine whether chronic pretreatment would suffice, by itself, to disrupt LI, we compared three conditions: amphetamine pretreatment (14 d) and vehicle in the preexposure-conditioning session; amphetamine pretreatment and amphetamine in the PE–conditioning session; and vehicle throughout.

LI was abolished if amphetamine was present during the LI procedure. However, animals given amphetamine for 2 wk, but preexposed and conditioned without the drug, showed a normal LI effect. This result demonstrated that amphetamine-induced abolition of LI is not the result of some nonspecific neurochemical effect of the drug, but of a specific action on the attentional processes underlying the development of LI (Weiner et al., 1984).

Although our initial results indicated that chronic amphetamine administration was necessary for disrupting LI, our later experiment showed that LI could be abolished by two injections of the drug (1.5 mg/kg *dl*-amphetamine or 1 mg/kg D-amphetamine), one before preexposure and one before conditioning (Weiner et al., 1987d,1988). However, such abolition was obtained only when the two stages were separated by 24 h, and not when they were given in one session. Figure 2 presents the results of an experiment that compared the effects of amphetamine on LI at two PE–conditioning delays: 0 and 24 h. In both conditions, two drug injections were given; in the 24-h delay, the drug was administered 15 min before preexposure and 15 min before conditioning; in the 0-delay condition, one injection was given 15 min before the start of the PE–conditioning session and the second 5 min before the start of conditioning. The test was carried out 24 h later, without drugs, in both conditions. As can be seen, LI was not affected in the immediate condition, but was absent in the delayed condition (Weiner et al., 1988).

3.1.2. The Locus of Drug Action

Having the information that LI is disrupted by amphetamine, we asked, What is the behavioral mechanism of this action? Our first aim was to determine whether amphetamine abolishes LI by disrupting animals' ability to learn that a stimulus is irrelevant. If this were the case, then the action of the drug should be confined to the preexposure stage, in which learning to ignore takes place. Consequently, we tested the effects of amphetamine administration confined to the preexposure stage on LI. The effects of both chronic and acute administration were assessed. We compared three conditions: acute amphetamine in PE (14 d vehicle + AMPH in PE); chronic amphetamine in PE (14 d AMPH + AMPH in PE); and vehicle throughout. The preexposure, conditioning, and test stages were given 24 h apart. The results clearly demonstrated that neither acute nor chronic amphetamine administration, confined to the preexposure stage, disrupts LI (Weiner et al., 1984). In fact, amphetamine treatment tended to produce a larger LI effect. Next, again using the 24-h-apart procedure, we investigated the effects of amphetamine (1.5 mg/kg *dl*-amphetamine) in the "drug–no-drug" design. Animals were assigned to one of 16 experimental groups in a 2 × 2 × 2 × 2 factorial design, consisting of PE–NPE, drug–no drug in preexposure, drug–no drug in conditioning, and drug–no drug in test. Since statistical analysis revealed no significant effects of drug in the test stage, data were analyzed collapsed over the factor of drug in test. The results of this experiment are shown in Fig. 3. As can be seen, the administration of the drug in either the preexposure or the conditioning stage did not affect LI. In contrast, when both stages were conducted under the drug, LI was abolished (Weiner et al., 1988).

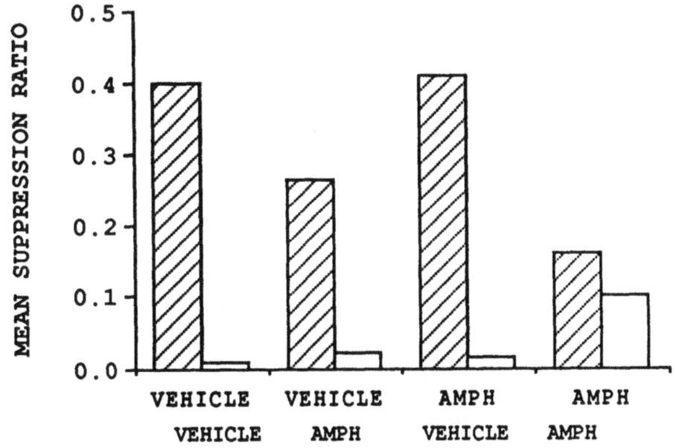

Fig. 3. Mean suppression ratios of the preexposed (PE; ▨) and nonpreexposed (NPE; □) groups under four drug conditions in preexposure and conditioning: vehicle–vehicle, vehicle–amphetamine (1.5 mg/kg *dl*-), amphetamine–vehicle, and amphetamine–amphetamine. Preexposure, conditioning, and test were given 24 h apart (from Weiner et al., 1988).

3.1.3. Drug Dose

In order to determine the effects of a higher dose of amphetamine on LI, acute and repeated (8 d) administration of 6 mg/kg *dl*-amphetamine were used. Acute administration, in the "drug–no drug" design did not disrupt LI. Figure 4 shows the results following 5 d of amphetamine pretreatment with drug administration in all three stages of the procedure. (The drug was also administered in test, because our previous results showed that this dose produced state dependency. Clearly, LI is not disrupted by chronic administration of 6 mg/kg *dl*-amphetamine (Weiner et al., 1987d).

3.2. The Avoidance Procedure

Figure 5 presents the results obtained using 1.5 mg/kg *dl*-amphetamine in the "drug–no drug" design. As can be seen, these results are identical to those obtained in the CER procedure, i.e., LI is disrupted only when the drug is given in both the preexposure and conditioning stages (Weiner et al., 1988). Two points are worth noting: First, there is a clear dissociation between the effects of amphetamine on avoidance learning and on LI. The administration of the drug in the conditioning stage

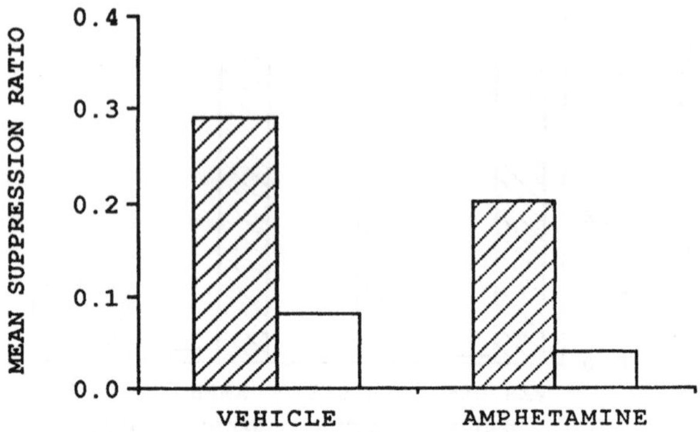

Fig. 4. Mean suppression ratios of the preexposed (PE; ▨) and nonpreexposed (NPE; □) groups under two drug conditions: vehicle and chronic (8-d) treatment with 6 mg/kg *dl*-amphetamine. Preexposure, conditioning, and test were given 24 h apart, and all stages were conducted under drug (from Weiner et al., 1987d).

markedly enhanced avoidance acquisition in both the placebo–drug and drug–drug conditions, yet LI was disrupted only in the latter condition. Thus, the disruption of LI was independent of the facilitatory effect of the drug on avoidance. Second, the abolition of LI in the CER and avoidance procedures was reflected in opposite behavioral effects: In the former, suppression of behavior (increased suppression of licking); in the latter, enhancement of behavior (facilitated avoidance). This double dissociation shows again that amphetamine-induced disruption of LI does not stem from some nonspecific behavioral action of this drug, but from a specific action on attentional processes underlying the development of LI.

3.3. Reversal of Amphetamine-Induced Disruption of LI by DA Antagonists

In support of the dopaminergic mediation of amphetamine-induced disruption of LI, the latter is prevented by a concomittant administration of the DA antagonists haloperidol (Weiner et al., in press) and chlorpromazine (Solomon et al., 1981).

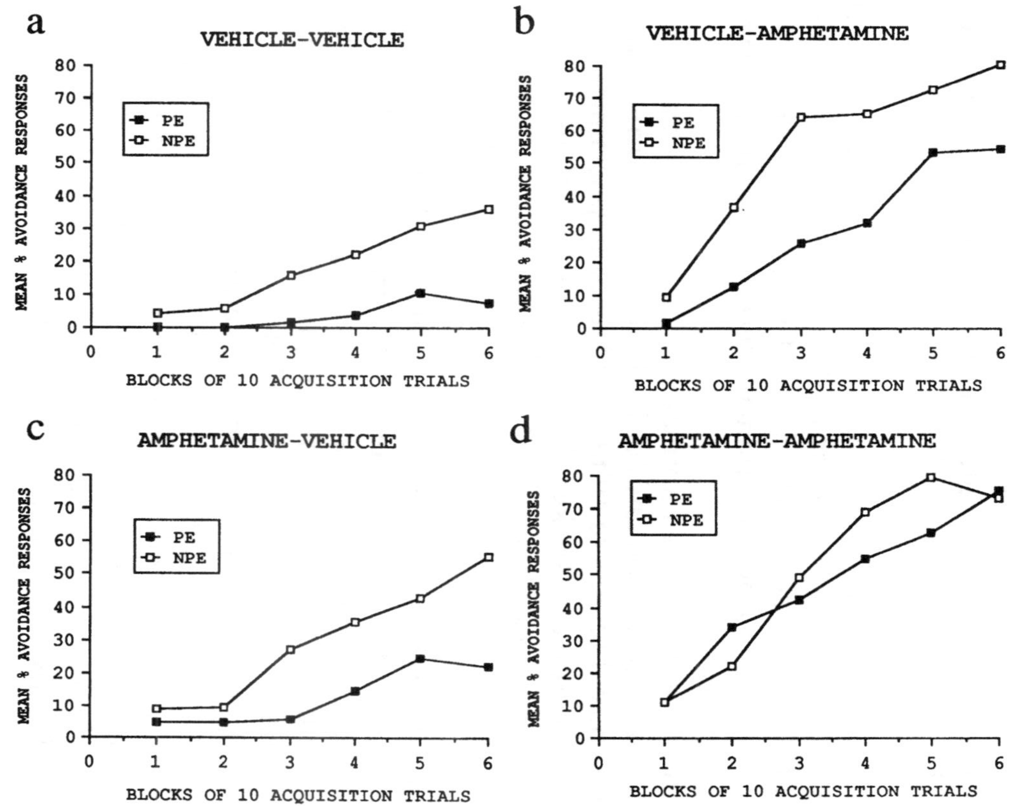

Fig. 5. Mean percent of avoidance responses over six blocks of 10 trials for the preexposed (PE; ■) and nonpreexposed (NPE; □) groups under four drug conditions in preexposure and conditioning: (a) vehicle–vehicle, (b) vehicle–amphetamine (1.5 mg/kg *dl*-), (c) amphetamine–vehicle, and (d) amphetamine–amphetamine (from Weiner et al., 1988).

3.4. Conclusions

3.4.1. Optimal Drug Administration Regimen for an Animal Model of Attentional Deficit

LI is not affected by high doses of amphetamine. It is disrupted by chronic administration of low doses of amphetamine, provided that both the preexposure and conditioning stages are conducted under the drug. In addition, acute administration of low doses prior to preexposure and prior to conditioning disrupts LI if the two stages are given 24 h apart. Obviously, the latter regimen provides the optimal injection paradigm for producing LI disruption.

Amphetamine-based animal models conventionally use repeated administration and/or high doses of the drug. However, in a recent review of animal models of amphetamine psychosis, Robinson and Becker (1986) concluded that, in order to produce behavioral sensitization (which, according to these authors, provides a suitable animal analog of psychosis), relatively low doses of amphetamine are more appropriate than high doses, and there is no need for long-term treatment. In fact, a single injection is sufficient for this purpose. Furthermore, they emphasized that a critical variable for obtaining behavioral sensitization is the interval between drug treatments. To produce robust behavioral sensitization, amphetamine must be given intermittently, with injections given relatively far apart being more efficacious than those given at short intervals. Our finding that two amphetamine injections, one given before preexposure and one before conditioning, with an interval of 30 min, did not disrupt LI, whereas the same two injections given 24 h apart did disrupt LI, provides strong support for their claim.

Robinson and Becker's (1986) model of behavioral sensitization focuses on amphetamine-induced motor effects, i.e., locomotion and stereotypy. Our results extend the model to attentional processes, demonstrating that the behavioral-sensitization paradigm produces, in addition to motor effects, an attentional deficit in the form of an inability to ignore irrelevant stimuli.

3.4.2. The Effects of Low vs High Doses of Amphetamine on LI: Neural Mechanisms

The differential effects of high and low doses of amphetamine on LI parallel those existing for the unconditioned effects of such doses and provide important clues regarding the involvement of brain DA mechanisms in LI and its disruption. Low doses of amphetamine produce locomotor hyperactivity, whereas high doses produce stereotyped behaviors (e.g., Groves and Rebec, 1976; Joyce and Iversen, 1984). There is abundant evidence that the locomotor enhancement is mediated by mesolimbic DA mechanisms, whereas stereotypy is principally mediated by striatal DA mechanisms (e.g., Creese and Iversen, 1975; Kelly et al., 1975; Pijnenburg et al., 1975; Staton and Solomon, 1984). This suggests that both the locomotor enhancement and the abolition of LI produced by low doses of amphetamine are mediated by the mesolimbic DA system. In contrast, high doses of the drug, which produce stereotypy via the striatal DA system, do not affect LI.

There is evidence that the two DA systems are activated differently by low and high doses of amphetamine. Porrino et al. (1984) demonstrated that, at the dose of 1 mg/kg, *d*-amphetamine metabolic activation (as measured by radiolabeled 2-deoxyglucose utilization) of the nucleus accumbens (NAcc) was maximal. In contrast, the administration of 5 mg/kg of the drug had no effect on the metabolic activity of the NAcc, but increased glucose utilization in the mesostriatal system. Hitzemann et al. (1980) reported that the administration of a high dose (6 mg/kg) of amphetamine increased the sensitivity of the mesostriatal system, but decreased the sensitivity of the mesolimbic system to amphetamine: Microinjections of the drug into the caudate produced enhanced stereotypy, whereas microinjections into the NAcc induced an attenuated motor response compared to controls. Electrophysiological studies show that at low to moderate doses of amphetamine there are qualitatively similar alterations in the neural activity in the NAcc and the striatum: Activity shows an initial brief increment followed by a long-lasting decline and subsequent recovery. In contrast, the effects of high doses differ: In the neostriatum, there is a sustained increase in firing, but in the NAcc, decline of neuronal activity predominates (Bashore et al., 1978; Groves and Tepper, 1983; Rebec and Zimmerman, 1980). These data demonstrate that with increased activation of the striatum following high amphetamine doses, the activation of the NAcc is blocked, suggesting a competitive relationship between the two systems. Additional support for the competitive relationship between the two DA systems was provided by Joyce and Iversen (1984). They showed that, following striatal DA depletion, locomotor activity emerged at high doses of amphetamine. Joyce and Iversen (1984) concluded that, at any given dose of amphetamine, the mesolimbic and the striatal systems compete for output pathways, with the mesostriatal system preventing the full expression of the mesolimbic activity. The competitive NAcc–striatum relationship explains the presence, or the reappearance, of LI with high doses of amphetamine: With the high dose, striatal activation blocks the effects of NAcc activation.

In direct support of the differential involvement of the mesolimbic and mesostriatal DA systems in LI, Solomon and Staton (1982) demonstrated that microinjections of *d*-amphetamine into the NAcc, but not into the caudate nucleus, eliminated LI. This dissociation is in line with the findings that LI is disrupted by low, but not high, doses of amphetamine and supports the conclusion that activation of the mesostriatal DA system leaves the LI effect intact, whereas the activation of the mesolimbic DA system disrupts LI.

3.4.3. The Locus of Amphetamine Action in LI: Behavioral Mechanisms

The process underlying amphetamine-induced disruption of LI does not take place in the preexposure stage: The administration of low doses of amphetamine in preexposure only, whether acutely or preceded by 14 d of drug treatment, leaves the LI effect intact. These results imply that low doses of amphetamine do not disrupt animals' ability to learn to ignore an irrelevant stimulus, but, instead, disrupt their ability to continue to respond to such a stimulus as irrelevant when it is followed by reinforcement. As we elaborated elsewhere (Ohad et al., 1987; Weiner et al., 1984; Weiner, 1990), in the LI paradigm the animal is successively exposed to two opposite environmental contingencies. In the preexposure stage, the target stimulus is consistently followed by nonreinforcement; in the conditioning stage, this stimulus is followed by reinforcement. Thus, in conditioning, the same stimulus signals conflicting information of relevance and irrelevance. In order to display LI, animals must continue to respond to the stimulus as irrelevant in spite of the fact that it comes to signal a significant outcome, i.e., reinforcement. Indeed, normal animals are under the control of their previous learning of irrelevance, rather than the new reinforcement contingency. In contrast, amphetamine-treated animals are under the control of the immediate reinforcement contingency, rather than their previous learning of irrelevance. In other words, they exhibit a rapid switch of responding upon the introduction of reinforcement in the conditioning stage.

Low doses of amphetamine (0.1–3.2 mg/kg) increase behavioral switching (Koob et al., 1978; Oades, 1985; Robbins and Everitt, 1982; Robbins and Sahakian, 1983). Moreover, switching appears to be mediated by the NAcc. Enhancement of DA activity in the NAcc promotes response switching, whereas DA depletion in this structure eliminates switching and gives rise to rigid and perseverative behavior. In contrast, DA depletion of the caudate gives rise to increased switching (Koob et al., 1978; Robbins and Everitt, 1982). These results suggest that the striatum subserves the mechanisms responsible for behavioral perseveration (the continued execution of behavioral sequences), whereas the NAcc subserves the mechanism of behavioral switching (shifting between different, or conflicting, behavioral sequences). We suggest that the activation of the NAcc (via systemic administration of low doses of amphetamine or microinjection into the NAcc) disrupts LI by promoting a rapid switch of association and consequent responding according to the changed contingency of reinforcement in the conditioning stage (Weiner, 1990). Indeed, Taylor and Robbins (1984,1986) pointed out that increased switching may be closely related to enhanced detection of environmental contingencies. These authors also

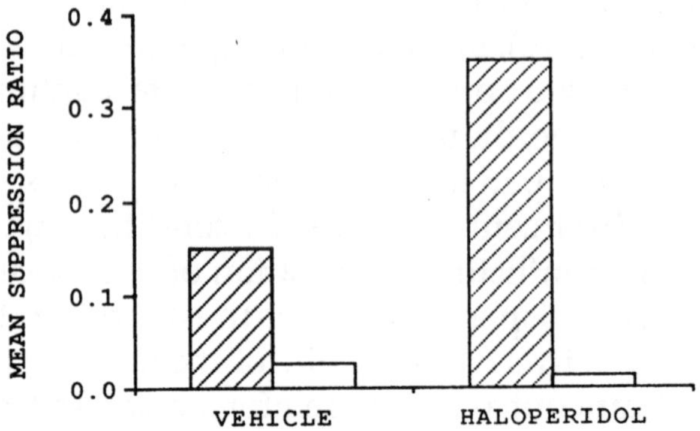

Fig. 6. Mean suppression ratios of the preexposed (PE; ▨) and nonpreexposed (NPE; □) groups under two drug conditions: vehicle and 0.1 mg/kg haloperidol. Preexposure consisted of 40 tone presentations. Drugs were given in both preexposure and conditioning (from Weiner and Feldon, 1987).

showed that enhanced contingency detection under amphetamine is dependent on the activation of the NAcc, but not the caudate nucleus. In the conditioning stage of LI, increased sensitivity to interevent contingencies following NAcc activation should be accentuated by the fact that the critical change in this stage involves the introduction of a highly salient event, i.e., reinforcement.

4. The Effects of Neuroleptics on LI

In all the experiments, the neuroleptic drug was administered ip 45 min before the behavioral procedure. The stages of the LI procedure were conducted 24 h (or more, as specified) apart. The CER test was conducted without drugs.

4.1. The CER Procedure

4.1.1. Number of Preexposures

Using 40 nonreinforced tone preexposures, we found that the administration of 0.1 mg/kg haloperidol before preexposure and before conditioning facilitated LI (Weiner and Feldon, 1987). This facilitatory effect is depicted in Fig. 6. As can be seen, learning of tone–shock association in the NPE animals was intact under haloperidol, i.e., these animals exhibited maximal sup-

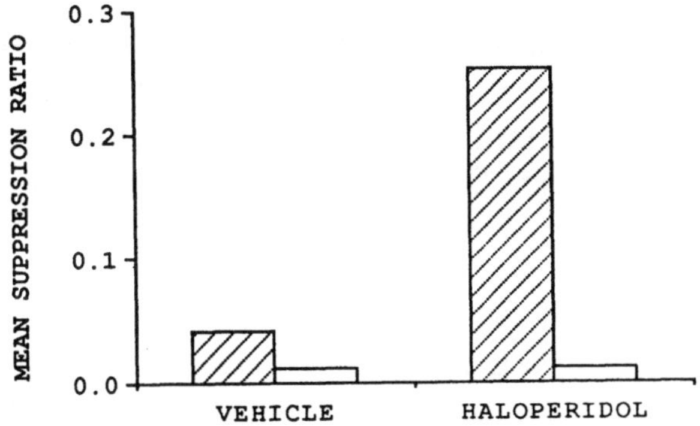

Fig. 7. Mean suppression ratios of the preexposed (PE; ▨) and nonpreexposed (NPE; □) groups under two drug conditions: vehicle and 0.1 mg/kg haloperidol. Preexposure consisted of 10 tone presentations. Drugs were given in both preexposure and conditioning (from Weiner and Feldon, 1987).

pression to the tone during the test, similar to that of vehicle-treated animals. This result is in line with other evidence that neuroleptics do not disrupt stimulus–stimulus associations (e.g., Beninger, 1983). In contrast, haloperidol exerted a marked influence on the PE animals, the PE–haloperidol group showing significantly less suppression to the tone than the PE–vehicle group, thus resulting in an enhanced LI effect in haloperidol-treated animals.

In order to test further the facilitatory effect of haloperidol on LI, we used a procedure that does not yield LI in normal animals, i.e., 10 nonreinforced stimulus preexposures (Weiner and Feldon, 1987). The results obtained with this procedure are shown in Fig. 7. As can be seen, normal animals do not show LI under these conditions. However, haloperidol-treated animals exhibit a robust LI effect. Again, this facilitation is caused exclusively by the action of the drug on PE animals, with NPE–haloperidol-treated animals responding in a manner identical to the NPE–vehicle-treated animals.

4.1.2. Drug Dose

Using 10 stimulus preexposures, we tested the effects of 0.02, 0.1, and 0.5 mg/kg haloperidol. Both 0.1 and 0.5 mg/kg haloperidol produced LI (Feldon and Weiner, in press). However, whereas 0.1 mg/kg affected the PE group exclusively, 0.5

mg/kg decreased suppression in both the PE and NPE groups, i.e., had a general disinhibitory effect. Haloperidol was reported to exert disinhibitory effects in experimental models of anxiety in rats (Pich and Samanin, 1986). Our results with CER may reflect a similar influence. Stimuli that were previously paired with footshock cause in rodents a selective activation of dopaminergic neurons that project to the frontal cortex and/or the NAcc (Deutch et al., 1985; Herman et al., 1982). Possibly haloperidol at the dose of 0.5 mg/kg blocks such stress-induced activation, leading to a decrease in response suppression. However, it is important to note that although 0.5 mg/kg haloperidol decreased suppression in both the NPE and the PE groups, the PE group was still significantly less suppressed than the NPE group, leading to the LI effect. Thus, the development of LI in haloperidol-treated rats was independent of the drug's effects on CER conditioning.

4.1.3. Repeated Drug Administration

Christison et al. (1988) demonstrated using the CER procedure with several levels of stimulus preexposure (0, 10, 13, 15, 20, and 40), LI facilitation following 5 daily injections of 0.3 mg/kg haloperidol.

4.1.4. Atypical Neuroleptic—Sulpiride

Most of the existing behavioral screening tests for antipsychotic drug action are insensitive to the atypical neuroleptic, sulpiride (Worms et al., 1983). Consequently, it was important to test the effects of this drug on LI, in order to determine the sensitivity of this model to atypical neuroleptics. We found that 100 mg/kg of sulpiride exerted an influence identical to that of haloperidol, i.e., produced LI following 10 stimulus preexposures (Feldon and Weiner, in press).

4.2. The Two-Way Avoidance Procedure

4.2.1. The Locus of Drug Action

Our CER results showed that haloperidol, in direct contrast to amphetamine, enhances animals' ability to ignore irrelevant stimuli. Following the logic adopted in amphetamine experi-

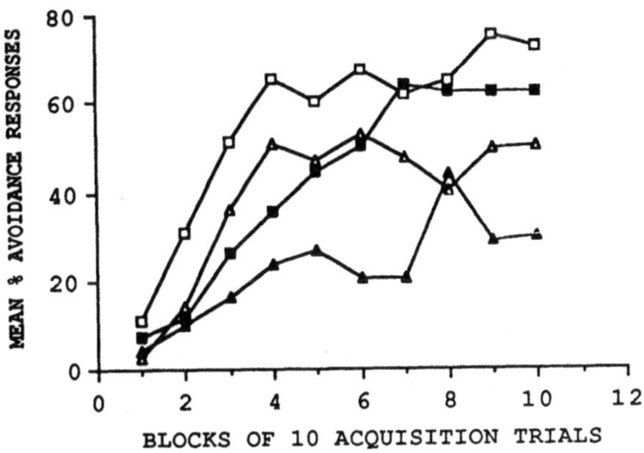

Fig. 8. Mean percent of avoidance responses over 10 blocks of 10 trials of the preexposed (PE) and nonpreexposed (NPE) groups in two drug conditions: vehicle (PE, ■; NPE, □) and 0.1 mg/kg haloperidol (PE, ▲; NPE, Δ). Avoidance conditioning was given 24 h after preexposure. Drugs were administered in preexposure only (from Weiner et al., 1987a).

ments, we tested whether such facilitation could be obtained with the administration of haloperidol confined to the preexposure stage, during which learning to ignore the irrelevant stimulus takes place (Weiner et al., 1987a).

In the first experiment, haloperidol was injected prior to preexposure, and avoidance training was given 24 h later without drugs. As can be seen in Fig. 8, the administration of haloperidol 24 h before avoidance conditioning led to poorer avoidance performance in both the PE and NPE conditions, consistent with numerous reports on the disruptive effects of neuroleptics on avoidance. In addition, the LI effect, i.e., poorer avoidance acquisition in the PE compared to the NPE groups, was evident in both the vehicle and the haloperidol conditions. However, there was no evidence for a facilitatory effect of haloperidol on LI. In the next experiment, we delayed avoidance training until 72 h after preexposure, in order to prevent the disruptive effects of haloperidol on avoidance learning, and again administered the drug only in PE. As expected, 72 h after haloperidol administration, there was no trace of the drug-produced interference of avoidance. LI was obtained in both the vehicle and haloperidol conditions, but again, there were no

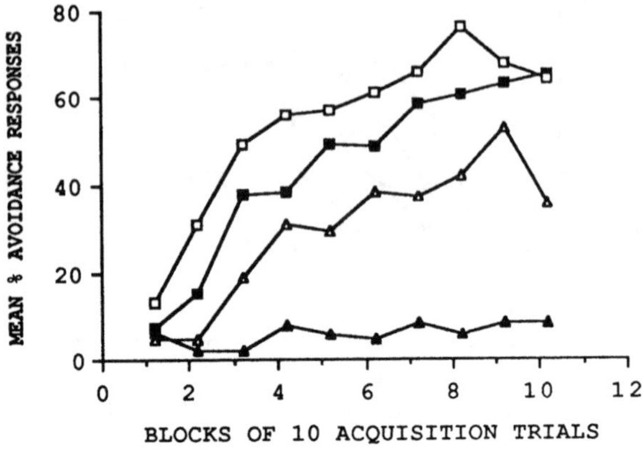

Fig. 9. Mean percent of avoidance responses over 10 blocks of 10 trials of the preexposed (PE) and nonpreexposed (NPE) groups in two drug conditions: vehicle (PE, ■; NPE,□)and 0.1 mg/kg haloperidol (PE, ▲; NPE, Δ). Drugs were administered in both the preexposure and conditioning stages (from Weiner, et al., 1987a).

appreciable differences between the magnitudes of the effect in the two conditions.

The results of these experiments showed conclusively that, when haloperidol was administered only in the preexposure stage, the LI effect was not enhanced. This result markedly contrasted with the dramatic facilitation of LI obtained in our CER experiments in which the drug was administered in both the preexposure and conditioning stages. Consequently, such administration was used in the third experiment, in order to determine whether the facilitatory effect of the drug depends on its presence throughout the LI procedure. In addition, 24 h after avoidance conditioning, animals were retested in the absence of the drug.

The results are presented in Fig. 9. It can be seen that, as expected, haloperidol interfered with avoidance; again, LI was present in both vehicle and haloperidol conditions. However, with haloperidol present in both preexposure and conditioning, the difference between the PE and the NPE group (i.e., the magnitude of LI) was much larger in the haloperidol than in the vehicle condition. This larger LI effect continued to be present in retest given 24 h later, without drugs.

In addition to LI facilitation, the above results demonstrated that the development of LI in haloperidol-treated animals was independent of the drug effects on avoidance performance. Thus, in the first and third experiments, haloperidol impaired avoidance acquisition in both the NPE and the PE groups, but the impairment was much greater in the PE groups, resulting in the LI effect. In the second experiment, when avoidance conditioning was conducted 72 h following preexposure, avoidance performance of the NPE–haloperidol group did not differ from that of the NPE–vehicle group, yet the PE–haloperidol group continued to exhibit impaired avoidance. Likewise, in the third experiment, when animals were retested in the absence of the drug, avoidance performance of the NPE-haloperidol group was similar to that of the NPE–vehicle group, whereas the PE–haloperidol group still showed pronounced retardation. It will be recalled that we demonstrated a similar dissociation between the effects of haloperidol on LI and CER acquisition. This dissociaton indicates that haloperidol exerts a specific effect on attentional processes, which can be dissociated from its motivational/motor effects on behavior.

4.3. Conclusions

4.3.1. The Effects of Sulpiride: Underlying DA Mechanisms

Sulpiride has been reported to act preferentially on the mesolimbic, rather than mesostriatal, system (e.g., White and Wang, 1983). Consequently, the fact that sulpiride facilitates LI provides further evidence for the involvement of the mesolimbic system in LI. Since the latter system has been repeatedly suggested to constitute the site of the antipsychotic action of neuroleptic drugs (e.g., Iversen, 1987), the possibility that the LI-facilitating effect of these drugs is mediated via this system, has important implications for the LI model.

4.3.2. The Locus of Haloperidol Action in LI: Behavioral Mechanisms

Haloperidol dramatically facilitates LI when given in both the preexposure and the conditioning stages. In contrast, animals receiving the nonreinforced preexposure under haloperidol, but conditioned without the drug, show a normal, nonfacilitated LI effect. Thus haloperidol, like amphetamine, does not affect animals' ability to learn that a stimulus is irrelevant. Instead, and in direct contrast to amphetamine, this drug enhances animals' capacity to continue to ignore such a stimulus when it signals a change in reinforcement contingencies. In other words, whereas amphetamine enhances behavioral switching, haloperidol reduces or blocks such switching.

4.3.3. Neuroleptic-Induced Facilitation of LI: A Unique Effect

The neuroleptic-induced facilitation of LI is unique in several respects:

1. It provides the first demonstration of a facilitatory influence of neuroleptics on learning;
2. It is specific to neuroleptics, since all other pharmacological and physiological manipulations tested to date have been found either to abolish LI or to leave it intact (Weiner, 1990);
3. The facilitatory effect on attentional processes is highly relevant to their clinical use.

The LI paradigm possesses two additional advantages:

1. It provides an instance of learning in the absence of reinforcement, thus enabling one to determine the effects of neuroleptics on learning and attentional processes unconfounded with either the motivational or the motor effects of these drugs.
2. It does not involve drug–drug interactions.

5. The LI Model: Summary and Implications

5.1. The Nature of the Attentional Deficit

Animals treated with a low dose of amphetamine retain their capability to learn that a given stimulus is irrelevant. However, this learning fails to control their subsequent response to this stimulus when it appears in a different constellation, i.e., is followed by reinforcement. Whereas normal animals continue to ignore such a stimulus, amphetamine-treated animals respond to it as if it were new. This failure to ignore irrelevant stimuli reults from enhanced detection of interevent contingencies and a rapid switching of associations, mediated by increased activation of the NAcc (Weiner, 1990). In LI, such activation speeds up switching from the old to the new reinforcement contingency. In general, it is expected to result in rapid switching of cognitive activity and ongoing informational processes (Swerdlow and Koob, 1987).

The above description emphasizes dopaminergic involvement in output processes, rather than input processes, underlying the LI phenomenon. In animals with hyperactive mesolimbic DA mechanism, learning to ignore a nonreinforced stimulus is intact, but the expression of this learning (i.e., continuing to ignore the irrelevant stimulus under changed reinforcement contingencies) is disrupted. Further support for the contention that DA activity does not affect animals' capacity to learn to ignore a nonreinforced stimulus, but affects the subsequent expression of this learning, comes from our results with haloperidol. This drug enhances LI only when given in both the preexposure and conditioning stages, but not when given in preexposure alone. Thus, neither the enhancement of DA transmission by amphetamine nor the blockade of DA transmission by haloperidol affect animals' capacity to learn to ignore irrelevant stimuli. Instead, the two manipulations affect the subsequent reactions to such stimuli, so that haloperidol enhances and amphetamine disrupts the animals' ability to continue to respond to a stimulus as irrelevant under changed environmental contingencies.

Here a comment is in order on the issue of input vs output processes and attention. Attentional processes are still often equated with input processes. Likewise, attentional mechanisms underlying the development of LI have been conventionally postulated to be confined to the nonreinforced preexposure stage, i.e., identified as input processes. The present approach shifts the emphasis to attentional processes taking place in conditioning, i.e., output processes. This is in line with modern theories of information processing that view attention as a control mechanism of information processing that can be deployed at any stage of such processing (e.g., Erdelyi, 1974; Underwood, 1978). In fact, these theories stress the crucial role of attention in the output, or response-organization, stages. For example, Shiffrin (1976) argued that attention is deployed very late in the information-processing sequence and that "attentional deficits arise not in the input...but in the retrieval and decision processes." According to Shiffrin and Schneider (1977), attentional processing is activated whenever the input provides conflicting information or gives rise to conflicting responses. Likewise, according to LaBerge (1976), attention may be deployed at any level of the information-processing sequence, but its main function lies in the processing for output. It is the output function of attention in LI, i.e., effective response choice in the face of conflicting contingencies signaled by the target stimulus, that is emphasized here and is disrupted by NAcc activation.

Output deficits have been emphasized in both schizophrenia and ADHD.

> Most clinicians and experimental researchers will agree that the impairments of schizophrenics are most conspicuous in tasks that could be described as "effortful" and "attention-demanding," or as requiring "controlled processes" or "attentional capacity". ...All these terms refer primarily to later stages of information processing on the way from sensory input to some response system. ...Deficiencies of schizophrenics in tasks requiring these more demanding modes of later processing are...ubiquitous" (Cohen and Borst, 1987).

Recently, Sergeant and van der Meere (1989) suggested that "output systems should be the primary candidates for future experimental diagnostics" of ADHD (p. 163). A similar position was taken by Sagvolden et al. (1989). This author argued that stimuli that have no reinforcing properties for normal children do have such properties for ADHD children and, consequently, acquire control over their behavior. The similarity of this view to ours is obvious. However, Sagvolden et al. (1989) treat this phenomenon as an output deficit and, therefore, as distinct from an attentional deficit, since the latter, in their view, reflects malfunctional input mechanisms. We would like to reiterate that output processes depend on attentional control as much as, and more than, input processes. In LI, this aspect of attentional control is expressed in the selection of past contingency (stimulus–no consequence) rather than present contingency (stimulus–consequence) for the control of behavior. In an animal with a hyperactive NAcc, such control is lost. Consequently, stimuli that ordinarily do not acquire reinforcing properties, and are thus ignored, rapidly acquire such properties and control behavior.

5.2. Neural Mechanisms and Drug Treatment

The DA hypotheses of both schizophrenia and ADHD are derived from the response of these syndromes to drug therapy, i.e., the amelioration of ADHD with DA agonists and the amelioration of schizophrenic symptoms with DA antagonists. However, the picture is far from being clear. First, although amphetamine is known to exacerbate schizophrenic symptoms, the drug has also been reported to improve some symptoms (e.g., van Kammen and Boronow, 1988). Second, it is believed that neuroleptics do not improve negative, or defect, symptoms. Third, quite paradoxically at first sight, ADHD can also be treated with neuroleptics (e.g., Werry and Aman, 1975), although this is not a preferred mode of treatment. For example, it has been reported that, in children with ADHD who had a partial response to stimulant treatment, a neuroleptic–stimulant combination was superior to a placebo–stimulant combination (Weitzman et al., 1984). Finally, in children with ADHD, stimulants may improve behavior on some tasks, but be without an effect or impair behavior on others.

Our results suggest the following: A certain level of NAcc activation, such as that experimentally produced in animals by systemic administration of low doses of amphetamine or by intraaccumbal injection, leads to an inability to ignore irrelevant stimuli, which is attributable to enhanced switching. Blockade of such activation, either by neuroleptic drugs or by higher doses of amphetamine, restores or even enhances the ability to ignore irrelevant stimuli by reducing or eliminating switching. High doses of amphetamine apparently exert this effect via the activation of the mesostriatal DA system, which in turn blocks the activitation of the mesolimbic DA system (*see* Section 3.4.2.). Neuroleptics most probably produce this effect via direct action on the mesolimbic DA system.

The relevance of the latter scheme in the clinic critically depends on the correspondence between the drug dosages used in our model and those used in the clinic. It is impossible to evaluate such a correspondence, because there are no data on the optimal drug regimens in either ADHD or schizophrenia (e.g., Gittelman, 1983; Gittelman-Klein, 1987; Rifkin and Siris, 1987) and because intraspecies comparisons are very difficult to make (e.g., Robbins and Sahakian, 1979). Consequently, we can only relate our comparisons and suggestions to the actual behavioral outcomes of those treatments in the clinic. Our results show how amphetamine can normalize the ability to ignore irrelevant stimuli since with higher doses of this drug, normal LI is obtained. Although it is experimentally nonsensical, we would expect that if our low-dose-amphetamine-treated animals were treated with a higher dose, their capacity to develop LI would be restored. The applicability of this finding to the clinical use of stimulants in ADHD rests on the assumption that these children receive doses that are comparable to our higher doses. This is likely to be the case, since the most conspicuous result of stimulant treatment that normalizes attention is a marked reduction of activity. Although this "calming" effect of stimulants has been considered for years "paradoxical," it is now clear that stimulants exert exactly the same actions, i.e., improve attention and reduce activity, in normal children (Rapoport et al., 1980). Robbins and Sahakian (1979) and Robbins et al. (1989) suggested that stimulant medication in children with ADHD may produce mild stereotypies that result in increased focusing of attention. Dyme et al. (1982) showed experimentally that this indeed may be the case.

In addition, our results show that neuroleptic treatment normalizes, and in fact enhances, the ability to ignore irrelevant stimuli. This is in line with numerous reports demonstrating the effectiveness of neuroleptic drugs in the amelioration of attention in schizophrenic individuals (e.g., Asarnow et al., 1988; Braff and Saccuzzo, 1982; Oltmanns et al., 1978; Spohn et al., 1977). In direct support of this suggestion are the recent findings that LI is absent in acute schizophrenics tested within the first week of their psychotic breakdown, but is restored in chronic schizophrenics maintained for at least six weeks on neuroleptic medication (Lubow et al., 1987; Baruch et al., 1988).

Finally, our results explain the seemingly paradoxical fact that children with ADHD may benefit from both stimulant and neuroleptic treatment.

We are left with the question of why these drugs are not beneficial under some conditions. The answer is rather simple: The influence of these drugs is dependent on the behavior we observe or want to affect and on the base state of the process on which the drugs are supposed to act. This can again be illustrated (and, hopefully, demonstrated empirically) with the LI paradigm.

When we perform drug experiments in LI (and in general), drug effects are not judged in isolation, but in comparison to undrugged controls. Thus, we create *a priori* conditions that lead to the etablishment of LI in undrugged animals by using an optimal combination of critical parameters, such as the number of stimulus preexposures and the number of CS–US pairings. Under these conditions, low doses of amphetamine disrupt LI, high doses leave it intact, and neuroleptics enhance it. But normal animals do not invariably show LI following nonreinforced stimulus preexposure. One condition that we described, in which LI does not develop, is a low number of preexposures (10). Under this type of condition, neuroleptics produce LI. Although this effect is important for demonstrating the dramatic facilitatory influence of neuroleptics on LI, it must be realized that the behavior of the neuroleptic-treated rats in this case is abnormal. Normal animals *do not* ignore the preexposed stimulus, rather they *do* switch associations under these conditions. Thus, neuroleptics "normalize" attention only within a certain range of stimulus preexposures (which produces LI in normal animals). Outside of this range, these drugs impair the capacity to switch associations. An additional way to demonstrate such impairment in LI is to increase the number of stimulus–reinforcement pairings in the conditioning stage to a level at which normal animals cease to show LI and, instead, switch to responding according to the new reinforcement contingency. Under these conditions haloperidol-treated animals should show LI, in contrast to undrugged controls, because they are incapable of switching associations. Again, although this effect can be seen as LI facilitation, it is abnormal. In principle, the same outcomes would be expected with higher doses of amphetamine.

It follows from these examples that, if we use a task that requires behavioral switching, we may find that both neuroleptics and relatively high doses of amphetamine impair, rather than improve, performance. This is probably what happens when stimulants are found ineffective in children with ADHD. Thus, with a child with ADHD,

> the decision has to be made: would stimulants at some optimal dose help him toward more normal and effective behavioral control? Or is the problem such that giving the child stimulants might actually impair his functioning by putting him into a state we designated as "overfocused" (Kinsbourne, 1983).

Robbins and Sahakian (1979) and Robbins et al. (1989) have also argued that stimulant medication may impair switching of attention. In support of this suggestion, they reported that 1 mg/kg of methylphenidate, although improving performance on a battery of tasks, impaired performance on a task that required attentional switching.

A similar explanation can be given for the ineffectiveness of neuroleptics in schizophrenic patients with negative symptoms. These patients, like those with positive symptoms, experience abnormal capture of attention by unimportant stimuli. However, in contrast to schizophrenics with positive symptoms in whom such capture leads to enhanced switching in patients with negative symptoms, such capture is apparently accompanied by an inability to switch (Anscombe, 1987; Cornblatt et al., 1983). Since neuroleptics reduce or prevent switching, they cannot be expected to be of benefit in such patients. However, amphetamine should be effective in low doses, since it enhances switching. Indeed, if we return to our haloperidol-treated animals showing LI under conditions in which normal animals do not show LI (e.g., after many CS–US pairings), we would expect to "normalize" these animals with low doses of amphetamine, which would restore their switching capacity. This is what happens, according to our model, in amphetamine-treated schizophrenic patients with negative symptoms. Again, this assumes that these patients are treated with doses comparable to our "low" doses.

5.3. A Single DA-Mediated Attentional Deficit?

The attentional dysfunction described here is linked to NAcc overactivity; however, the primary source of such overactivity in the afflicted brain is not necessarily dopaminergic, although it could be (Carlsson, 1988). LI is also disrupted by hippocampal and medial raphe nucleus (the origin of the mesolimbic serotonergic system) lesions. In a recent review on the neural substrates of LI (Weiner, 1990), we concluded that LI is critically dependent on the hippocampal input into the NAcc, and that the disruption of such input leads to enhanced activation of the switching mechanism of the NAcc, resulting in the abolition of LI. However, additional major inputs to NAcc, i.e., those from prefrontal cortex and the amygdala, are also likely to play a crucial role, as well as the mesolimbic serotonergic system (Weiner, 1990). There are intimate interconnections between all these brain sites, and lesions in each of these regions increase NAcc activity. Thus, although we believe that an attentional deficit of the type described here, i.e., enhanced switching, is always related to NAcc hyperactivation, such a deficit may assume different forms, depending on the primary source of such hyperactivation. NAcc hyperfunction associated with frontal–cortical dysfunction will present a different picture from NAcc hyperfunction associated with hippocampal dysfunction, and so forth. Therefore, in various neuropsychiatric conditions associated with an attentional deficit, such a deficit will be only one symptom in a wider syndrome, the characteristics of which will be determined by affected brain areas other than the NAcc.

The existing data on the pharmacology and physiology of LI make it a highly promising tool for investigating the pharmacological and neural mechanisms underlying organisms' inability to effectively ignore unimportant aspects of their environment—the very core of many psychopathological syndromes.

6. Developmental Model of Attention Deficit: Early Nonhandling in Male Rats

It has been often suggested that ADHD reflects a developmental abnormality (e.g., Rutter, 1989; Varley, 1984). In line with such suggestions, several developmental animal models of ADHD exist, i.e., the neonatal 6-hydroxydopamine-depleted rat, the spontaneously hypertensive rat, and the isolated rat (Oades, 1987, 1989; Robbins et al., 1989). All these treatments produce hyperactivity that can be reduced by amphetamine and some learning impairments. However, attentional deficits have not been assessed within these models.

Our intention was to produce disruption of LI following some developmental manipulation. We looked for a purely behavioral, rather than lesion- or drug-induced, treatment that is given in infancy and produces deleterious effects on adult behavior. Our choice was based on data from the vast literature on early handling, albeit with expectations different from those typically described in this literature.

"Early handling" consists of a brief daily removal of the pups from their mothers, between birth and weaning (d 22). In the early-handling literature, handling constitutes the experimental condition, the effects of which are assessed in comparison to the control condition of nonhandling, which consists of leaving the pups completely undisturbed from birth to weaning. The resulting outcome is a beneficial effect of handling on adult behavior. However, from the survey of the literature, we were impressed by "the other side of the coin," i.e., the deleterious effects of nonhandling, particularly in male rats. We therefore set out to test whether early handling or nonhandling would affect the development of LI in adult animals. As we shall see, the results are quite intriguing.

6.1. Early Handling/Nonhandling Procedure

- Early experience treatments: Approximately 15 d before giving birth, pregnant females are placed in individual maternity cages, 35 × 29 × 16 cm in size, made of opaque plastic, with wood shavings as bedding. At birth, litters are culled to eight pups, comprising, as nearly as possible, four males and four females, and randomly assigned to handled or nonhandled conditions. For the handled groups, on each of the 20-d between d 2 and weaning at d 22, the mother is removed from the litters to a holding cage and the pups are individually placed into a 5- × 5- × 5-cm cardboard box with wood shaving as bedding, for 3 min. The mother and pups are then returned to the maternity cages. The nonhandled litters are left entirely undisturbed until weaning. At weaning, animals are sexed, marked for identification, and housed in hanging cages with 2–4 littermates of the same sex.

Adult testing: At 75 d of age, animals are assigned to the experimental groups, with the provision that only one animal per litter is included in any of the experimental groups. The animals are rehoused individually in regular plastic cages, and run in the LI experiments between 85 and 100 d of age.

6.2. The Effects of Early Handling/Nonhandling on LI

The experimental procedures used for testing LI were CER and avoidance, described above. In line with earlier reports, handled animals exhibited less suppression of response to the tone and better avoidance than nonhandled animals. In both procedures, LI was present in handled males and females, as well as in nonhandled females. In contrast, nonhandled males failed to exhibit LI in both CER (Weiner et al., 1987b) and avoidance (Weiner et al., 1985).

Two points must be emphasized: First, there was a clear dissociation between the effects of early handling/nonhandling on CER and avoidance learning and the effects on LI. Early handling improved avoidance performance and reduced conditioned suppression. (These effects are attributable to the well-documented lower emotionality of handled animals.) However, in spite of the opposite behavioral effects in avoidance (enhancement) and conditioned suppression (reduction), clear LI was obtained in handled animals in both procedures.

More important, nonhandling led to poorer avoidance and increased conditioned suppression in both males and females; yet, in both procedures, LI was obtained in nonhandled females, whereas nonhandled males failed to develop the LI effect.

Second, early nonhandling led to different outcomes in males and females. Sex differences in response to early handling have been demonstrated in shock-induced fighting (Erskine et al., 1975), avoidance learning (Weinberg and Levine, 1977), exploration (Wells, 1976; Weinberg et al., 1978), and the partial reinforcement-extinction effect (PREE) (Weiner et al., 1987c). Weinberg et al. (1978) concluded that "...early experience differentially affects males and females and that females have greater flexibility than males in responding to novel or aversive situations." The findings with LI demonstrated sex differences following early experience in attentional processes: Nonhandled males, but not females, develop an attentional deficit consisting of an inability to ignore irrelevant stimuli.

6.3. Reversal of Attentional Deficit by Haloperidol

Figure 10 depicts the effects of 0.1 mg/kg haloperidol on LI in handled and nonhandled males and females in the CER procedure. As expected, haloperidol facilitated LI in all animals. However, it exerted a dramatic effect on nonhandled males: In this condition, LI was reinstated (Feldon and Weiner, 1988).

6.4. The Effects of Amphetamine on Locomotor Activity of Nonhandled Males

Nonhandled males exhibited an attenuated locomotor response to *d*-amphetamine (0.3, 1, and 2.5 mg/kg). These rats did exhibit increased activity with increasing drug doses, but the response at each dose was smaller than that seen in handled males. Reduced responsiveness to *d*-amphetamine in nonhandled males was also found by Schreiber et al. (1978). In their experiment, nonhandled males, compared to handled males, showed a decline in rearings induced by 2 mg/kg *d*-amphetamine.

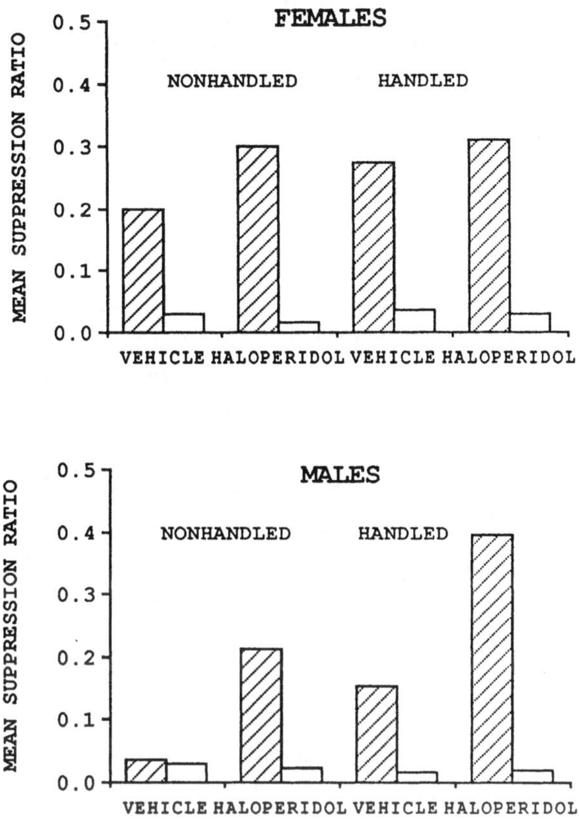

Fig. 10. Mean suppression ratios of the preexposed (PE; ▨) and nonpreexposed (NPE; □) groups in eight conditions: nonhandled females (NHF)–vehicle; NHF–haloperidol; handled females (HF)–vehicle; HF–haloperidol; nonhandled males (NHM)–vehicle; NHM–haloperidol; handled males (HM)–vehicle; HM–haloperidol (from Feldon and Weiner, 1988).

6.5. Conclusions

The early-nonhandling model is at the beginning of its development, but it already has several noteworthy features:

1. It takes into account developmental variables.
2. It does not involve drug- or lesion-induced changes.
3. It is based on an attentional deficit.
4. The attentional deficit is related to DA function. Nonhandled males exhibit an LI deficit that is identical to that found in

normal animals treated with low doses of amphetamine, and this deficit is reversed with haloperidol. They also exhibit reduced locomotor activation in response to amphetamine. These findings suggest that nonhandling in males produces some functional alterations in the mesolimbic DA system. The possibility of mesolimbic involvement is further strengthened by the remarkable resemblance between the performance of nonhandled males and that of animals with hippocampal lesions. Hippocampal lesions abolish latent inhibition (Ackil et al., 1969; Solomon and Moore, 1975) and the PREE (Rawlins et al., 1980), but leave the partial punishment effect (PPE) intact (Brookes et al., 1983). Likewise, nonhandled males fail to develop LI (Weiner et al., 1985,1987b) and the PREE (Weiner et al., 1987c), but show a normal PPE (Weiner et al., 1987c).

5. We believe that the most impressive (and exclusive) feature of this model is its "built-in" sex difference: Only nonhandled males develop an attentional deficit. The parallel of this outcome to the male predominance in ADHD (e.g., Jacklin, 1989; Rutter, 1989; Taylor, 1986) is quite striking. (Interestingly enough, male predominance is apparently found also in schizophrenia [e.g., Lewine et al., 1984]).

Before concluding, one point must be elaborated on because of its methodological importance. We have often been asked to explain why, in our hands, nonhandled males fail to show LI, whereas all other LI experiments using what appear to be nonhandled males, show normal LI. The importance of our finding that nonhandled males fail to develop LI derives precisely from the fact that

> conventionally, in the early handling literature the nonhandling condition serves as the control group against which the effects of handling are evaluated. Our results with (both) LI (and PREE) indicate that nonhandling can produce clear deficits in adult behavior, whereas handling appears to exert no beneficial effects. This questions the suitability of the nonhandled condition as a baseline against which "facilitative" effects of handling are assessed (Weiner et al., 1987b).

As we pointed out in our previous work, the condition of nonhandling is characterized by a radical absence of external stimulation, not only in comparison to the handling condition, but also in comparison to normal laboratory rearing conditions. Infant rats raised under normal laboratory conditions are exposed to a variety of external stimulation, such as that elicited by the common maintenance procedures of cage cleaning and woodshavings changing. During these procedures, pups are in fact handled, at least to some extent (Greenberg and Bursdal, 1982.) It was shown that short periods of handling are sufficient to result in the long-term physiological and behavioral effects found following conventional periods of handling (from birth to weaning) (Denenberg, 1962; Levine and Lewis, 1959; Schaefer, 1963). Thus, it is the handling, rather than the nonhandling, condition that resembles the conditions of normal laboratory rearing. Indeed, Levine pointed out in 1960 that the nonhandled group should be viewed as the experimental condition. As detailed above, in several studies nonhandled males indeed emerged as the most impaired group, as indicated by both physiological and behavioral measures (e.g., Erskine et al., 1975; Weinberg and Levine, 1977; Wells, 1976).

7. Evaluation of the LI Model

McKinney and Bunney (1969) proposed a set of criteria for a valid animal model of psychopathology, which has been very influential:

1. Similarity of inducing conditions;
2. Similarity of behavioral symptoms;
3. Common underlying neurobiological mechanisms; and
4. Reversal of the abnormal behavior by the clinically effective treatment techniques.

A survey of the existing animal models of psychopathology reveals that these objectives have not been fulfilled, and this fact has often been interpreted as a failure of animal models. However, at least two of these criteria, 1 and 3, simply cannot be fulfilled. The inducing conditions of most neuropsychiatric disorders are still largely unknown, so there is nothing to model. The same is true, although to a lesser extent, for the underlying neurobiological mechanisms; at best, for each psychopathology, we have several candidates that can be modeled.

Recently, McKinney (1988) recognized these problems and proposed an alternative approach to the evaluation of animal models. According to McKinney (1988), such models should be evaluated according to the specific purpose for which they were developed. There are four general kinds of animal models:

1. Behavioral-similarity models, designed to produce in animals certain specific aspects, or symptoms, of the human disorder;
2. Theory-driven models, designed to evaluate either a specific theory of a given form of psychopathology or a general theory about the importance of certain variables in a given psychopathology;
3. Mechanistic models, designed to study underlying mechanisms, behavioral and/or neurobiological; and
4. Empirical-validity models, designed to evaluate or screen drugs in order to predict their clinical effects.

The LI model combines all four of the above functions:

1. It produces in animals a specific symptom of the human disorder, namely, an inability to ignore irrelevant stimuli.
2. It evaluates the DA hypothesis.
3. It allows the elucidation of both the behavioral and the neural mechanisms of the phenomena under study.
4. It provides a novel screening technique for detecting drugs with attention-normalizing potential.

Four additional points are worth elaborating: First, point 1 above satisfies the criterion for face validity of the model (Willner, this volume). Second, the LI paradigm has been shown to be directly applicable to schizophrenic patients (*see* Section 5.2.). This lends the LI model construct validity (Willner, this volume), rarely attained in animal models of psychopathology, and fulfills yet another requirement of McKinney: "Ultimately, we will have to develop hypotheses based on experimental work with animals and test them directly in humans, rather than reasoning directly from results in one species to another" (p. 32). The use of the LI paradigm in schizophrenic patients has been a direct outcome of animal studies, and the results obtained were in line with expectations derived from the animal laboratory. We believe that the LI paradigm can be used to yield further fruitful avenues of clinical investigation.

Third, although the LI model as presented here, is related to the DA hypothesis, it can and should be used for studying the involvement of other neural systems in attention. Indeed, quite a lot of work has been done already (Weiner, 1990), but more is necessary.

Fourth, as a screening test for neuroleptic drugs, the LI model possesses several unique advantages (*see* Section 4.3.3.), and fulfills the requirements for predictive validity (Willner, this volume).

We would like to end with the following quotation:

> One of the most exciting areas of schizophrenia research is in the area of arousal or attention. It is phrased in various ways, but it appears that a definite subgroup of schizophrenic patients suffers from the inability to filter out irrelevant stimuli. They appear to be in a state of chronic arousal and are unable to selectively attend. This may be a trait-related risk factor. Could it be studied experimentally in animal models? If so, then we would have an experimental system that could help unravel the pathogenesis of certain aspects of schizophrenia. Again, no one claims to have a model for all aspects of schizophrenia, but this phenomenon does seem to have held up as a core defect. Experimental analogues in animals would permit the study of this phenomenon in relation to genetic, developmental, and neurobiological factors in ways impossible to do in humans (McKinney, 1988, p. 140).

Acknowledgments

Our research cited in this chapter was supported by the Israeli Academy of Sciences and Humanities—Basic Research Foundation, and the Israeli Ministry of Health—Chief Scientist's Office.

References

Ackil J., Mellgren R. L., Halgren C., and Frommer, S. P. (1969) Effects of CS preexposure on avoidance learning in rats with hippocampal lesions. *J. Comp. Physiol. Psychol.* **69,** 739–747.

Anscombe F. (1987) The disorder of consciousness in schizophrenia. *Schiz. Bull.* **2,** 241–260.

Asarnow R. F., Marder S. R., Mintz J., Van Putten T., and Zimmerman K. E. (1988) Differential effect of low and conventional doses of fluphenazine on schizophrenic outpatients with good or poor information-processing abilities. *Arch. Gen. Psychiat.* **45,** 822–826.

Baruch I., Hemsley D., and Gray J. A. (1988) Differential performance of acute and chronic schizophrenics in a latent inhibition task. *J. Nerv. Ment. Dis.* **176,** 598–606.

Bashore T., Rebec G. V., and Groves P. M. (1978) Alterations of spontaneous neuronal activity in the caudate-putamen, nucleus accumbens, and amygdaloid comple:of rat produced by *d*-amphetamine. *Pharmacol. Biochem. Behav.* **8,** 467–474.

Beninger R. J. (1983) The role of dopamine in locomotor activity and learning. *Brain Res. Rev.* **6,** 173–196.

Braff D. L. and Saccuzzo D. P. (1982) Effect of antipsychotic medication on speed of information processing in schizophrenic patients. *Am. J. Psychiat.* **139,** 1127–1130.

Brookes S., Rawlins J. N. P., and Gray J. A (1983) Hippocampal lesions do not alter the partial punishment effect. *Exp. Brain Res.* **52,** 34–40.

Carlsson A. (1988) The current status of the dopamine hypothesis of schizophrenia. *Neuropsychopharmacology* **1,** 179–186.

Christison G. W., Atwater G. E., Dunn L. A., and Kilts C. D. (1988) Haloperidol enhancement of latent inhibition: Relation to therapeutic action? *Biol. Psychiat.* **23,** 746–749

Cohen P. and Borst V. (1987) Psychological models of schizophrenic impairments, in *Search for the Causes of Schizophrenia* (Hafner H., Gattaz W. F., and Janzavik W., eds.), Springer-Verlag, Berlin.

Cornblatt B. A., Lenzenweger M. F., Dworkin P. H., and Erlenmeyer-Kimling L. (1983) Positive and negative schizophrenic symptoms, attention, and information processing. *Schiz. Bull.* **11,** 397–408.

Creese I. and Iversen S. D. (1975) The pharmacological and anatomical substrates of the amphetamine response in the rat. *Brain Res.* **83,** 419–436.

Cutting J. (1985) *The Psychology of Schizophrenia.* Church and Livingstone, Edinburgh.

Denenberg V. H. (1962) An attempt to isolate critical periods of development in the rat. *J. Comp. Physiol. Psychol.* **55,** 813–815.

Deutch A. J., Tam S-Y, and Roth R. H. (1985) Footshock and conditioned stress increase 3,4-dihydroxyphenylactic acid (DOPAC) in the ventral tegmental area but not the substantia nigra. *Brain Res.* **333,** 143–146.

Dyme I. Z., Sahakian. J., Golinko B., and Rabe E. (1982) Perseveration induced by methylphenidate in children: Preliminary findings. *Prog. Neuropsychopharmacol. Biol. Psychiat.* **6,** 269–273.

Erdelyi M. H. (1974) A new look at the new look: Perceptual defense and vigilance. *Psychol. Rev.* **81,** 1–25

Erskine M. S., Stern J. M., and Levine S. (1975) Effects of prepubertal handling on shock-induced fighting and ACTH in male and female rats. *Physiol. Behav.* **14,** 413–420

Feldon J. and Weiner I. (1988) Long term attentional deficit in nonhandled males: Possible involvement of the dopaminergic sytem. *Psychopharmacology* **95,** 231–236.

Feldon J. and Weiner I. (1989) Abolition of the acquisition but not the expression of latent inhibition by chlordiazepoxide in rats. *Pharmacol. Biochem. Behav.* **32,** 123–127.

Feldon J. and Weiner I. (in press) The latent inhibition model of schizophrenic attention disorder: Haloperidol and sulpiride enhance rats' ability to ignore irrelevant stimuli. *Biol. Psychiatry.*

Gittelman P. (1983) Experimental and clinical studies of stimulant use in hyperactive children and children with other behavioral disorders, in *Stimulants: Neurochemical, Behavioral and Clinical Perspectives* (Creese I., ed.), Raven, New York.

Gittelman-Klein R. (1987) Pharmaco-therapy of childhood hyperactivity: An update, in *Psychopharmacology: The Third Generation of Progress* (Meltzer H. Y., ed.), Raven, New York.

Greenberg G. and Bursdal C. (1982) Animal colony practices in North American academic institutions: A survey. *J. Gen. Psychol.* **106,** 165–173.

Groves P. M. and Rebec G. V. (1976) Biochemistry and behavior: Some central actions of amphetamines and antipsychotic drugs. *Ann. Rev. Psychol.* **27,** 91–127.

Groves P. M. and Tepper J. M. (1983) Neuronal mechanisms of action of amphetamine, in *Stimulants: Neurochemical, Behavioral and Clinical Perspectives* (Creese I., ed.), Raven, New York.

Herman J. P., Guillonneau D., Dantzer R., Scatton B., Smerdijan-Rouquier L., and Le Moal M. (1982) Different effects of inescapable footshocks and stimuli previously paired with footshocks on dopamine turnover in cortical and limbic areas of the rat. *Life Sci.* **30,** 2207–2214.

Hitzemann R., Wu J., Hom D., and Loh H. (1980) Brain locations controlling the behavioral effects of chronic amphetamine intoxication. *Psychopharmacology* **72,** 92–101.

Iversen S. D. (1987) Is it possible to model psychotic states in animals? *J. Psychopharmacol.* **1,** 154–176.

Jacklin C. N. (1989) Female and male: Issues of gender. *Am. Psychol.* **44,** 127–133.

Joyce E. M. and Iversen S. D. (1984) Dissociable effects of 6-OHDA-induced lesions of neostriatum on anorexia, locomotor activity and stereotypy: The role of behavioural competition. *Psychopharmacology* **83,** 363–366.

Kelly P. H., Seviour P. W., and Iversen S. D. (1975) Amphetamine and apomorphine responses in the rat following 6-OHDA lesions of the nucleus accumbens septi and corpus striatum. *Brain Res.* **94,** 507–522.
Kinsbourne M. (1983) Toward a model for the attention deficit disorder, in *The Minnesota Symposia on Child Psychology* (Perlmutter M., ed.), Erlbaum, Hillsdale, NJ.
Koob G. F., Riley S. J., Smith C., and Robbins T. W. (1978) Effects of 6-hydroxydopamine lesions of the nucleus accumbens septi and olfactory tubercle on feeding, locomotor activity and amphetamine anorexia in the rat. *J. Comp. Physiol. Psychol.* **92,** 917–927.
Kornetzky C. (1972) The use of simple test of attention as a measure of drug effects in schizophrenic patients. *Psychopharmacologia* **24,** 99–106.
LaBerge D. (1976) Perceptual learning and attention, in *Handbook of Learning and Cognitive Processes* vol. 4 (Estes W. K., ed.), Erlbaum, Hillsdale, NJ.
Levine S. (1960) Stimulation in infancy. *Sci. Am.* **202,** 80–86.
Levine S. and Lewis G. W. (1959) The relative importance of experimenter contact in an effect produced by extra-stimulation in infancy. *J. Comp. Physiol. Psychol.* **52,** 368,369.
Lewine R. L., Burbach D., and Meltzer H. Y. (1984) Effect of diagnostic criteria on the ratio of male to female schizophrenic patients. *Am. J. Psychiat.* **141,** 84–87.
Lubow R. E. (1973) Latent inhibition. *Psychol. Bull.* **79,** 398–407.
Lubow R. E. (1989) *Latent Inhibition and Conditioned Attention Theory,* Cambridge University Press, New York.
Lubow R. E., Weiner I., and Feldon J. (1982) An animal model of attention, in *Behavioral Models and the Analysis of Drug Action* (Spiegelstein M. Y. and Levy A., eds.), Elsevier, New York.
Lubow R. E., Weiner I., and Schnur P. (1981) Conditioned attention theory, in *The Psycholoqy of Learning and Motivation* vol. 15 (Bower G. H., ed.), Academic, New York.
Lubow R. E., Weiner I., Schlossberg A., and Baruch I. (1987) Latent inhibition and schizophrenia. *Bull. Psychon. Soc.* **25,** 464–467.
Mackintosh N. J. (1973) Stimulus selection: Learning to ignore stimuli that predict no change in reinforcement, in *Constraints on Learninq: Limitations and Predispositions* (Hinde R. A. and Hinde J. S., eds.), Academic, Cambridge.
Mackintosh N. J. (1975) A theory of attention: Variations in the associability of stimuli with reinforcement. *Psychol. Rev.* **82,** 276–298.
Mackintosh N. J. (1983) *Conditioning and Associative Learning.* Oxford University Press, New York.
McKinney W. T. (1988) *Models of Mental Disorders: A New Comparative Psychiatry.* Plenum, New York.
McKinney W. T. and Bunney W. F. (1969) Animal models of depression. I. Review of evidence: Implications for research. *Arch. Gen. Psychiat.* **21,** 240–248.

Moore J. W. (1979) Brain processes and conditioning, in *Mechanisms of Learning and Motivation: A Memorial Volume for Jerzy Konorski* (Dickinson A. and Boakes R. A., eds.), Erlbaum, Hillsdale, NJ.

Moore J. W. and Stickney K. J. (1980) Formation of attentional-associative networks in real time: Role of the hippocampus and implications for conditioning. *Physiol. Psychol.* **8,** 207–217.

Oades R. D. (1985) The role of noradrenaline in tuning and dopamine in switching between signals in the CNS. *Neurosci. Biobehav. Rev.* **9,** 261–282.

Oades R. D. (1987) Attention deficit disorder with hyperactivity (ADDH). The contribution of catecholaminergic activity. *Prog. Neurobiol.* **29,** 365–391.

Oades R. D. (1989) Attention deficit disorder and hyperkinetic syndrome: Biological perspectives, in *Attention Deficit Disorder: Clinical and Basic Research* (Sagvolden T. and Archer T., eds.), Erlbaum, Hillsdale, NJ.

Ohad D., Lubow R. E., Weiner I., and Feldon J. (1987) The effects of amphetamine on blocking. *Psychobiology* **15,** 137–143.

Oltmanns T. F., Ohayon J., and Neale J. M. (1978) The effect of antipsychotic medication and diagnostic criteria on distractability in schizophrenia. *J. Psychiatr. Res.* **14,** 81–91.

Pich E. M. and Samanin R. (1986) Disinhibitory effect of buspirone and low doses of sulpiride and haloperidol in two experimental anxiety models in rats: Possible role of dopamine. *Psychopharmacology* **89,** 125–130.

Pijnenburg A. J. J., Honig W. M. M., and van Rossum J. M. (1975) Inhibition of *d*-amphetamine-induced locomotor activity by injection of haloperidol into the nucleus accumbens of the rat. *Psychopharmacology* **41,** 87–95.

Porrino L. J., Lucignani G., Dow-Edwards D., and Sokoloff L. (1984) Correlation of dose-dependent effects of acute amphetamine administration on behavior and local cerebral metabolism in rats. *Brain Res.* **307,** 311–320.

Rapoport J. L., Buchsbaum M. S., Weingartner H., Zahn T. P., Ludlow C. M., and Mikkelsen E. J. (1980) Dextroamphetamine: Its cognitive and behavioral effects in normal and hyperactive boys and normal men. *Arch. Gen. Psychiat.* **37,** 933–943.

Rawlins J. N. P., Feldon J., and Gray J. A. (1980) The effects of hippocampectomy and of fimbia section upon the partial reinforcement extinction effect in rats. *Exp. Brain Res.* **38,** 273–283.

Rebec C. V. and Zimmerman K. S. (1980) Opposite effects of *d*-amphetamine on spontaneous neuronal activity in the neostriatum and nucleus accumbens. *Brain Res.* **201,** 485–491.

Reiss S. and Wagner A. (1972) CS habituation produces a "latent inhibition effect" but no active "conditioned inhibition." *Learn. Motiv.* **3,** 237–245.

Rescorla R. A. (1969) Pavlovian conditioned inhibition. *Psychol. Bull.* **72,** 77–94.

Rescorla R. A. (1971) Summation and retardation tests of latent inhibition. *J. Comp. Physiol. Psychol.* **75,** 77–81.

Rifkin A. and Siris S. (1987) Drug treament of acute schizophrenia, in *Psychopharmacology: The Third Generation of Progress* (Meltzer H. Y., ed.), Raven, New York.

Robbins T. W. and Everitt B. J. (1982) Functional studies of the central catecholamines. *Int. Rev. Neurobiol.* **23,** 303–365.

Robbins T. W., Jones G. H., and Sahakian B. J. (1989) Central stimulants, transmitters and attentional disorder: A perspective from animal studies, in *Attention Deficit Disorder: Clinical and Basic Research* (Sagvolden T. and Archer T., eds.), Erlbaum, Hillsdale, NJ.

Robbins T. W. and Sahakian B. J. (1979) "Paradoxical" effects of psychomotor stimulant drugs in hyperactive children from the standpoint of behavioral pharmacology. *Neuropharmacology* **18,** 931–950.

Robbins T. W. and Sahakian B. J. (1983) Behavioral effects of psychomotor stimulant drugs: Clinical and neuropsychological implications, in *Stimulants: Neurochemical, Behavioral and Clinical Perspectives* (Creese I., ed.), Raven, New York.

Robinson T. E. and Becker J. B. (1986) Enduring changes in brain and behavior produced by chronic amphetamine administration: A review and evaluation of animal models of amphetamine psychosis. *Brain Res. Rev.* **11,** 157–198.

Rutter M. (1989) Attention deficit disorder/hyperkinetic syndrome: Conceptual and research issues regarding diagnosis and classification, in *Attention Deficit Disorder: Clinical and Basic Research* (Sagvolden T. and Archer T., eds.), Erlbaum, Hillsdale, NJ.

Sagvolden T., Wultz B., Moser E. I., Moser M. B., and Morkrid L. (1989) Results from a comparative neuropsychological research program indicate altered reinforcement mechanisms in children with ADD, in *Attention Deficit Disorder: Clinical and Basic Research* (Sagvolden T. and Archer T., eds.), Erlbaum, Hillsdale, NJ.

Schaefer T. (1963) Early experience in its effects in later behavioral processes in rats: II: A critical factor in the early handling phenomenon. *Trans. NY Acad. Sci.* **25,** 871–889.

Schmajuk N. A. and Moore J. W. (1985) Real-time attentional models for classical conditioning and the hippocampus. *Physiol. Psychol.* **13,** 278–290.

Schmajuk N. A. and Moore J. W. (1988) The hippocampus and the classically conditioned nictitating membrane response: A real-time attentional–associative model. *Psychobiology* **16,** 20–35.

Schreiber H., Bell R., Wood G., Carlson R., Wright L., Kufner M., and Villescas R. (1978) Early handling and maternal behavior: Effect on *d*-amphetamine responsiveness in rats. *Pharmacol. Biochem. Behav.* **9,** 785–789.

Sergeant J. and van der Meere J. J. (1989) The diagnostic significance of attentional processing: Its significance for ADDH classification—a future DSM, in *Attention Deficit Disorder: Clinical and Basic Research* (Sagvolden T. and Archer T., eds.), Erlbaum, Hillsdale, NJ.

Shiffrin R. M. (1976) Capacity limitations in information processing, attention and memory, in *Handbook of Learning and Cognitive Processes* vol. 4 (Estes W. K., ed.), Erlbaum, Hillsdale, NJ.

Shiffrin R. M. and Schneider W. (1977) Controlled and automatic human information processing: II. Perceptual learning, automatic attending and a general theory. *Psychol. Rev.* **84,** 127–190.

Solomon P. R. (1980) Temporal versus spatial information processing theories of hippocampal function. *Psychol. Bull.* **86,** 1272–1279.

Solomon P. and Moore J. W. (1975) Latent inhibition and stimulus generalization of the classically conditioned nictitating membrane response in rabbits *(Oryctolagus cuniculus)* following dorsal hippocampal ablation. *J. Comp. Physiol. Psychol.* **89,** 1192–1203.

Solomon P. and Staton D. M. (1982) Differential effects of microinjections of *d*-amphetamine into the nucleus accumbens or the caudate putamen on the rat's ability to ignore an irelevant stimulus. *Biol. Psychiat.* **17,** 743–756.

Solomon P., Lohr C., and Moore J. W. (1974) Latent inhibition of the rabbit's nictitating response: Summation tests for active inhibition as a function of a number of CS preexposures. *Bull. Psychon. Soc.* **4,** 557–559.

Solomon F., Crider., Winkelman J. W., Turi A., Kamer R. M., and Kaplan L. J. (1981) Disrupted latent inhibition in the rat with chronic amphetamine or haloperidol-induced supersensitivity: Relationship to schizophrenic attention disorder. *Biol. Psychiat.* **16,** 519–537.

Spohn H. E., Lacoursiere R. B., Thompson K., and Coyne L. (1977) Phenothiazine effects on psychological and psychopharmacological dysfunction in chronic schizophrenics. *Arch. Gen. Psychiat.* **34,** 633–644.

Staton D. M. and Solomon P. (1984) Microinjections of *d*-amphetamine into the nucleus accumbens and caudate-putamen differentially affect stereotypy and locomotion in the rat. *Physiol. Psychol.* **12,** 159–162.

Swerdlow N. W. and Koob G. F. (1987) Dopamine, schizophrenia, mania and depression: Toward a unified hypothesis of cortico-striato-pallido-thalamic function. *Behav. Brain Sci.* **10,** 215–247.

Taylor E. A. (1986) The causes and development of hyperactive behavior, in *The Overactive Child* (Taylor A., ed.), MacKeith, London.

Taylor E. A. (1989) On the epidemiology of hyperactivity, in *Attention Deficit Disorder: Clinical and Basic Research* (Sagvolden T. and Archer T., eds.), Erlbaum, Hillsdale, NJ.

Taylor J. and Robbins T. W. (1984) Enhanced behavioral control by conditioned reinforcers following micronjections of *d*-amphetamine into the nucleus accumbens. *Psychopharmacology* **84,** 405–412.

Taylor J. and Robbins T. W. (1986) 6-Hydroxydopamine lesions of the nucleus accumbens but not of the caudate nucleus, attenuate enhanced responding with reward-related stimuli produced by intra-accumbens *d*-amphetamine. *Psychopharmacology* **90,** 390–397.

Underwood G. (1978) Attentional selectivity and behavioral control, in *Strategies of Information Processing* (Underwood G., ed.), Academic, London.

van Kammen D. P. and Boronow J. J. (1988) Dextro-amphetamine diminishes negative symptoms in schizophrenia. *Int. Clin. Psychopharmacol.* **3,** 111–121.

Varley C. K. (1984) Attention deficit disorder (the hyperactivity syndrome): A review of selected issues. *J. Dev. Behav. Pediatr.* **5,** 254–258.

Weinberg J. and Levine S. (1977) Early handling influences in behavioral and physiological responses during active avoidance. *Dev. Psychobiol.* **10,** 161–169.

Weinberg J., Krahn E. A., and Levine S. (1978) Differential effects of handling on exploration in male and female rats. *Dev. Psychobiol.* **11,** 251–259.

Weiner I. (1990) Neural substrates of latent inhibition: The switching model. *Psychol. Bull.* **108,** 442–461.

Weiner I. and Feldon J. (1987) Facilitation of latent inhibition by haloperidol. *Psychopharmacology* **91,** 248–253.

Weiner I., Feldon J., and Katz Y. (1987a) Facilitation of the expression but not the acquisition of latent inhibition by haloperidol in rats. *Pharmacol. Biochem. Behav.* **26,** 241–246.

Weiner I., Feldon J., and Ziv-Harris D. (1987b) Early handling and latent inhibition in the conditioned suppression paradigm. *Dev. Psychobiol.* **20,** 233–240.

Weiner I., Israeli-Telerant A., and Feldon J. (1987d) Latent inhibition is not affected by acute or chronic administration of 6 mg/kg *dl*-amphetamine. *Psychopharmacology* **91,** 345–351.

Weiner I., Lubow R. E., and Feldon J. (1984) Abolition of the expression but not the acquisition of latent inhibition by chronic amphetamine in rats. *Psychopharmacology* **83,** 194–199.

Weiner I., Lubow R.E., and Feldon J. (1988) Disruption of latent inhibition by acute administration of low doses of amphetamine. *Pharmacol. Biochem. Behav.* **30,** 871–878.

Weiner I., Halevy G., Alroy G., and Feldon J. (1987c) The effects of early handling on the partial reinforcement extinction effect and the partial punishment effect in male and female rats. *Q. J. Exp. Psychol.* **39B,** 245–263.

Weiner I., Schnabel I., Lubow R. E., and Feldon J. (1985) The effects of early handling on latent inhibition in male and female fats. *Dev. Psychobiol.* **18,** 291–298.

Weiner I., Shofel A., and Feldon J. (in press) Disruption of latent inhibition by low dose of amphetamine is antagonized by haloperidol and apomorphine. *J. Psychopharmacol.*

Weitzman A., Weitz R., Szekely G., Tiano S., and Belmaker R. H. (1984) Combination of neuroleptic and stimulant treatment in attention deficit disorder with hyperactivity. *J. Am. Acad. Child Psychiat.* **23,** 295–298.

Wells P. A. (1976) Sex difference in response to early handling in the rat. *J. Psychosom. Res.* **20,** 259–266.

Werry J. S. and Aman M. G. (1975) Methylphenidate and haloperidol in children. *Arch. Gen. Psychiat.* **32,** 790–795.

White F. J. and Wang R. Y. (1983) Differential effects of classical and atypical antipsychotic drugs on A9 and A10 dopamine neurons. *Science* **221,** 1054–1056.

Worms P., Broekkamp C. L. E., and Lloyd K. G. (1983) Behavioral effects of neuroleptics, in *Neuroleptics: Neurochemical, Behavioral and Clinical Perspectives* (Coyle J. T. and Enna S. J., eds.), Raven, New York.

A Computerized Methodology for the Study of Neuroleptic-Induced Oral Dyskinesias

Gaylord Ellison and Ronald E. See

1. Introduction

Although there is a dramatic need for animal models related to schizophrenic symptomatology, the literature has been disappointing in this regard. This is also true for models of the side effects of antipsychotic drugs, particularly with regard to rodent models. Considerable controversy exists as to whether the oral movements induced by antipsychotics in rodents represent an acute-dystonia-like effect or a tardive dyskinesia-like effect.

We have developed a highly novel methodology for measuring oral dyskinesias in rats administered neuroleptics. Using this computerized technique, which is based on rapid detection of UV-sensitive ink spots placed on the rat's upper and lower jaws, we have found several distinctive syndromes that gradually develop in rats administered chronic neuroleptics. A syndrome similar to human tardive dyskinesia (TD) develops with continuous typical neuroleptics. This syndrome is characterized by oral movements of the same energy spectrum as those seen in human TD patients, and it worsens upon drug withdrawal. Weekly injections of neuroleptics induce a distinctively different syndrome, with very large, fast, oral movements with a much higher energy spectrum. Both drug regimens, however, also gradually induce very small oral movements, the nature of which

From: *Neuromethods, Vol. 18: Animal Models in Psychiatry I*
Eds: A. Boulton, G. Baker, and M. Martin-Iverson

remains poorly understood. Atypical neuroleptics do not induce the TD-like syndrome in rats, although other unexpected alterations in oral behaviors occur. Only through measurements of the exact form of oral movements could these observations have been made, and the computerized television-based procedures described here represent the only extant technology for making such observations.

2. Forms of Neuroleptic-Induced Movement Disorders in Humans

Compared to other psychopharmacological treatments, such as antidepressants and antianxiety agents, relatively little progress is being made in the development, and then progression into widespread use, of novel antipsychotic medications. One reason for this is the difficulties encountered in assessment of favorable drug effects with antipsychotic drugs. Schizophrenia often is a chronic disorder; consequently, many schizophrenic patients have already, inconsistently, received substantial quantities of drug over prolonged time periods. Futhermore, drug treatment often includes various drug prescribed under a variety of drug regimens. The complex drug-treatment histories of such patients has made it particularly difficult to assess causal factors of the extrapyramidal side effects (EPS) that often accompany neuroleptic administration. Because most, or perhaps all, currently available antipsychotic drugs may ultimately be proved capable of producing motor disorders to some extent, these side effects have become the limiting factor in the prescription of these highly efficacious compounds.

However, the lack of progress may also be attributable to the paucity of good models. There is a great need to develop and utilize optimal animal models in order to understand etiological factors of, and treatment approaches to, schizophrenia. At present, there are few convincing animal models related to schizophrenia, largely because of the difficulty of modeling complex cognitive dysfunctions in animals. Yet neuroleptic-induced motor side effects would seem to be perfectly suited to animal models, since (a) these side effects of neuroleptics are believed to be exerted through extrapyramidal motor regions, particularly the nigrostriatal dopamine (DA) pathway, rather than cortical DA pathways (Marsden and Jenner, 1980; Creese, 1983) and (b) the symptoms are so overt and distinctive.

Neuroleptics induce three reversible EPS (i.e., ones that generally disappear when the drug is withdrawn); these are pseudo-Parkinsonism, akathisia, and acute dystonia. Chronic neuroleptic treatment can also induce TD, which appears only after prolonged drug therapy and which can be irreversible (Jeste and Wyatt, 1982). Some side effects of neuroleptics have proved quite amenable to animal modeling, most notably pseudo-Parkinsonism. Thus, the akinesia and muscular stiffness induced by acute neuroleptics in humans are mimicked quite well by the neuroleptic-induced catalepsy seen in animals. However, animal models of other side effects, particularly TD, have proven to be more controversial.

2.1. Acute EPS

Acute EPS can be characterized in several forms. Neuroleptic-induced Parkinsonism is similar to idiopathic Parkinsonism with symptomatic movements of hypokinesia, rigidity, and tremor (McGeer et al.,1961). Dystonia is a common acute reaction occurring within a few days of treatment (Ayd,1961). Dystonias induced by neuroleptic drugs have been recently reviewed by Rupniak et al. (1986). Although acute dystonias, which appear shortly after the onset of drug administration, are the most common, tardive dystonias have also been reported (Burke et al., 1982). The dystonias consist of muscle spasms and abnormal postures of the cranial, neck, and trunk musculature; abnormal movements of the arms and legs; facial grimacing; and ocular spasms. Acute dystonias are best treated with anticholinergics. Finally, there is akathisia ("inability to sit still"), which consists of subjective feelings of motor restlessness, often accompanied by objective signs of incessant motor movement (Adler et al., 1989). Although often defined by a subjective experience (a highly unpleasant inner restlessness), it can be behaviorally diagnosed by watching for incessant movement, especially of the lower extremities (such as tapping movements of the feet, shifting in place, shuffling from foot to foot, or pacing).

2.2. Tardive Dyskinesia

Tardive dyskinesia affects about 20–40% of patients receiving chronic antipschotic medication (Casey, 1985a). Tardive dyskinesia is classically defined as a syndrome that primarily consists of abnormal, stereotyped involuntary movements, usually of a choreoathetoid nature, which typically affect the mouth and face (the "buccolinguo-masticatory triad," or BLM syndrome), but sometimes involves the limbs and trunk as well. More recent accounts of this disorder have discussed how the disorder can vary, and this has led to the suggestion of distinct subsyndromes of TD (Gardos et al., 1987). It has also been proposed that TD represents the end stage of a progression that begins with acute akathisia and then progresses, through stages, to the BLM syndrome and TD (Barnes and Braude, 1985). It is clear, however, that the BLM syndrome typically appears relatively late in the course of treatment involving chronic antipsychotic drugs and that drug treatment is a necessary factor in the etiology of this syndrome (Jeste and Wyatt, 1982). Tardive dyskinesia is considered the most undesirable of the neuroleptic-induced EPS, because the orofacial dyskinesias that are a predominant feature of this syndrome sometimes persist long after discontinuation of the drugs, or may even be permanent (Crane, 1973; Casey, 1985b).

The literature on this syndrome is full of conflicts as to which procedures minimize, and which induce, TD. This is largely in consequence of several complicating variables. A typical patient who develops TD has been exposed to a variety of neuroleptics, often given in differing dosage regimens. Tardive dyskinesia patients often have other impairments, including ventricular enlargement and cognitive dysfunctions (Waddington et al., 1985). Finally, it has been well established that age is a critical indicator of the presence of orofacial dyskinesia (Kane and Smith, 1982; Guy et al., 1985). Consequently, retrospective clinical studies have not been convincing as to what induces or worsens TD. Treatment is also not well understood, for although various strategies have been developed for reducing the incidence of or treating TD (such as reducing the dopaminergic excess, or enhancing cholinergic or GABAergic tone), each has generally fallen into disfavor after further study (Jeste et al., 1988).

One obvious strategy is to develop new antipsychotic medications that might not produce severe motor side effects. However, historically there has been a variety of promising candidates that led to falsely raised hopes. Jeste and Wyatt comment (1982, p. 86),

> A number of new neuroleptics, which were initially claimed to have a low risk of inducing TD, were later found to be as likely to produce dyskinesia as the older drugs. Probably all neuroleptics, when given in equivalent amounts and for similar periods to similar patients, carry a comparable risk of causing TD.

The most effective antipsychotic in use that appears not to induce TD is the dibenzoxazepine clozapine (Pi and Simpson, 1983; Lindstrom, 1988). In addition, neuroleptics of the substituted benzamide class, most notably sulpiride, have also been reported to show a low incidence of TD (Gerlach and Casey, 1984). It is possible that these and other newly developed antipsychotic drugs may eventually replace presently used neuroleptics that do cause TD.

A second strategy has been to attempt to minimize these undesirable side effects through the use of optimal dosing strategies. One such strategy involves giving the patient periodic drug-free periods ("drug holidays"). This practice has fallen into disfavor because of the considerable controversy about what is an optimal drug holiday, how best to discontinue the drug, and whether this practice is actually deleterious (Goldman and Luchins, 1984). Another drug-regimen issue that has been discussed is whether continuous, steady levels of neuroleptics (which can be most clearly achieved with intramuscular depot injections) have different long-term effects than more fluctuating neuroleptic levels (as are obtained with oral regimens). Because a prolonged drug exposure is necessary, the role of drug regimen may be especially important for the development of TD. Evidence from dopamine-agonist studies indicate that dopaminergic systems have entirely different responses to chronic fluctuating vs chronic steady-state levels of stimulation (Post, 1980; Ellison and Morris, 1981). Evidence for such differences with neuroleptics remains controversial (Csernansky et al., 1981; Beresford and Ward, 1987).

3. Animal Models of Neuroleptic-Induced Side Effects

3.1. Primate Studies

Initial research into the effects of chronic neuroleptics on nonhuman primates led some authors to conclude that a variety of TD-like symptomatic movements, including orobuccolingual dyskinesias, clearly develop in primates after prolonged neuroleptic treatment (Deneau and Crane, 1969; Gunne and Barany, 1976; reviewed by Domino, 1985) and that these are excellent animal models of TD. However, as pointed out by Rupniak et al. (1986), it is very important to distinguish between acute dystonic effects of neuroleptics and TD-like effects. These two reactions probably reflect very different mechanisms, yet are often confused or intermixed. Acute dystonic reactions are Parkinsonian-like: They often occur shortly after the drug is administered and they are diminished by anticholinergics. TDs, conversely, are usually maximal when the drug is wearing off (Gardos et al., 1978), and (although this is controversial) sometimes respond well to cholinomimetics (Moore and Bowers, 1980). However, this confusion of the two side effects has led to very controversial practices and interpretations of primate models. Many pharmaceutical companies use the propensity of a drug to induce acute dystonic reactions in "primed" monkeys as an indication of whether a new drug will induce extrapyramidal dysfunction (Neale et al., 1984). However, it has not been demonstrated that these acute dystonic reactions are accurate predictors of TD development. Many such laboratories maintain a colony of monkeys that have been primed, typically by receiving large injections of neuroleptics once each week in a nondepot form (*see* Liebman and Neale, 1980; Kovacic et al., 1986). Upon injection of the neuroleptic, these monkeys immediately show acute dystonic-like reactions, such as writhing, abnormal posture, rigid limb extension, and flinging themselves against the wall (Weiss and Santell, 1978; Neale et al., 1984). The highly fluctuating (once a week) regimen clearly contributes to the progressive appearance of this syndrome, and it is noteworthy that humans are never given such an extremely fluctuating drug regimen. However, if treated long enough with high enough doses, all cebus, squirrel, and green vervet monkeys develop dystonias; only some macaques and baboons do (Rupniak et al., 1986).

This reaction does not fit the pattern of TD. Pharmacologically, it behaves in an opposite fashion to TD, for it disappears when drug levels decline. Yet, because it is not convenient to administer neuroleptics several times daily for many months to monkeys, this is rarely the practiced drug regimen. Domino and Kovacic (1983), in fact, concluded that, of 121 monkeys given neuroleptics in several published studies, most developed only the acute dystonic or Parkinsonism syndrome, and only 10 developed TD (some of the monkeys of Gunne and Barany, 1976; Bedard et al., 1977; McKinney et al., 1980). Kovacic and Domino (1984), Johansson (1989), and Gunne et al. (1984) subsequently increased this sample, but the number of documented cases is clearly low.

3.2. Rodent Studies

Many studies have purported to show TD-like behaviors in rats after chronic neuroleptic administration, which they have put forth as viable rodent models of TD. Initial studies with repeated neuroleptic exposure in rats and guinea pigs found increased oral stereotypies in neuroleptic-treated animals in response to a DA agonist, such as apomorphine (Klawans, 1973; Tarsy and Baldessarini, 1974). However, these studies do not represent true animal models of TD, since the periods of drug treatment were relatively short (three weeks) and no spontaneous changes in oral movements were noted. Later studies have used longer periods of neuroleptic treatment (3–18 months) and have focused on spontaneous changes in oral activity. Some reports noted an increase in "spontaneous stereotypies" (Sahakian et al., 1976) and "spontaneous mouth movements" (Iversen et al., 1980) after repeated neuroleptic treatment. Waddington et al. (1983) gave a variety of oral neuroleptics, as well as intramuscular injections of fluphenazine decanoate, to groups of rats. After six months of drug administration, spontaneous increases in nondirected oral movements could be observed in many of the drugged animals when the rats were placed in an observation cage. Such late-onset increases in oral movements have also been reported by several other laboratories using various neuroleptics (Gunne and Haggstrom, 1983; Johansson et al., 1986; Mithani et al., 1987).

However, several important characteristics of this type of observer-rated vacuous chewing movements are controversial. Other investigators have failed to replicate these findings, and still others argue that these vacuous oral movements in rats are not truly models of TD. Clow et al. (1979,1980) described spontaneous mouthing movements in rats after 12 months of neuroleptics, but reported that these movements disappeared relatively rapidly after drug withdrawal, unlike TD. Rodriguez et al. (1986) failed to observe "vacuous chewing" after prolonged neuroleptic treatment and conclude that this is a "controversial animal model of TD." Furthermore, Gunne et al. (1986) and Rupniak et al. (1983) noted that, whereas cholinergic agents often ameliorate TD in humans (and monkeys) and anticholinergics exacerbate them, the vacuous chewing movements in rats induced by chronic haloperidol, trifluoperazine, or sulpiride are increased by cholinergic agents and decreased by anticholinergics. An even more severe problem is that several laboratories have reported that "vacuous oral movements" occur in rodents as soon as 24 hours after an acute injection (Glassman and Glassman, 1980; Rosengarten et al., 1983; Rupniak et al., 1985). Thus, in some situations, they are clearly *not* tardive in appearance. This entire issue has been reviewed by Rupniak et al. (1986) and Stewart et al. (1988), who conclude that vacuous oral movements actually represent acute dystonic reactions in rats, rather than TD-like movements. A related problem is that, although these oral movements are often labeled as being dyskinetic or dystonic, they are often reported to seem normal in appearance. As stated by Stewart et al.,

> However, it must be stressed again that the effect of neuroleptic treatment on the rodent is to increase the incidence of part of the normal repertoire of behavior of the animal. The purposeless chewing movements induced by neuroleptic drugs are not abnormal (p. 352).

Thus, it is clear that the time-course of any enhancement of oral movements produced by neuroleptics in rodents, and even the basic phenomenon, has been inconsistent, with some researchers observing enhanced oral activity early in the neuroleptic-exposure period or even upon acute drug treatment; others observing only a delayed enhancement of oral movements, occasionally not apparent until drug withdrawal; and still other studies demonstrating elevated oral activity during long-term neuroleptic treatment, which rapidly returns to control levels upon drug discontinuation. The definition of "opening the mouth" in all these studies is very vague, and attempts to categorize "oral movements" more precisely (such as including only those oral movements accompanied by audible tooth grinding, and so forth) have not proved successful in reconciling the discrepancies noted above.

One might question the basic premise of all these attempts at developing rodent models, for the basic definition of dystonias and dyskinesias in humans is that of movement that is abnormal in form, not just any movement, albeit at an increased frequency. The direct counting of oral movements in humans with TD has been attempted and that this method yields data which correlates very poorly with all other TD scales, including the AIMS scale (Chien et al., 1980), whereas the intercorrelations between other TD scales is high. Indeed, the patients are often severely tranquilized, and often demonstrate markedly diminished organized oral movements, including speech. Rather, it is the form of the movements that characterize the syndrome, not just the frequency of oral (or any other) movement *per se.*

4. An Artifact in Assessing Oral Activity in Rats in the Open-Cage Test

The discrepancies seen among these various attempts at modeling TD may be caused by several factors, including rat strain, observer criteria, and drug regimen. One particularly important factor for the interpretation of these results is the observational technique used to assess oral activity. Typically, rodents are administered chronic neuroleptics, intermittently removed from their home cage, and placed in a novel observation cage or open field for a few minutes. A human observer stands close to the cage and counts, or rates, the incidences of spontaneous oral movements (sometimes called "vacuous chewing movements").

We have reported a possible confound in the open-cage test in three separate experiments (Levy et al., 1987). In these experiments, groups of rats were chronically administered neuroleptics in a variety of ways (chronic injections, subcutaneous implants, and decanoate injections) and then examined for oral movements in two different tests: (a) in an open cage using a human observer and (b) in a Plexiglas™ tube enclosure, where oral movements were monitored by both a human observer and a computerized video-analysis system. These two testing methods showed dramatically different effects of neuroleptic administration. In one experiment, when the rats were tested in the tubes, oral activity was depressed during haloperidol administration and elevated after treatment was discontinued, but when these animals were then observed in the open-cage test, there was no sign of increased oral movements. In another experiment, we found that oral activity was elevated in the open cages during neuroleptic exposure, but that this effect rapidly became attenuated upon drug withdrawal. Conversely, in the same animals, oral activity was enhanced following withdrawal when tested in the observation tube.

These two studies clearly demonstrated that neuroleptic-induced oral activity varies considerably, depending on the testing environment, and a third experiment provided a possible explanation for these data. Rats were examined for oral behavior in the observation cages and in the tubes in rapid succession. Oral movements were elevated in the cages immediately following haloperidol injections (at a time when motor activity was drastically reduced), but there was no sign of these increased oral movements when the animals were placed, immediately afterward, into the observation tubes. Several days after haloperidol injections (and in the continuous haloperidol group), when motor behavior was only moderately suppressed, there was no alteration in oral activity in either testing situation.

These data suggest that an artifact related to activity levels can exist in measurements of oral movements in the open-cage test. It is possible that this is attributable in part to the difficulty involved in observing the mouth of an animal that is extremely active, but we have also observed that oral activity in rats is quite dramatically inhibited during bouts of exploratory behavior, when the animals are rearing, sniffing, and moving about the cage. If all that is recorded during a test is the total number of oral movements, this can lead to misleading results, since animals injected with neuroleptics appear to show inflated OM (oral-movement) scores simply because they are less exploratory. This appears to be the result of an intrinsic incompatibility between the expression of gross skeletal activity and that of spontaneous, "vacuous" oral activity (rats who are rearing and sniffing move their mouths less often; tranquilized rats rarely rear and sniff).

This effect can be seen in several different tests. We have consistently observed that when rats (either neuroleptic-treated or drug-naive) are placed into novel cages, exploration over the next 30 minutes progressively decreases, whereas oral-movement scores correspondingly increase. Another situation in which this effect may occur is in acute drug tests with tranquilizers, for we have confirmed the observations of others (Rupniak et al., 1985) that, when tested in observation cages, rats can sometimes be observed to show elevated oral movements very soon after the beginning of neuroleptic administration (Rupniak et al., 1984) or even following acute injections (Rosengarten et al., 1983,1986). Similarly, the elevated oral movements observed in the open-cage test following chronic neuroleptics returned to control levels soon after drug discontinuation. This may be because haloperidol withdrawal leads to hyperactivity (Owen et al.,1980), resulting in a suppression of the expression of oral movements caused by the mechanism described above. When tested in the observation tubes, a different trend is often observed, especially in the initial stages of drug administration and testing. It is only after the animals have been receiving drugs and tested repeatedly in both conditions for months that the two measures begin to show some concordance (See et al., 1988). From these data it seems clear that simply counting normal-appearing oral activity will not develop into a noncontroversial model of a disorder that is fundamentally a dyskinesia. Future models must also demonstrate that the behavior measured is abnormal in form.

5. Computerized Assessment of Oral Activity

It seems obvious that, at this stage in technology, with the rapid development of inexpensive and powerful computing techniques, the computer should be able to act as a highly sophisticated observer of behavior. An immense number of devices that can store television images ("frame-grabbers") are now available, and the information contained within each image can be analyzed and compared with the information obtained from television images of the animal just before and just after. However, one soon learns that this apparently easy solution has a severe limitation, that of information overload for the computer. A typical frame-grabber might store the television image in a matrix of 480 × 600 pixels. Each pixel might contain, for example, 24 shades of gray, and, if a color image is being analyzed, there will be three such matrices, each one corresponding to the red, blue, or green gun of the television screen. If the computer is then asked to analyze this immense amount of information to detect position, orientation, and movement of the animal, even the most powerful microprocessor can only analyze each frame very slowly, at speeds on the order of 10 seconds to several minutes, depending on the amount of data stored and the type of information required. This may be sufficient for some kinds of simple measurements, but it is obviously entirely inappropriate for the measurement of rapid movements and, especially, for tasks that require an analysis of the precise form of the movement. Yet one clearly cannot derive conclusions about dyskinesias without actually determining the form of the movement in question. One solution to this is to turn to dedicated video processors, which can perform on-line analyses of aspects of the video image. Yet such systems become quite expensive, and are so complex that most available systems are proprietary, and the code of the analysis performed is not available to the experimenter, meaning that only extremely general data outputs are available.

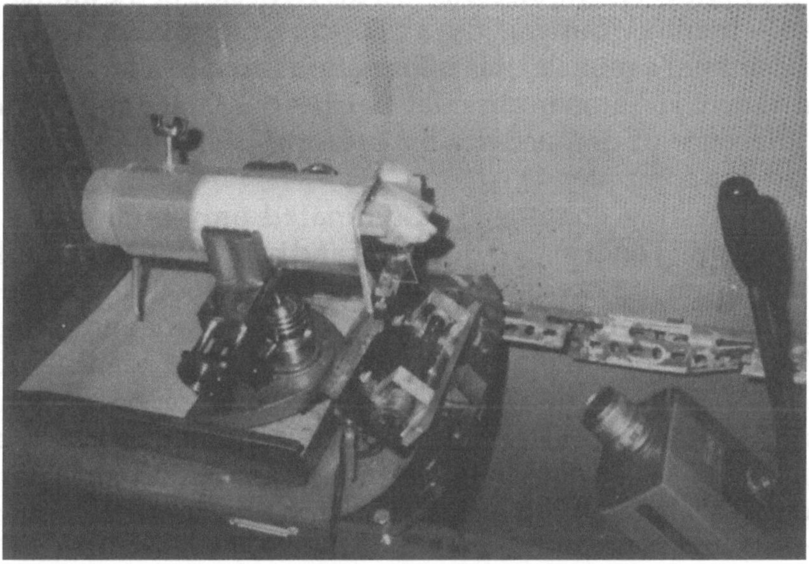

Fig. 1. View of the rat in the recording tube.

We have adopted a different kind of solution to this problem by developing a procedure by which oral movements in rats can be precisely and rapidly quantified. It is based on an attempt to simplify the video image to its pure essentials by giving the closed-circuit television camera monitoring the animal only an extremely simple image, building a special board that breaks this image down into a very simple and rapid quantification on-line, thereby giving the computer an extraordinarily easy task.

Using close-up television images, the oral regions of rats can be observed while the animals are in round Plexiglas™ tubes (Sant and Ellison, 1984; Gunne et al., 1986). We have added a much more precise quantification of oral movements with direct computerized measurements of jaw movements using a fluorescent dye marking that can be simultaneously verified by human obsevational scoring. Rats are initially habituated to being placed for brief periods in loosely fitting Plexiglas™ tubes that rest inside a soundproof chamber (Fig. 1). At one end of the tube is a hole through which the rat's head protrudes. The rat is not tightly confined in this tube, and struggling is minimal during the six-minute testing sessions. On the right side of the animal, a closed-circuit TV camera with a close-up lens provides a picture of the animal's muzzle; this information is fed to a large television screen. The human observer watches this image and records the presence of oral activity by pressing keys on a keyboard linked to a computer.

The testing chamber is illuminated only by a 6-W UV ("black-light") bulb placed in front of and below the rat's muzzle. Before the rat is placed in the tube, two small spots are painted, one on the upper and one on the lower jaw of the rat, using a UV-sensitive dye ("Black-Ray Swimming-pool readmission ink—#A801 green color" from Willard Marking Devices, Santa Ana, CA). If too weak, this ink can be concentrated by evaporating some of the alcohol in it, using a stream of nitrogen gas. A second TV camera with a close-up lens is positioned 22 cm in front of and beneath the rat's muzzle. This camera has a UV filter in front of the lens (03FIB008 from Melles Griot, Irvine, CA), so that it detects only the two fluorescent dots and not the background illumination. The output from this camera is adjusted so that only the two spots are visible (Fig. 2), and the resulting digital signal is fed to a computer with a movement-detection circuit (Biotronic Designs,Tarzana, CA). This circuit contains a counter that stores the number of TV rasters that occur from the bottom of the top spot until the top of the bottom spot is detected. The actual hardware involved consists of a "one-shot" that is fired when the first spot is detected. This one-shot remains on during a time that is set to coincide with one complete horizontal sweep of the television signal. If the signal is still present at the end of this time, the one-shot immediately fires again. This process continues until the bottom of the first spot is reached, at that time the one-shot finally "times out" and the counter is enabled. This counter then begins to increment upon each horizontal synch pulse, continuing until the second spot is detected. The counter then ceases incrementing, and, when the vertical synch pulse occurs at the end of the TV frame, a hardware interrupt is fired, which, through a subroutine written in assembler language, places the counter value into computer memory and resets the counter. The net result is that an integer

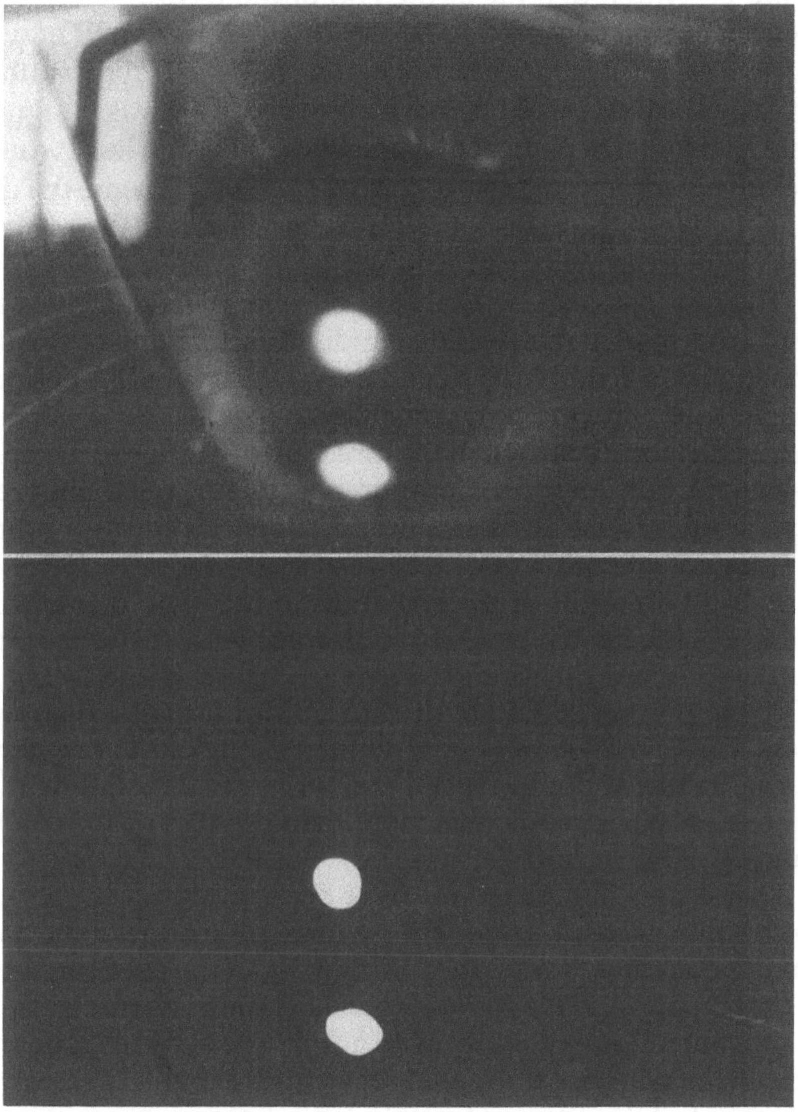

Fig. 2. View of the rat with UV-sensitive ink spots on the muzzle, as seen by the recording TV camera before digitizing (above), and as the digitized input into the computer (below).

value, which corresponds to the number of TV rasters between the two spots, is stored 60 times/second.

This distance represents the amount of mouth opening, and it, together with the human observer's reports every 60th of a second, is written to a file at the end of the recording session. This system was designed to be maximally sensitive to mouth movements, for it is the distance between the two spots, rather than their absolute position, that is recorded. This was designed so that horizontal or vertical movements of the entire head would be relatively undetected (since during such movements the difference between the two spots would change minimally relative to the change during jaw movements).

The raw data can be analyzed in several ways. One basic unit of analysis is a unidirectional individual opening or closing of the mouth, a "computer-scored movelet," or "CSM," defined to begin each time the dots come closer together or go farther apart, and to terminate when the direction of change in distance between the dots reverses. Computer-scored movelets are presently sorted into different amplitude categories corresponding to number of television rasters crossed by the movelet: 2, 3, 4–5, 6–9,or $\geq$ 10 (in terms of the actual distance of jaw movement: 0.6, 0.9, 1.2–1.5, 1.8–2.7, or >3.0 mm, respectively, of jaw opening or closing). In our earlier experiments, because of interlacing of the TV picture, we could not discern CSMs of smaller than two rasters. However, we have now switched to television cameras that can be run at double speed and will provide a full field of 512 rasers at 60/seconds; this more than doubles our previous technological precision.

Data-inspection programs have been written that present, on a computer screen, a moment-by-moment picture of the data from the data files. Simultaneous viewing of the observer's behavioral reports and the computerized files indicate that isolated oral movements are typically of an amplitude between four and nine rasters, with rise times that are neither extremely slow nor extremely fast. Sudden, extremely large head movements can be identified by their large amplitudes, steep slopes, and nonrepetitive nature. These two types of behavior can be sorted at least partially using restrictions ("filters") on amplitude, distance between CSMs, degree of repetitiveness, and waveform. The scoring programs also calculate the slope (amplitude divided by duration) of each CSM, determining within each amplitude category the average slope for each animal. More recently, we have written programs to analyze the energy spectrum of oral activity using fast-Fourier transforms, and this has proved an incredibly useful way to detect alterations in the form of oral movements.

This fluorescent-dye technique may become a useful tool in behavioral analyses, because it is inexpensive, it permits precise recordings of movement without restraining wires, and highly simplified data can be collected at a rapid rate (compared with relatively slow analyses of the entire video image, as is the case with typical freeze-frame methods). The data-collection boards can be purchased for the same price as a typical frame-grabber. Single spots can be used, with the X and Y coordinates being stored, to analyze, limb movements, tracking, and so forth.

6. Effects of Prolonged Neuroleptics on Oral Movements as Assessed by Computerized Measurement

In an initial validation study (Ellison et al., 1987), we described the results obtained when slow-release silastic implants filled with haloperidol (HAL) were implanted (release rate of 0.45 mg/kg/d) and oral activity measured (for construction details of these implants, *see* Appendix 1). From what we had read in the literature, we had expected to find greatly heightened oral movements gradually developing in the HAL-implanted animals during drug administration, but in general, we did not see such an increase. Instead, the files of the human observer indicated only a very small increase in the total duration of observed oral movements during eight months of continuous HAL, and no increase in OM frequency. The computerized two-spot files collected at six and eight months correlated well with the behavioral reports by the observer, with observer reports of oral movements often preceded, by about 100 ms (reflecting the reaction time of the observer), by CSMs of amplitudes between four and nine rasters. These CSMs all showed a typical form of opening and closing of the jaws, but each individual animal had a characteristic waveform. Observer reports of head movements were accompanied by rapid and large-amplitude shifts in distance between the two dots. The computerized device also added a number of further interesting details about the oral movements in the chronic HAL-implanted animals. Two kinds of differences were noted between the control animals and the animals who had received HAL for six to eight months. There were acute drug effects to which tolerance did not develop, such as fewer large-amplitude oral movements in the HAL-implanted animals, with those oral movements that did occur being sluggish in form (i.e., they had slow rise times or small slopes). However, when we tested drug-naive rats administered acute injections of HAL, similar effects were noted, thus suggesting that these effects reflect persisting sedative effects of HAL that do not show tolerance.

However, there were also distinctive changes in the HAL-implanted animals, for a totally unexpected result was also obtained. Although the computer-analyzed files indicated that CSMs of the largest amplitudes were chronically depressed in the HAL-implanted animals, they also indicated that CSMs of the smallest amplitude (2 rasters) were increased in these animals, an effect that clearly did not occur with acute injections of HAL. The meaning of this finding became more clear when the implants were removed, for total oral activity in the chronic-HAL rats dramatically increased to above control levels, and the amplitude distribution also shifted dramatically as what had been isolated, small oral movements in the HAL-implanted rats developed into repetitive trains of larger mouth movements. We replicated these finding in a separate study (Levin et al., 1987). These results led us to rethink our model of TD in rats, for our computer analysis had told us two things: (a) large-amplitude oral movements were not dramatically increased in chronic HAL-implanted animals, but were rather chronically suppressed in many ways; and (b) tiny tremorous movements, which are completely undetected by the human eye under conditions used by all other laboratories, appeared to be our best indicator of "late-stage" neuroleptic effects.

Since these results were strikingly different from what we had expected based on the literature, we speculated that perhaps our observers were not seeing dramatically increased large-amplitude oral movements because our silastic implants were not releasing sufficient amounts of drug. Consequently, we conducted a long-term study using quite sizable depot injections (See et al., 1988). Rats were chronically administered either haloperidol or fluphenazine decanoate via depot injections for eight months. These were administered via im (21 mg/kg) injections in the upper thigh muscle, given once every three weeks, giving an average daily dose of 1 mg/kg/day. Since decanoate neuroleptics can persist for long times after drug withdrawal, the animals were then prepared for final drug withdrawal by giving these same drug (nondecanoate form) in their drinking water for the next two months, and then the drugs were withdrawn. Throughout the experiment the animals were tested repeatedly in the computerized tube device, which measured CSMs, and, in the latter half of the experiment, were also scored for observed oral movements in the tube, as well as in an open cage, by a human observer. In the tube, the animals in both neuroleptic-treated groups showed initial decreases in the number of CSMs and made sluggish CSMs; these effects were generally larger in the fluphenazine-treated animals. After six months of chronic neuroleptics, the haloperidol-treated animals showed slightly increased oral movements, as reported both by the human observer and in CSMs of all amplitudes. This effect of increased large oral movements became significant only upon drug withdrawal (Fig. 3a). Fluphenazine-treated animals showed a more persistent depression of both oral movements and CSMs of large amplitudes during and after drug treatment. However, the behavior most characteristic of both neuroleptic-treated groups was again the gradual development of increases in CSMs of the smallest amplitudes measurable (Fig. 3b).

We began to believe that these tiny CSMs represented our best animal model of TD. Because rats sometimes show very small, very rapid repetitive oral tremors of the masseter muscles, we had initially surmised that the small, computer-detected CSMs represented tiny oral tremors, and that this was apparently what represented the analog of TD in rats. We therefore wrote computer programs that would allow us to conduct fast-Fourier transform (FFT) analyses of our data, expecting to find

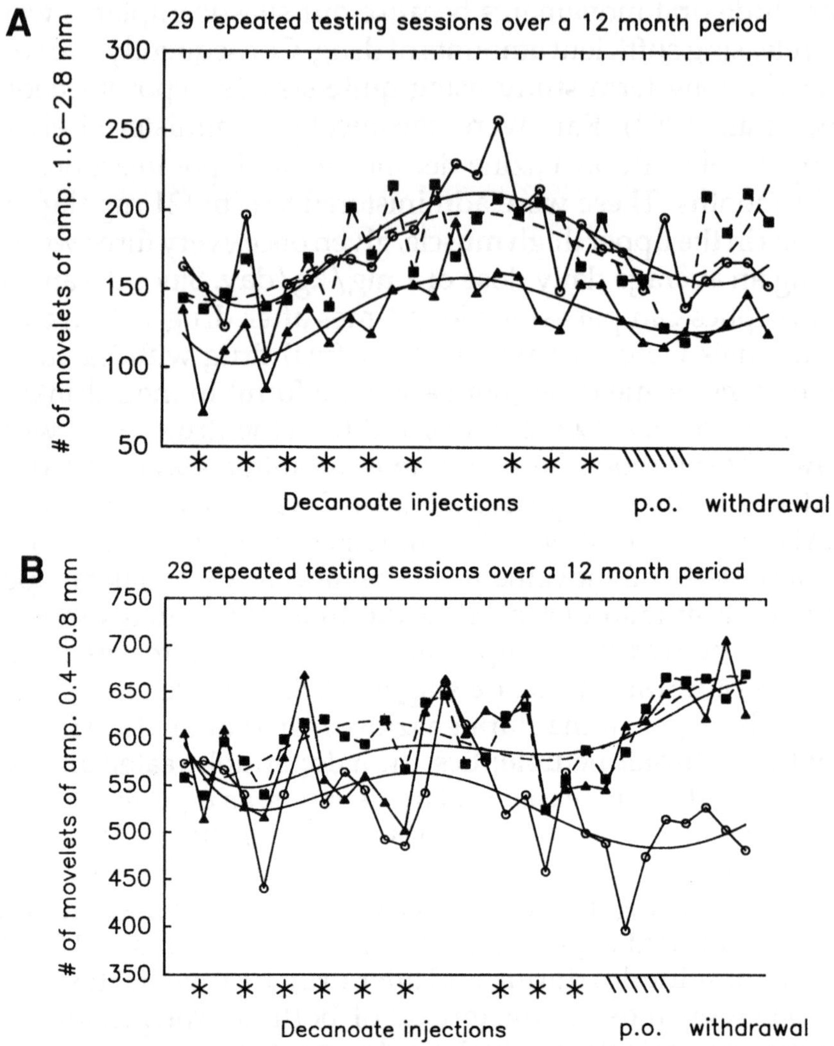

Fig. 3. **(A)** This shows that large-amplitude CSMs in rats administered chronic haloperidol (■—■) and fluphenazine (▲—▲) are not significantly elevated compared to control values (O—O) during drug administration, whereas the smallest CSMs **(B)** do become elevated in both drug groups, especially upon drug withdrawal.

increased energy at frequencies at which tremors occur (about 6–10 Hz) in chronic-neuroleptic-treated animals. When we actually analyzed our previous data (See et al., 1988) using FFT, we found that, when the animals were first started on the drug,

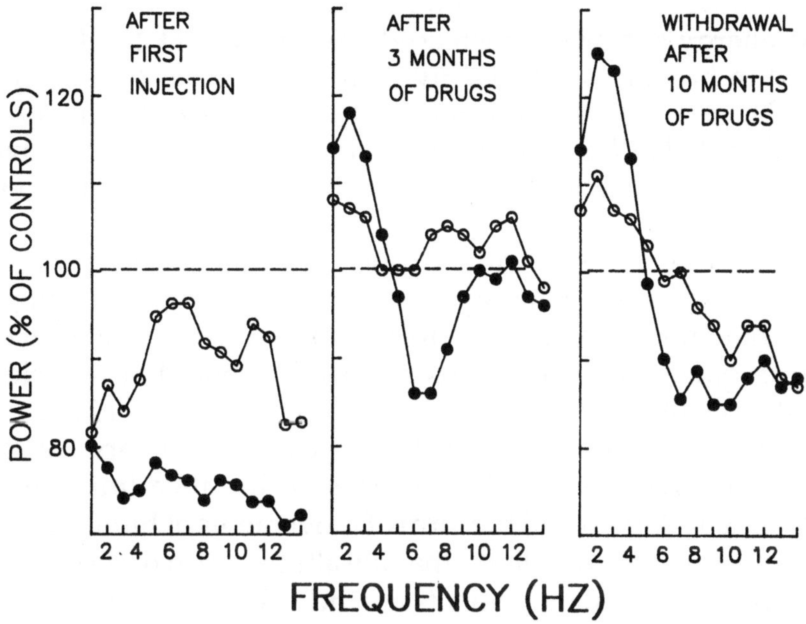

Fig. 4. FFT analyses of oral-movement energy in the two drug groups (expressed as percent of controls) at the various frequencies at (left) 5 days after the first decaoate injection, (center) just after the fifth monthly injection, and (right) 10 days after drug withdrawal following 10 months of drug administration: O, haloperidol; ●, fluphenazine.

there was an initial period of sedation (decreased energy at all frequencies). With continued administration, to our surprise, the drugged animals began to show increased oral movements of 1–3 Hz, an effect that then increased substantially upon drug withdrawal (Ellison and See, 1989; *see* Fig. 4). This increased energy was not caused by the very small CSMs, for it had a different time-course (appearing sooner in time) and different group characteristics (it was greater in the fluphenazine animals). Also, because the tiny CMS are so small in amplitude, they did not possess sufficient energy to cause these changes. What this meant, then, was that the oral movements of appreciable and readily observable size were not necessarily increasing in frequency with chronic neuroleptic, but were slowly changing in form, and that this effect showed the exact temporal characteristics expected of TD (late appearance and exaggeration upon drug withdrawal).

This was a remarkable finding, for this is precisely the altered energy spectrum (increased energy only at 1–2 Hz) observed in humans with TD. Five different studies (Caligivri et al., 1989; Lees, 1985; Rondot and Bathien, 1986;Alpert et al., 1976; Wirshing et al., 1989) have recorded the oral or extremity movements of humans with TD, using various devices, and all report the movements of TD showing exaggerated energy at the same place in the energy spectrum. In one other investigation that reported on the altered energy spectra of TD (Tryon and Pologe, 1987), not enough data were presented to determine at exactly what frequency the alterations occurred, although it was clear that the results were not incompatible with the 1–2 Hz result. This very close agreement between the altered form of the movements recorded in this experiment in rats administered chronic neuroleptics and the form of movements of humans with TD would appear to indicate that the altered form of these computer-recorded oral movements are indeed a quite valid animal model.

In a further study designed to determine how drug regimen would affect these results, rats were given HAL for eight months, either via highly fluctuating, weekly injections or via more continuous administration (via their drinking water and osmotic minipumps). The two groups received approx equivalent total doses of HAL. The results were quite striking, for they indicated that with time these two groups developed very distinctively different changes in oral movement form, as determined both by the computer system and by the human observer. Continuous administration again resulted in late-onset changes in oral activity at 1–3 Hz and withdrawal increases in CSMs, a pattern indicating TD. However, intermittent treatment (weekly large injections) produced a pattern more analogous to a primed dystonia-like reaction: large-amplitude CSMs that had steep onset slopes and a peak energy at 4–7 Hz (Fig. 5). These results demonstrate the importance of drug regimen in determining the type of neuroleptic-induced dyskinesias that develop with prolonged neuroleptic treatment in rodents, and validate our unique measuring device, for we can now distinguish between two forms of neuroleptic-induced motor dyskinesias (TD) and tardive

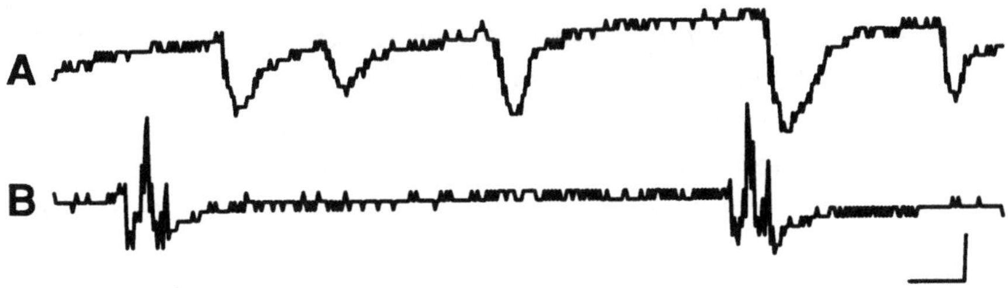

Fig. 5. Two representative tracings of mouth movements in animals 10 days after discontinuation of eight months of chronic haloperidol. **(a)** following continuous HAL (oral and minipump). The observer reported five oral movements during this segment. Smooth, repetitive movements with energy at 1–2 HZ are apparent. **(b)** Another animal following eight months of intermittent HAL at the same total dosage. Repetitive, sharp oral movements occur, each of which has a similar waveform. Calibration: 0.5 seconds and 1.5 mm (i.e., 5 rasters).

dystonia and can now study drugs that exacerbate or ameliorate these two different syndromes.

This experiment will go a long way toward resolving a great deal of controversy in this field. The intermittent-injection group appear to be showing the "primed dystonia" effect that has been demonstrated repeatedly in monkeys (Liebman and Neale, 1980; Kovacic et al., 1986).

We have also studied, using our rodent model, some promising atypical neuroleptics. These are drugs that have a low propensity for inducing TD (Tamminga and Gerlach, 1987). One such promising drug is clozapine, a nonselective antipsychotic that can be differentiated from typical neuroleptics by its noncataleptogenic nature (Niemegeers and Janssen, 1979) and its inability to block DA-agonist-induced stereotypies (Robertson and MacDonald, 1984). Clozapine has been shown to be a potent antipsychotic (Gerlach et al., 1974) that apparently does not induce TD (Lindstrom, 1988). A second direction in the search for atypical neuroleptics has been the development of compounds that are more specific for the dopamine D2 receptor, particularly the substituted benzamides, of which sulpiride is the most widely established in the treatment of schizophrenia (Tamminga and Gerlach, 1987). The substituted benzamides also appear to show less acute and tardive EPS (Gerlach and Casey,

1984). We chose to employ one of the most promising substituted benzamides, raclopride (Hall et al., 1988). In this experiment (See and Ellison, 1990b), rats were administered haloperidol, clozapine, raclopride, or no drug for either 28 days or eight months, and then withdrawn from drug treatment for three weeks. Four weeks of neuroleptic administration produced no changes in CSMs in any drug-treated group. Long-term administration induced distinctively different patterns of oral activity in the three drug groups, both in the number of CSMs and, the form of these movements. The oral movements that developed in the haloperidol-treated rats fit our previously described syndrome of late-onset oral dyskinesias that increased upon drug withdrawal. The clozapine- and raclopride-treated rats did not show the increased oral movements seen in the haloperidol-treated animals, but each exhibited uniquely different CSM characteristics compared to controls. In fact, Fig. 6 compares the form of oral movements, as determined by FFT, of rats administered eight months of chronic haloperidol (either continuously or intermittently) with that of rats administered clozapine or raclopride for the same length of time. This figure graphically demonstrates the radically different energy spectra of continuous- vs intermittent-haloperidol-treated animals, but it further demonstrates a novel finding. The clozapine-treated animals, who progressively developed a profound suppression of oral movements, correspondingly show decreased energy at all parts of the spectrum. However, although these animals showed profound differences from haloperidol-treated animals, it is also apparent that they eventually developed an altered form of oral activity, as determined by FFT, which in some ways showed characteristics similar to those of haloperidol-treated animals (i.e., peak energy at 1–3 Hz and decreased activity at 6–8 Hz). Raclopride clearly did not induce this altered profile of oral-movement form (although it is possible that the doses of raclopride given were somewhat lower than those of the other drugs on a scale of clinical effectiveness). These results provide a strong rodent model for assessing different oral behaviors following chronic administration of typical and atypical neuroleptics and imply that haloperidol, but not clozapine or

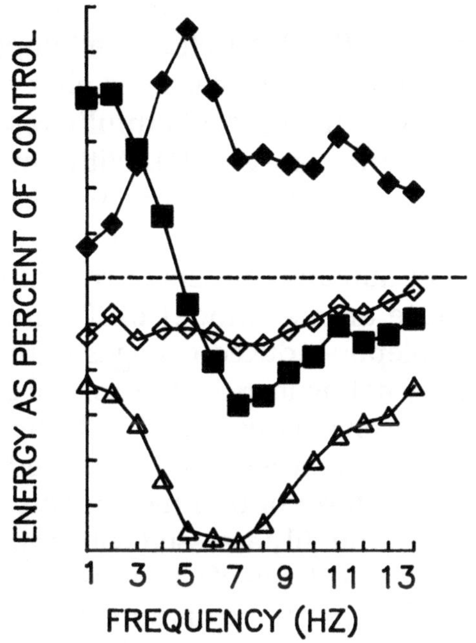

Fig. 6. FFT analysis (compared to controls) of oral movements 10 day after drug withdrawal in rats who had received, respectively, 8 months of continuous haloperidol (■—■), intermittent haloperidol (◆—◆), clozapine (△–△), or raclopride (◇—◇). Abcissa: percent of control levels, going from 40 to 160% in steps of 10%; the dashed line represents control levels. The distinctly different waveforms of oral movements in animals pretreated with continuous vs intermittent haloperidol are clear. Although raclopride-treated animals show no distinctive alteration in energy, the clozapine-treated animals appear to have a waveform similar to that of continuous haloperidol treatment, but decreased energy at all parts of the spectrum, reflecting the dramatically decreased number of oral movements that progressively developed in these animals. Note that the controls for the animals intermittently treated with haloperidol received weekly saline injections, whereas the controls for all other groups received nondrugged, flavored drinking water and sham minipump implants.

raclopride, produces in rats late-onset oral dyskinesias that fit the pattern expected for TD.

7. Conclusions

A comprehensive rodent model of TD would greatly facilitate our understanding of this disorder and provide a means of assessing treatment approaches, including the development of novel neuroleptics. A difficult question in attempting to develop animal models of human pathologies is always the extent to which the syndrome in the model faithfully represents the pathology seen in humans (Casey, 1984). With a movement disorder, such as TD, this can only be answered by showing some kind of correspondence in the actual form of the altered movements. Since TD in humans consists of truly abnormal movements, an accurate animal model of TD must be able to demonstrate neuroleptic-induced changes in the shape of oral movements. Simply noting increases in normal chewing behavior fails to provide an analogous model of the human condition.

The rodent model that we have presented here indicates that alterations in the form of oral movements in neuroleptic-treated rats can be thoroughly assessed with appropriate testing procedures. Three distinct types of alterations in oral movements with chronic neuroleptics can be discerned. With continuously presented typical neuroleptics, such as fluphenazine and haloperidol, normal-sized oral movements gradually become altered in form, and this effect increases dramatically upon drug withdrawal. The close agreement between the altered forms of these movements recorded in rats administered chronic neuroleptics and those of humans with TD would appear to indicate that these computer-recorded oral movements with increased energy at 1–2 Hz are indeed a quite valid animal model of TD.

A second type of altered oral syndrome develops when animals are given highly fluctuating neuroleptics, for example, one large injection weekly. In this case, large, rapid, and repetitive gaping movements develop; these appear to be similar to dystonic behaviors.

A third type of altered behavior is the very small, computer-detected oral movements that reliably develop very late in the administration of neuroleptics. These very small CSMs develop using both types of drug regimen discussed above, and their clinical correlate remains a mystery.

Several avenues of study remain for further extension of this rodent model of TD. Responsivity of these oral movements to various pharmacological probes is needed to assess those drugs that may ameliorate or exacerbate the symptomatic movements. This includes drugs that have previously been found to modulate oral activity in humans and animals, including cholinergic, dopaminergic, and GABAergic drugs. Further study of the role of drug regimen is needed, particularly in light of our findings of distinctly different oral syndromes caused by interminent and continuous haloperidol (See and Ellison, 1990a). Finally, techniques of assessing neurotransmitter function can be applied to rats exhibiting oral dyskinesias, in order to ascertain possible biochemical correlates of such behavior. We have already found that rats showing the "TD-like" and "dystonia-like" syndromes show alterations in D2 and GABA receptor binding in caudate putamen and substantia nigra, but that the changes are remarkably similar in the two groups even though the behavioral syndromes are completely different. It is now extremely important to determine, using this model, where in the brain the altered pattern generator for these oral dyskinesias lies, for this will shed considerable light on the mechanisms underlying TD.

References

Adler L. A., Angrist B., Reiter S., and Rotrosen J. (1989) Neuroleptic-induced akathisia: A review. *Psychopharmacology* **97**, 1–11.

Alpert M., Diamond F., and Friedhoff A. J. (1976) Tremographic studies in tardive dyskinesia. *Psychopharmacol. Bull.* **12(2)**, 5.

Ayd F. J. (1961) A survey of drug-induced extrapyramidal reactions. *JAMA* **175**, 1054–1060.

Barnes T. R. and Braude W. M. (1985) Akathesia variants and tardive dyskinesia. *Arch. Gen. Psychiat.* **42**, 874–878.

Bedard P., Delean J., Lafleur J., and Larochelle L. (1977) Haloperidol-induced dyskinesias in the monkey. *J. Can. Sci.* **4**, 197–201.

Beresford R. and Ward A. (1987) Haloperidol decanoate: A preliminary review of its pharmacodynamic and pharmacokinetic properties and therapeutic use in psychosis. *Drugs* **33**, 31–49.

Burke R. E., Fahn S., Jankovic J., Marsden C. D., Lang A. E., Gollomp S., and Ilson J. (1982) Tardive dystonia: Late-onset and persistent dystonia caused by antipsychotic drugs. *Neurology* **32**, 1335–1346.

Caligiuri M., Jeste D., and Harris M. (1989) Instrumental assessment of lingual motor instability in tardive dyskinesia. *Neuropsychopharmacology* **2**, 309–312.

Casey D. E. (1984) Tardive dyskinesia–animal models. *Psychopharmacol. Bull.* **20,** 376–379.

Casey D. E. (1985a) Tardive dyskinesia: Epidemiologic factors as a guide for prevention and management, in *Chronic Treatments in Neuropsychiatry* (Kemali D. and Racagni G., eds.), Raven, NY, pp. 15–24.

Casey D. E. (1985b) Tardive dyskinesia: Reversible and irreversible, in *Dyskinesia—Research and Treatment* (Casey D. E., Chase T. N., Christensen A. V., and Gerlach J., eds.), Springer, Heidelberg, pp. 88–97.

Chien C., Jung K., and Ross-Townsend A. (1980) Methodological approach to the measurement of tardive dyskinesia: Piezoelectric recording and concurrent validity test on give give clinical rating scales. *Tardive Dyskinesia: Research and TreatmentI* (Frann W., Smith R., Davis J., and Domino E., eds.). Spectrum Publications, New York, NY, pp. 233–241.

Clow A., Jenner P., and Marsden C. D. (1979) Changes in dopamine-mediated behaviour during one year's neuroleptic administration. *Eur. J. Pharmacol.* **57,** 365–375.

Clow A., Theodorou A., Jenner P., and Marsden C. D. (1980) Cerebral dopamine function in rats following withdrawal from one year of continuous neuroleptic administration. *Eur. J. Pharmacol.* **63,** 145–157.

Crane G. E. (1973) Persistent dyskinesia. *Br. J. Psychiat.* **122,** 395–405.

Creese I. (1983) Receptor interactions of neuroleptics, in *Neuroleptics: Neurochemical, Behavioral, and Clinical Perspectives* (Coyle J. T. and Enna S. J., eds.), Raven, NY, pp. 183–223.

Csernansky J. G., Grabowski K., Cervantes J., Kaplan J., and Yesavage J. A. (1981) Fluphenazine decanoate and tardive dyskinesia: A possible association. *Am. J. Psychiat.* **138(10),** 1362–1365.

Deneau G. and Crane G. (1969) Dyskinesia in rhesus monkeys tested with high doses of chlorpromazine, in *Psychotropic Drugs and Dysfunction of the Basal Ganglia* (Crane G. and Gardner R., eds.), US Public Health Service, Washington, DC, pp. 12–14.

Domino E. and Kovacic B. (1983) Monkey models of tardive dyskinesia. *Mod. Probl. PharmacoPsychiat.* **21,** 21–33.

Domino E. F. (1985) Induction of tardive dyskinesia in Cebus apella and macaca speciosa monkeys: A review, in *Dyskinesia—Research and Treatment* (Casey D. E., Chase T. N., Christensen A. V., and Gerlach J., eds.), Springer, Heidelberg, pp. 217–223.

Ellison G. and Morris W. (1981) Opposed stages of continuous amphetamine administration: Parallel alterations in motor stereotypies and in vivo spiroperidol accumulation. *Eur. J. Pharmacol.* **74,** 207–214.

Ellison G. D. and See R. E. (1989) Rats administered chronic neuroleptics develop oral movements which are similar in form to those in humans with tardive dyskinesia. *Psychopharmacology* **98,** 564–566.

Ellison G., Staugaitis S., and Crane P. (1981) A silicone delivery system for producing binge and continuous ethanol intoxication in rats. *Pharmacol. Biochem. Behav.* **14,** 207–211.

Ellison G., Eison M., Huberman H., and Daniel F. (1978) Structural and biochemical alterations in dopaminergic innervation of the caudate nucleus following continuous amphetamine administration. *Science* **201,** 276–278.

Ellison G. D., See R. E., Levin E. D., and Kinney J. (1987) Tremorous mouth movements in rats administered chronic neuroleptics. *Psychopharmacology* **92,** 122–126.

Fann W. E., Stafford J., Malone R., Frost J., and Richman B. (1977) Clinical research techniques in tardive dyskinesia. *Am. J. Psychiat.* **134,** 759.

Gardos G., Cole J. O., and Tarsy D. (1978) Withdrawal syndromes associated with antipsycholic drugs. *Am. J. Psychiat.* **135,** 1321–1324.

Gardos G., Cole J. O., Salomon M., and Schnielbolk S. (1987) Clinical forms of severe tardive dyskinesia. *Am. J. Psychiat.* **144,** 895–902.

Gerlach J. and Casey D. E. (1984) Sulpiride in tardive dyskinesia. *Acta Psychiatr. Scand. (Suppl.)* **311,** 93–102.

Gerlach J., Koppelhus P., Helweg E., and Monrad A. V. (1974) Clozapine and haloperidol in single-blind cross-over trial: Therapeutic and biochemical aspects in the treatment of schizophrenia. *Acta Psychiatr. Scand.* **50,** 410–424.

Glassman R. and Glassman H. (1980) Oral dyskinesias in brain-damaged rats withdrawn from a neuroleptic: Implications for models of tardive dyskinesia. *Psychopharmacology* **69,** 19–25.

Goldman M. B. and Luchins D. J. (1984) Intermittent neuroleptic therapy and tardive dyskinegia: A literature review. *Hosp. Comm. Psychiat.* **35,** 1215–1219.

Gunne L. M. and Barany S. (1976) Haloperidol-induced tardive dyskinesia in monkeys. *Psychopharmacology* **50,** 237–240.

Gunne L. M. and Haggstrom J. E. (1983) Reduction of nigral glutamic acid decarboxylase in rats with neuroleptic-induced oral dyskinesia. *Psychopharmacology* **81,** 191–194.

Gunne L. M., Haggstrom J. E., and Syoquist B. (1984) Association with persistent neuroleptic-induced dyskinesia of regional changes in brain GABA synthesis. *Nature* **309(24),** 347–349.

Gunne L. M., Andersson U., Bondesson U., and Johansson P. (1986) Spontaneous chewing movements in rats during acute and chronic antipsychotic drug administration. *Pharmacol. Biochem. Behav.* **25,** 897–901.

Guy W., Ban T. A., and Wilson W. H. (1985) An international survey of tardive dyskinesia. *Prog. Neuropsychopharmacol. Biol. Psychiat.* **9,** 401–405.

Hall H., Kohler C., Gawell L., Farde L., and Sedvall G. (1988) Raclopride, a new selective ligand for the dopamine-D2 receptors. *Prog. Neuropsycopharmacol. Biol. Psychiat.* **12,** 559–568.

Iversen S. D., Howells R. B., and Hughes R. P. (1980) Behavioral consequences of long-term treatment with neuroleptic drugs. *Adv. Biochem. Psychopharmacol.* **24,** 305–313.

Jeste D. V., Lohr J. B., Clark K., and Wyatt R. J. W. (1988) Pharmacological treatments of tardive dyskinesia in the 1980s. *J. Clin. Psychopharmacol.* **8,** 385–485.

Jeste D. V. and Wyatt R. J. W. (1982) Therapeutic strategies against tardive dyskinesia: Two decades of experience. *Arch. Gen. Psychiat.* **39,** 803–816.

Johansson P. (1989) Characterization and application of animal models for tardive dyskinesia. Uppsala, Sweden, *Acta Univ. Upsaliensis* **222,** pp. 1–46.

Johansson P., Casey D. E., and Gunne L. M. (1986) Dose-dependent increases in rat spontaneaus chewing rates during long-term administration of haloperidol but not clozapine. *Psychopharmacol. Bull.* **22,** 1017–1019.

Kane J. M. and Smith J. M. (1982) Tardive dyskinesia: Prevalence and risk factors, 1959–1979. *Arch. Gen. Psychiat.* **39,** 473–481.

Klawans H. L. (1973) The pharmacology of tardive dyskinesias. *Am. J. Psychiat.* **130,** 82–86.

Kovacic B. and Domino E. F. (1984) Fluphenazine-induced acute and tardive dyskinesia in monkeys. *Psychopharmacology* **84,** 310–314.

Kovacic B., Ruffing D., and Stanley M. (1986) Effect of neuroleptics and of potential new antipsychotic agents (MJ13859-1 and MJ13980-1) on a monkey model of tardive dyskinesia. *J. Neural. Transm.* **165,** 39–49.

Lees A. J. (1985) *Tics and Related Disorders* (Churchill Livingstone, Edingburgh), pp. 191–234.

Levin E. D., Galen D., and Ellison G. D. (1987) Chronic haloperidol effects on radial-arm maze performance and oral movements in rats. *Pharmacol. Biochem. Behav.* **26,** 1–6.

Levy A. D., See R. E., Levin E. D., and Ellison G. D. (1987) Neuroleptic-induced oral movements in rats: Methodological issues. *Life Sci.* **41,** 1499–1506.

Liebman J. and Neale R. (1980) Neuroleptic-induced acute dyskinesias in squirrel monkeys: Correlation with propensity to cause extrapyramidal side effects. *Psychopharmacology* **68,** 25–29.

Lindstrom L. H. (1988) The effect of long-term treatment with clozapine in schizophrenia: A retrospective study in 96 patients treated with clozapine for up to 13 years. *Acta Psychiatr. Scand.* **77,** 524–529.

McGeer P. L., Boulding J. E., Gibson W. C., and Foulkes R. G. (1961) Drug-induced extrapyramidal reactions. *J. Am. Med. Assoc.* **177,** 665–670.

McKinney W. T., Moran E. C., Kraemer G. W., and Prange A. J. (1980) Long-term chlorpromazine in rhesus monkeys: Production of dyskinesias and changes in social behavior. *Psychopharmacology* **72,** 35–39.

Marsden C. D. and Jenner P. (1980) The pathophysiology of extrapyramidal side effects of neuroleptic drugs. *Psychol. Med.* **10,** 55–72.

Mithani S., Atmadja S., Baimbridge K. G., and Fibiger H. C. (1987) Neuroleptic-induced oral dyskinesias, effects of progabide and lack of correlation with regional changes in glutamic acid decarboxylase and choline acetyltransferase activities. *Psychopharmacology* **93,** 94–100.

Moore D. C. and Bowers M. B. (1980) Identification for a subgroup of tardive dyskinesia patients by pharmacologic probes. *Am. J. Psychiat.* **137,** 1202–1205.

Neale R., Gerhardt S., and Liebman J. M. (1984) Effects of dopamine agonists, catecholamine depletors, and cholinergic and GABAergic drugs on acute dyskinesias in squirrel monkeys. *Psychopharmacology* **82,** 20–26.

Niemegeers C. J. E. and Janssen P. A. J. (1979) A systematic review of the pharmacological activities of dopamine antagonists. *Life Sci.* **24,** 2201–2216.

Owen F., Cross A. J., Waddington J. L., Poulter M., Gamble S. J., and Crow T. J. (1980) Dopamine-mediated behaviour and 3H-spiperone binding to striatal membranes in rats after nine months haloperidol administration. *Life Sci.* **26(1),** 55–59.

Pi E. H. and Simpson G. M. (1983) Atypical neuroleptics: Clozapine and the benzamides in the prevention and treatment of tardive dyskinesia. *Mod. Probl. PharmacoPsychiat.* **21,** 80–86.

Post R. M. (1980) Intermittent versus continuous stimulation: Effect of time interval on the development of sensitization or tolerance. *Life Sci.* **26,** 1275–1282.

Potthoff A. D., Ellison G., and Nelson L. (1983) Ethanol intake increases during continuous administration of amphetamine and nicotine, but not several other drugs. *Pharmacol. Biochem. Behav.* **18,** 489–495.

Robertson A. and MacDonald C. (1984) Atypical neuroleptics clozapine and thioridazine enhance amphetamine-induced stereotypy. *Pharmacol. Biochem. Behav.* **21,** 97–101.

Rodriguez L. A., Moss D. E., Reyes E., and Camarena M. L. (1986) Perioral behaviors induced by cholinesterase inhibitors: A controversial animal model. *Pharmacol. Biochem. Behav.* **25,** 1217–1221.

Rondot P. and Bathien N. (1986) Movement disorders in patients with coexistent neuroleptic-induced tremor and tardive dyskinesia: EMG and pharmacologic study. *Adv. Neurol.* **45,** 361.

Rosengarten H., Schweitzer J. W., and Friedhoff A. J. (1983) Induction of oral dyskinesias in naive rats by D1 stimulation. *Life Sci.* **33,** 2479–2482.

Rosengarten H., Schweitzer J. W., and Friedhoff A. J. (1986) Selective dopamine D2 receptor reduction enhances a D1 mediated oral dyskinesia in rats. *Life Sci.* **39,** 29–35.

Rupniak N. M. J., Jenner P., and Marsden C. D. (1983) Cholinergic manipulation of perioral behaviour induced by chronic neuroleptic administration to rats. *Psychopharmacology* **79,** 226–230.

Rupniak N. M. J., Jenner P., and Marsden C. D. (1985) Pharmacological characterisation of spontaneous or drug-associated purposeless chewing movements in rats. *Psychopharmacology* **85,** 71–79.

Rupniak N. M. J., Jenner P., and Marsden C. D. (1986) Acute dystonia induced by neuroleptic drug. *Psychopharmacology* **88,** 403–419.

Rupniak N. M. J., Mann S., Hall M. D., Fleminger S., Kilpatrick G., Jenner P., and Marsden C. D. (1984) Differential effects of continuous administration for 1 year of haloperidol or sulpiride on striatal dopamine function in the rat. *Psychopharmacology* **84,** 503.

Sahakian B. J., Robbins T. W., and Iversen S. D. (1976) Fluphenthixol-induced hyperactivity by chronic dosing in rats. *Eur. J. Pharmacol.* **37,** 169–178.

Sant W. W. and Ellison G. (1984) Drug holidays alter onset of oral movements in rats following chronic haloperidol. *Biol. Psychiat.* **19,** 95–99.

See R. E. and Ellison G. (1990a) Intermittent and continuous haloperidol regimens produce different types of oral dyskinesias in rats. *Psychopharmacology* **100,** 404–412.

See R. E. and Ellison G. (1990b) Comparison of chronic administration of haloperidol and the atypical neuroleptics, clozapine and raclopride, in an animal model of tardive dyskinesia. *Eur. J. Pharmacol.* **181,** 175–186.

See R. E., Levin E. D., and Ellison G. D. (1988) Characteristics of oral movements in rats during and after chronic haloperidol and fluphenazine administration. *Psychopharmacology* **94,** 421–427.

See R., Sant W. W., and Ellison G. (1987) Recording oral activity in rats reveals a long-lasting subsensitivity to haloperidol as a function of duration of previous haloperidol treatment. *Pharmacol. Biochem. Behav.* **28,** 175–178.

Stewart B. R., Rupniak N. M. J., Jenner P., and Marsden C. D. (1988) Animal models of neuroleptic-induced acute dystonia, in *Advances in Neurology,* vol. 50: *Dystonia* 2 (Fahn S., Marsden C. D., and Calne D. B., eds.), Raven, New York., pp. 343-359.

Tamminga C. A. and Gerlach J. (1987) New neuroleptics and experimental antipsychotics in schizophrenia, in *Psychopharmacology—The Third Generation of Progress* (Meltzer H. Y., ed.), Raven, NY, pp. 1129–1140.

Tarsy D. and Baldessarini R. (1974) Behavioral supersensitivity to apomorphine following chronic treatment with drugs which interfere with synaptic function of catecholamines. *Neuropharmacology* **13,** 927,928.

Tryon W. W. and Pologe B. (1987) Accelerometric assessment of tardive dyskinesia. *Am. J. Psychiat.***144,** 1584–1587.

Waddington J. L., Cross A. J., Gamble S. J., and Bourne R. C. (1983) Spontaneous orofacial dyskinesia and dopaminergic function in rats after six months of neuroleptic treatment. *Science* **220,** 530–532.

Waddington J. L., Youssef H. A., Molloy A. G., O'Boyle K. M., and Pugh M. T. (1985) Association of intellectual impairment, negative symptoral movements, and aging with tardive dyskinesia: Clinical and animal studies. *J. Clin. Psychiat.* **46,** 29–33.

Weiss B. and Santelli S. (1978) Dyskinesias evoked in monkeys by weekly administration of haloperidol. *Science* **200,** 799–801.

Wirshing W. C., Cummings J. L., Lathers P., and Engel J. (1989) The machine measured characteristics of tardive dyskinesia. *Schizophr. Res.* **2,** 240.

Appendix 1: Construction of Slow-Release Haloperidol Pillows

Chronic drug administration always presents a problem, and this is particularly true when drugs must be administered for at least six months before the desired behavioral results are obtained, as is the case with chronic neuroleptic studies. One solution is to give repeated depot injections of the decanoate form of the drug (i.e., See et al., 1988); however, this can be undesirable because sudden drug withdrawal becomes difficult or impossible, yet withdrawal effects are one of the cardinal characteristics of TD. Another solution is to give the drug via osmotic minipumps, but this becomes prohibitively expensive with very long-term studies. Repeated injections of the drug in the nondepot form induce an entirely different syndrome and induce a drug regimen completely unlike that of humans given neuroleptics. Spiking the animal's drinking water with the neuroleptic is probably the method most commonly employed; however, this has several undesirable features. One is that different animals invariably consume different quantities of the drug, inducing uncontrolled variability. Another is that different animals consume the drug in different ways. We have recorded the drinking patterns of rats being administered neuroleptics in their drinking water and found that some rats, for example, will go virtually the entire 12-hour lights-on period without any drug consumption, whereas others regularly consume some drug every 60 minutes throughout the day–night cycle. It is apparent that this induces different drug regimens in different animals.

A novel solution that we have developed is to administer the drug using slow-release silicone implants. We originally developed these for short-term studies of the effects of continuous amphetamines (Ellison et al., 1978), other psychoactive compounds (Potthoff et al., 1983), or even ethanol (Ellison et al., 1981), but they are particularly useful in chronic neuroleptic studies, for they are tolerated very well by the animals, are extraordinarily steady in their release rate, and can be constructed so that they will continue to release effective drug doses for periods approaching six months.

The "haloperidol pillows" that we have developed are constructed from two 2.5- × 5-cm pieces of silastic medical-grade sheeting (Dow Corning). The release rate is controlled by the thickness of the sheeting chosen; typically we have used either thick (0.02-in.) or thin (.005-in.) sheeting. For most purposes, only the thinner sheeting yields doses that are sufficient to induce appreciable behavioral changes, but a sturdy pillow can be made with one thick and one thin side (See et al., 1987).

The two pieces are fused along three outer edges using a silicone-type medical adhesive (Dow-Corning #891), and the glued edges are clamped together overnight to dry, usually using a standard paper clamp. Powdered HAL base (100 mg) (the salt form will not cross the silicone barrier) is then inserted into the pillow and distributed evenly about the pillow, and the final edge is sealed with the silicone adhesive. After drying, the HAL powder is spread evenly about the pillow by shaking, and spreading the HAL with force, using a blunt utensil, such as the back of a spoon. It is extremely important that the HAL be evenly distributed, with no clumps, for, as the drug begins to leave the capsule, output will decline dramatically over time if clumps exist. This is because regions with little HAL will become empty, and only the remaining clumps are left to distribute.

The edges of the pillow can be coated with Dow-Corning medical-grade Elastomer™ (or, because this product is no longer available from Dow-Corning, a comparable substitute, such as A-101 medical-grade Elastomer from Factor II, Lakeside, AR); this is done to remove rough edges, thereby reducing tissue irritation from thick pillows, and also to provide extra support for implantation of thin pillows. Then the "pillow" is implanted into the animal, using a general anaesthetic. We typically employ a volatile anaesthetic, such as ether or halothane, because a practiced surgeon can implant a pillow in perhaps 3–5 min. The pillow is dusted with an antibiotic powder (Neosporin™, Burroughs-Wellcome), and an area on the flank of the animal is shaved with a clipper. Then an incision equal to the width of the pillow is made with scissors, a large cavity is made by separating the skin from the underlying tissue using blunt dissection, and the pillow is inserted into the cavity using forceps with tips coated with rubber or silicone (so as to not puncture the pillow). The pillow is carefully smoothed under the skin so as to be lying flat, and the wound is closed with wound clips. A topical antibiotic is applied and im antibiotics injected. The wound clips should be removed within a week. Control implants can be made from an empty silastic container.

These pillows remain remarkably free from infection, and only rarely do fluid sacs develop. The animals should be periodically inspected and any tumors removed. The release rate for double thin-wall (0.05-in. thick) pillows, determined by HPLC assay over a 3-week period, is 0.23 mg/d (or 0.69 mg/kg/d for a 333-g rat). A complete analysis of release rates at various time-periods was made for pillows made of one thick and one thin wall, with half of the pillows being double-chambered (100 mg) and half being single-chambered (50 mg). Over 24,12, and six week, double-chambered pillows released 22.9,12.3, and 5.2 mg of HAL base, respectively; for single-chambered pillows, these values were 11.25, 7.45, and 3.9 mg (See et al., 1987).

Index

MIX
Papier aus verantwortungsvollen Quellen
Paper from responsible sources
FSC® C105338

If you have any concerns about our products,
you can contact us on
ProductSafety@springernature.com

In case Publisher is established outside the EU,
the EU authorized representative is:
Springer Nature Customer Service Center GmbH
Europaplatz 3, 69115 Heidelberg, Germany

Printed by Libri Plureos GmbH
in Hamburg, Germany